D0164458

ENVIRONMENTAL SCIENCE

FFA

ENVIRONMENTAL SCIENCE

FUNDAMENTALS AND APPLICATIONS

Edited by L. DeVere Burton

Australia • Brazil • Japan • Korea • Mexico • Singapore • Spain • United Kingdom • United States

Environmental Science: Fundamentals and Applications
L. DeVere Burton

Vice President, Career and Professional Editorial: Dave Garza

Director of Learning Solutions: Matthew Kane

Acquisitions Editor: David Rosenbaum

Managing Editor: Marah Bellegarde

Product Manager: Christina Gifford

Editorial Assistant: Scott Royael

Vice President, Career and Professional Marketing: Jennifer McAvey

Marketing Director: Deborah Yarnell

Marketing Coordinator: Jonathan Sheehan

Production Director: Carolyn Miller

Production Manager: Andrew Crouth

Content Project Manager: Elizabeth C. Hough

Art Director: Dave Arsenault

Technology Project Manager: Mary Colleen Liburdi

Production Technology Analyst: Tom Stover

Library of Congress Control Number: 2007940808

ISBN-13: 978-1-4180-5354-3

ISBN-10: 1-4180-5354-6

Delmar
5 Maxwell Drive
Clifton Park, NY 12065-2919
USA

Cengage Learning is a leading provider of customized learning solutions with office locations around the globe, including Singapore, the United Kingdom, Australia, Mexico, Brazil, and Japan. Locate your local office at:
international.cengage.com/region

Cengage Learning products are represented in Canada by Nelson Education, Ltd.

For your lifelong learning solutions, visit **delmar.cengage.com**

Visit our corporate website at **www.cengage.com**

Notice to the Reader
Publisher does not warrant or guarantee any of the products described herein or perform any independent analysis in connection with any of the product information contained herein. Publisher does not assume, and expressly disclaims, any obligation to obtain and include information other than that provided to it by the manufacturer. The reader is expressly warned to consider and adopt all safety precautions that might be indicated by the activities described herein and to avoid all potential hazards. By following the instructions contained herein, the reader willingly assumes all risks in connection with such instructions. The publisher makes no representations or warranties of any kind, including but not limited to, the warranties of fitness for particular purpose or merchantability, nor are any such representations implied with respect to the material set forth herein, and the publisher takes no responsibility with respect to such material. The publisher shall not be liable for any special, consequential, or exemplary damages resulting, in whole or part, from the readers' use of, or reliance upon, this material.

Printed in Canada
1 2 3 4 5 6 7 12 11 10 09 08

CONTENTS

PREFACE XV

ACKNOWLEDGMENTS XVII

ABOUT THE AUTHOR XIX

SECTION 1 Introduction to Environmental Science 3

CHAPTER 1 **Understanding Ecology** 5

Perceptions in Ecology 6
Principles of Ecology 7
Pollution 10
Conservation of Natural Resources 14
Looking Back 16
Self-Evaluation 16
Discussion and Essay 16
Multiple-Choice Questions 16
Learning Activities 17

CHAPTER 2 **The Science Behind Environmental Science** 19

Conservation of Matter 20
The Science of Ecosystems 21
The Ultimate Concept in Ecology 23
Laws of Energy 23
The First Law of Energy 24
The Second Law of Energy 24
Natural Cycles 25
The Carbon Cycle 25
The Nitrogen Cycle 27
The Water Cycle 28
Food Chains 31
Humans and the Food Pyramid 32
Biological Succession 33
Looking Back 38

Self-Evaluation 38
Essay Questions 38
Multiple-Choice Questions 38
Learning Activities 39

Chapter 3 **Ecosystem Management** 41

Resource Management 42
The Nature of Resources 44
Nonexhaustible Resources 44
Renewable Resources 45
Exhaustible Resources 45
Balance of Nature 46
Carrying Capacity 47
Human Population and Resource Use 48
Balancing Population Needs with Resources 48
Conservation 50
Preservation 51
Multiple Use 52
Reclaiming Damaged or Polluted Resources 53
Preserving Air Quality 55
Looking Back 55
Self-Evaluation 55
Essay Questions 55
Multiple-Choice Questions 56
Learning Activities 57

SECTION 2 Ecosystems and Natural Resources 59

Chapter 4 **Air** 61

Air Quality 62
Threats to Air Quality 62
The Greenhouse Effect and Global Warming 67
The Greenhouse Effect 67
Hole in the Ozone Layer 67
Global Warming: A Topic for Scientific Debate 68
Precautions Against Global Warming 71
Air and Living Organisms 72
Maintaining and Improving Air Quality 73
Regulating Air Quality 74
Looking Back 75
Student Activities 75
Self-Evaluation 76
Discussion and Essay 76
Multiple-Choice Questions 76
Matching 77
Learning Activities 77

CHAPTER 5 **Water** 79

The Nature of Water 81

Fresh Versus Salt Water 82

Aquatic Food Chains and Webs 83

The Universal Solvent 83

The Water Cycle 85

Relationships of Land and Water 86

Precipitation 86

Water and Watersheds 86

Marshes and Wetlands 88

Land as a Reservoir 90

Groundwater 90

Improving Water Quality 91

Water as a Living Environment 94

Water pH 94

Temperature 95

Salinity 95

Total Dissolved Solids 96

Major Dissolved Ionic Components 96

Trace Elements 98

Dissolved Gases 99

Organic Material and Its Breakdown Products 105

Conserving Water 107

Looking Back 108

Self-Evaluation 109

Essay Questions 109

Multiple-Choice Questions 109

Learning Activities 111

CHAPTER 6 **Land and Soil** 113

Soil Formation 114

Parent Material 114

Weathering 116

Soil Organic Matter 117

Characterizing Soils 118

The Soil Profile 118

Soil Physical Properties 119

Medium for Plant Growth 121

Anchorage 121

Water 121

Oxygen 121

Nutrients 122

Soil Uses 123

Cropland 123

Grazing Land 123

Urbanization 124

Foundations 125

Waste Disposal 125

Forest 126

Recreation 126

Soil as a Living Environment 127
Soil Conservation 128
How Erosion Occurs 129
Preventing and Controlling Erosion 134
Looking Back 137
Self-Analysis 137
Essay Questions 137
Multiple-Choice Questions 137
Learning Activities 138

Chapter 7 **Forests** 141

Forest Regions of North America 143
Northern Coniferous Forest 143
Northern Hardwoods Forest 144
Central Broad-Leaved Forest 145
Southern Forest 145
Bottomland Hardwoods Forests 145
Pacific Coast Forests 146
Rocky Mountain Forest 146
Tropical Forest 146
Hawaiian Forest 146
Important Types and Species of Trees in the United States 146
Softwoods 146
Hardwoods 149
Tree Growth and Physiology 152
Forest Management 154
Forest Reproduction and Regeneration 154
Woodlot Management 157
Timber Management 158
Sustained Yield 163
Environmental Impact Statements 163
The Forest as a Resource 165
Economic Value 165
Biological Value 165
Habitat Value 166
Climate Moderation Value 169
Energy Value 169
Livestock Value 170
Recreation Value 172
Watershed Value 172
Protecting Forests 173
Pests 173
Diseases 174
Insects 174
Fire 175
Forest Soils 176
Soil Orders 177
Erosion 177
Looking Back 178

Self-Evaluation 179
Essay Questions 179
Multiple Choice 179
Learning Activities 180

Chapter 8 **Grasslands and Wetlands** 183
Grasslands 185
History of the North American Grassland Ranges 187
Types of Grassland Vegetation 192
Range Management Techniques 193
Objectives of Range Management 193
Grazing Capacity 194
Grazing Management 194
Range Restoration 195
Wetlands 196
What Are Wetlands? 196
Characteristics 196
Two Definitions 197
History of Wetlands in the United States 198
Wetland Identification 200
Types of Wetlands 202
Marshes 202
Floodplains 203
Ponds 204
Rivers and Streams 204
Swamps 204
Bogs 205
Prairie Potholes 205
Vernal Pools 206
Status of Wetlands in the United States 206
Wetland Preservation 208
Wetlands Management 209
Other Government Programs 210
Looking Back 211
Self-Analysis 212
Essay Questions 212
Multiple-Choice Questions 212
Learning Activities 213

Chapter 9 **Wildlife Biology and Management** 215
Animal Behaviors and Habits 216
Food 217
Water 221
Shelter 223
Habitat 224
Space 224
Animal Growth 226
Animal Reproduction 227
Arrangement 228
Wildlife Relationships 229

Parasitism 229
Mutualism 229
Predation 230
Commensalism 230
Competition 230
Preserving and Restoring Wildlife Populations and Habitats 231
The U.S. Endangered Species Act 232
Human Impacts 233
Habitat Loss 234
The Principle of Stewardship 236
Extinction and Its Causes 236
Managing Endangered and Threatened Species 239
Approved Practices in Wildlife Management 239
Farm Wildlife 239
Forest Wildlife 240
Wetlands Wildlife 241
Stream Wildlife 242
Lakes and Ponds Wildlife 244
Looking Back 245
Self-Analysis 245
Essay Questions 245
Multiple-Choice Questions 246
Learning Activities 247

SECTION 3 The Human Impact 249

CHAPTER 10 **Agriculture and Sustainability** 251
Foundation for Agriculture 252
Farmers and Ranchers 253
Sustainable Agriculture 257
Principles of Sustainable Agriculture 258
Agriscience in a Growing World 261
Impact of Agriscience 262
Looking Back 264
Self-Evaluation 264
Essay Questions 264
Multiple-Choice Questions 265
Learning Activities 265

CHAPTER 11 **Integrated Pest Management** 267
History of Pest Management 269
The Need For Pest Management 270
Types of Pests 270
Weeds 271
Annual Weeds 271
Biennial Weeds 272
Perennial Weeds 272
Noxious Weeds 272
Insects 273

Insect Anatomy 273
Rodents 274
Plant Diseases 275
Environmental Concerns 276
Pest-Control Alternatives 276
Cultural Control 276
Biological Control 278
Genetic Control 278
Chemical Control 279
Principles and Concepts of Integrated Pest Management 282
Key Pests 282
Crop and Biology Ecosystem 284
Ecosystem Manipulation 284
Threshold Levels 285
Monitoring 286
Integrated Pest Management Program Components 286
Planning 286
Setting Action Thresholds 287
Site Monitoring 287
Pest Identification 287
Plan Implementation 287
Looking Back 287
Self-Analysis 288
Essay Questions 288
Multiple-Choice Questions 288
Learning Activities 289

CHAPTER 12 **Population Ecology** 291

Population Characteristics 292
Population Growth Factors 293
Extinction and Its Causes 296
Biomes 300
Freshwater Biomes 300
Marine Biome 305
Terrestrial Biomes 308
Looking Back 315
Self-Analysis 316
Essay Questions 316
Multiple-Choice Questions 316
Learning Activities 317

CHAPTER 13 **Waste Management** 319

Solid Wastes 320
Types of Solid Waste 324
Disposal of Solid Waste 326
What Is a Landfill? 327
Landfill Design 328
Natural Attenuation Landfills 328
Containment Landfills 330
Recycled Animal Wastes 331

Why Recycling? 333
Looking Back 333
Self-Analysis 333
Essay Questions 333
Multiple-Choice Questions 334
Learning Activities 335

Chapter 14 **Fossil Fuels** 337
Coal 338
Oil or Petroleum 341
Natural Gas 343
Oil Shale and Tar Sands 344
Fossil Fuel Conservation 345
Looking Back 346
Self-Analysis 346
Essay Questions 346
Multiple-Choice Questions 347
Learning Activities 347

Chapter 15 **Energy and Alternative Fuels** 349
Solar Energy 350
Nuclear Power 353
Geothermal Energy 355
Alcohol 356
Methane 358
Hydropower 360
Tidal Power 361
Wind 362
Wood 362
Looking Back 363
Self-Analysis 363
Essay Questions 363
Multiple-Choice Questions 364
Learning Activities 365

Chapter 16 **Toxic and Hazardous Substances** 367
Toxic and Hazardous Materials 368
Harm from Hazardous Materials 370
Assessing Risk 372
Toxic and Hazardous Substances Legislation 372
Looking Back 375
Self-Analysis 376
Essay Questions 376
Multiple-Choice Questions 376
Learning Activities 377

Chapter 17 **Careers in Environmental Science** 379
Selecting a Career 380
Career Options 381

Air Quality Control 382
Biologist: Aquatic 382
Biologist: Fish and Wildlife 382
Biologist: Marine 383
Botanist 383
Dendrology, Silviculture, and Forestry 383
Ecologist 384
Educator: Forestry 384
Entomologist 385
Environmental Analyst 385
Environmental Engineer 385
Environmental Quality Technician 386
Environmental Scientist 386
Forester 387
Game Bird Farm Manager 387
Herpetologist 388
Ichthyologist 388
Inspector: Environmental Safety 389
Mitigation Specialist 389
Oceanologist 389
Ornithologist 389
Predatory Animal Control Officer 389
Science Teacher 390
Soil Conservationist 390
Soil Specialist 390
Taxonomist 390
Technology Specialist 391
Water Treatment Specialist and Wastewater Treatment Specialist 391
Wildlife Conservation Officer 391
Wildlife Technician 392
Learning Activities 392

GLOSSARY 393

INDEX 407

PREFACE

A GENERAL PUBLIC awareness of environmental issues is increasing every day in the United States of America. This is as it should be, because all of us depend on a healthy environment. We not only live in it ourselves but also share space and resources with all other living organisms. The debate remains, however, about how to interpret data obtained from environmental research. There are also huge differences among people in the philosophies they express when discussing environmental needs and solutions.

Some of the polarization that occurs result from mistrust that is directed toward some of the environmental studies that are reported. Too many of them appear to be funded by special interest groups, which raises issues of objectivity. Others appear to be manipulated to produce the results that support a researcher's philosophy. The general public is often left somewhere in the middle of the debate trying to form educated opinions. Meanwhile, a whirlwind of sound information and misinformation assaults us on all sides.

Sound scientific research is needed in order to reduce the contention that exists among those who see little need to protect vulnerable species and those who are willing to sacrifice modern industries to restore habitat and displaced species. It is the belief of the author that citizens should seek to find "middle ground" in resolving environmental issues, and that it is wise to avoid radical positions on either side of most issues.

This textbook, *Environmental Science: Fundamentals and Applications*, seeks middle ground. A conscious attempt has been made to present both sides of the issues that are discussed in these chapters, allowing students to form their own opinions. After all, this is a key element of education and democracy: to gather information, evaluate the soundness of the data, and form opinions based on sound principles.

This textbook is organized into three general sections. The first three chapters work generally as an introduction to environmental sciences and examine our understanding of the sciences that affect the environment, including the principles of science that are related to life, growth, ecology, and human efforts to support healthy ecosystems. The second section of five chapters addresses relationships among natural resources such as air, water, soils, forests, grasslands, and wetlands in the ecosystems of North America. In the final section, eight chapters discuss human impacts on the environment, including agricultural practices, the sustainability of agricultural resources, integrated pest management, population ecology, waste management, fossil fuels, alternative fuels, and toxic and hazardous substances. The last chapter in this text discusses careers in environmental sciences and how students can prepare themselves now and in the near future to enter such fields if they choose.

The textbook is accompanied by a ClassMaster that includes the Instructor's Guide, which offers student activities and keys to questions for discussion and review that are listed at the end of each chapter.

Included in each chapter are such features as chapter-opening objectives and terms to know, Environmental Science and Interest Profiles, end-of-chapter reviews, self-analyses (including discussion, essay, and multiple-choice questions), and Learning Activities. Each chapter includes photographs and illustrations to aid students as they seek to understand the concepts that are presented.

A great deal of effort has been directed toward schools to integrate the principles of math and science in the context of how the knowledge of the subject will be used. This textbook is an example of how science can be integrated into a subject in such a way that the student can understand how the principles of science apply to living organisms in the ecosystems of North America.

ACKNOWLEDGMENTS

BOTH THE AUTHOR and publisher extend their appreciation to the individuals, agencies, associations, and organizations that have supplied the photographs and information needed to create this textbook. U.S. Deparment of Agriculture photo libraries have provided many photographs and images that have enhanced the finished product.

We also acknowledge the contributing authors to this textbook whose works have been used, sometimes separately, and sometimes blended together, to address the topics contained herein. This textbook presents the collective works of the lead author and the following individuals:

- Edward J. Plaster, *Soil Science and Management*, 4th edition;
- Elmer L. Cooper, coauthor, *Agriscience Fundamentals and Applications*, 4th edition;
- Kevin H. Deal, *Wildlife and Natural Resource Management*, 2nd edition;
- William G. Camp and Thomas B. Daugherty, *Managing Our Natural Resources*, 4th edition; and
- Rick Parker, *Aquaculture Science*, 2nd edition.

A special thanks is also extended to the reviewers who offered suggestions for the development of this new textbook in the agriscience series: Douglas G. Brown, Central Columbia High School (Bloomsburg, Pennsylvania), and John Busekist, Cattaraugus-Little Valley Central School District (Cattaraugus, New York).

L. DeVere Burton
Twin Falls, Idaho
January 2008

ABOUT THE AUTHOR

L. DeVere Burton, editor of *Environmental Science: Fundamentals and Applications*, is a lifelong educator who has been engaged in agricultural education since he was a high school student. He recently retired to write full-time. His career includes service as Dean of Instruction at the college of Southern Idaho, as well as Director of Research and State Supervisor of Agricultural Science and Technology with the Idaho State Division of Professional-Technical Education. He has served as President of the National Association of Supervisors of Agriculture Education and has participated as a member of several national curriculum-related task forces for the National Agricultural Education Council. He has also served as an adult consultant to the National Future Farmers of America Nominating Committee.

The author was a high school agriculture teacher for 15 years and has been involved as a professional educator in agricultural education since 1967. He has experienced teaching assignments in both large and small schools, and in both single- and multiple-teacher departments. He has taught at four different schools and at a major land grant university. He was involved in agriculture program supervision from 1987 to 1997. All of these experiences have contributed to his philosophy that "education must be fun and exciting for those who learn and for those who teach."

A wide range of experiences have prepared the author for his career as an educator in agriculture and natural resources. He was raised on a farm in western Wyoming that bordered on forest lands, and he experienced many pleasant hours in the canyons and along the streams that were parts of the forests of his youth. During his years as a university student, he worked in the forest industry as a logger and sawmill worker. Other jobs held by the author included testing milk for butterfat content; caring for livestock on a combination beef, swine, and trout ranch; maintenance and warehouse worker in a feed mill; manager of a dairy; finish carpenter; and animal research assistant. He has also worked in the food-processing, metal fabrication, and concrete construction industries and owned and managed a purebred sheep and row-crop farm for several years.

Dr. Burton earned his BS degree in agricultural education from Utah State University in 1967. He was awarded an MS degree in animal science from Brigham Young University in 1972. His PhD was earned at Iowa State University in 1987, where he was also an instructor in the Agricultural Engineering Department.

Dr. Burton has authored three textbooks and collaborated in revisions for another. They include *Agriscience and Technology, Fish and Wildlife: Principles of Zoology and Ecology, Introduction to Forestry Science,* and *Agriscience Fundamentals and Applications*. These textbooks have been written in a serious attempt to strengthen the science content and expand the breadth of the curriculum in the nation's agriculture and natural resource education programs.

ENVIRONMENTAL SCIENCE

section one

Introduction to Environmental Science

CHAPTER 1

Understanding Ecology

CHAPTER 2

The Science Behind Environmental Science

CHAPTER 3

Ecosystem Management

TERMS TO KNOW

- environmental science
- ecology
- ecologist
- environmentalist
- organism
- population
- community
- ecosystem
- biosphere
- ecosphere
- terrestrial
- aquatic
- pollution
- surface water
- industrial waste
- solid wastes
- pesticides
- insecticide
- herbicide
- rodenticide
- hazardous materials
- petroleum
- groundwater
- aquifer
- acid precipitation
- law of conservation of matter
- preservationists
- conservationists

CHAPTER 1

understanding ecology

Objectives

After completing this chapter, you should be able to

- define the terms *environmental science* and *ecology*
- describe the fundamental difference between an ecologist and an environmentalist
- identify similarities and differences among the terms *organism*, *population*, and *community*
- explain the makeup of an ecosystem
- specify the differences between terrestrial and aquatic organisms
- describe pollution of natural resources and its effects on living organisms
- identify some of the materials that make up industrial waste
- clarify differences between solid waste and hazardous materials
- classify the different forms of pesticides according to their use
- distinguish between surface water and groundwater
- list ways in which acid precipitation affects natural resources and the organisms that depend on them
- explain the law of conservation of matter
- describe some differences in philosophy between preservationists and conservationists

A simple definition for **environmental science** is the study of the interactions and relationships among living organisms and the physical and chemical features of the surroundings in which they live. Environmental science also deals with pollution and damage to the environment resulting from both natural events and human activities. Other environmental issues include conservation, contamination of groundwater and soil, waste management, air and noise pollution, and the uses of natural resources. Environmental science is based on principles of ecology.

Ecology is the branch of science that deals with the complex relationships among living things and their environments. Different organisms relate differently to their environments. They obtain nutrients in different ways. They benefit from different forms of shelter, and each organism requires a favorable environment in which to reproduce a new generation of its species.

Most plants and animals are also used as food by a variety of other organisms. All of these relationships are described through the science of ecology, which is concerned with all of the activities that affect the environment and the living organisms that depend on it for nutrients and shelter.

Perceptions in Ecology

ecologist versus environmentalist

It is important to note that ecology is a science. An **ecologist** is a scientist who studies the complex relationships and interactions among living organisms and their environment. *Environmentalism,* which is simply a strong concern for the environment, is not the same thing at all. Environmentalism is based on emotion, values, beliefs, and politics. An **environmentalist** is a political activist with a special interest in some aspect of the environment. Environmentalists concern themselves with what they perceive as right and wrong, good and bad, and moral or amoral.

As a science, ecology is based on the observation and objective interpretation of data. In the role of a scientist, an ecologist does not attempt to make decisions based on moral interpretations (value judgments). We say that science is *amoral,* which simply means that value judgments about good and bad should not be a part of science. Scientists make judgments based on data obtained though testing and measurement.

To illustrate the distinction, consider this situation: A person in a fishing boat catches a large shark just off a beach used by swimmers. Should he or she release the shark because the fish has a right to live? Should the shark be killed out of concern for the human swimmers? That sort of judgment is one of values. It is a moral question, not a question for science. An ecologist (while acting as a scientist) would simply look at the shark as a large predator and an important part of the food web. To the ecologist, the question of killing the shark or releasing it would probably be concerned with the place of the shark in the ecosystem and whether there were too many or too few sharks for the food web to support. The role of science in this situation would be amoral—that is, non-value-oriented.

This does not mean a scientist cannot also be an environmentalist. When a scientist stops using science and starts advocating environmentalist positions on political issues, he or she is not talking as a scientist but rather as an environmentalist—and perhaps a well-informed environmentalist.

Principles of Ecology

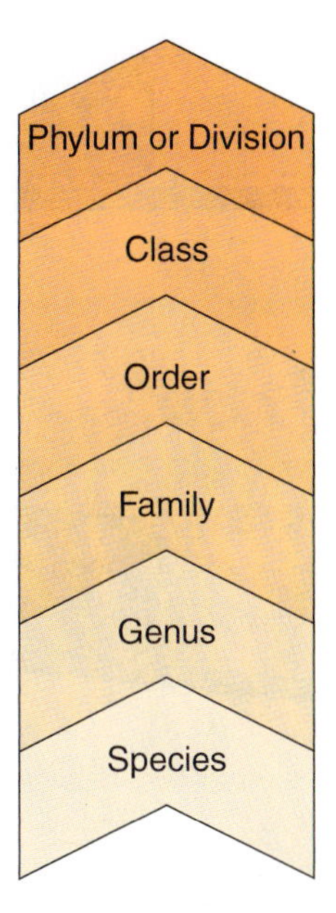

Figure 1–1 Taxonomy is a system that biologists use to organize living things according to their relationships to one another.

As students of ecology, it is important to understand the language and organization of this branch of science. All living organisms that are known to humans are classified into naturally-related groups in a systematic way. These are known as *taxonomic groups* and include the following divisions: species, genus, family, order, class, and phylum (animals) or division (plants) (see Figure 1–1). Animals or plants of the same species are closely related and exhibit many of the same characteristics. Those of the same order exhibit some of the same characteristics but are not considered to be closely related.

Scientists who study ecology organize living things into a different set of classifications. An **organism** is an individual plant, animal, or other living thing. A group of similar organisms that is found in a defined area is known as a **population** of plants or animals. A **community** includes all of the populations of organisms that live within a defined area such as a woodland, marsh, or cornfield.

An **ecosystem** is any partially self-contained environmental and living system—for example, a lake, a forest, a large valley, or a desert. It comprises the community of living organisms plus all of the nonliving features of the environment such as water, soil, rocks, and buildings. Each component of an ecological system or ecosystem affects other components of the ecosystem.

taxonomic groups
ecosystem, population, community

In a very real sense, we exist in an ecosystem. We depend on our environment for life itself. For thousands of years, that was no problem—our use of natural resources had little effect on our ecosystem. Today, however, our numbers are increasing, our technology is becoming tremendously powerful, and we have had ever-increasing impacts on our ecosystem.

Relationships also exist between two or more ecosystems. For example, soil erosion in Canada has an effect on the ecosystem from which the soil was lost, but it also affects survival rates of aquatic organisms in the river systems and in the freshwater and saltwater ecosystems where the silt is deposited. When all of the ecosystems of the Earth are considered as a whole, we call it a **biosphere** or **ecosphere** (Figure 1–2).

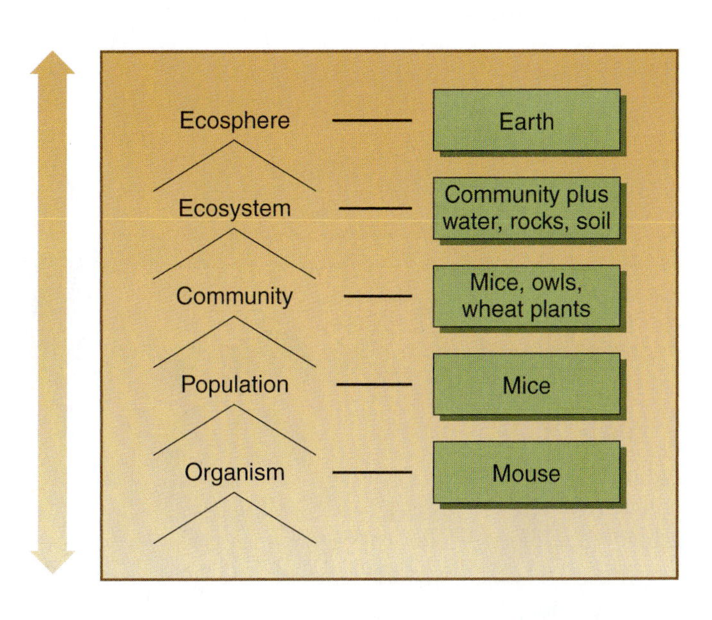

Figure 1–2 Ecologists organize living organisms and their environments into groups to make it easier to define and understand their relationships.

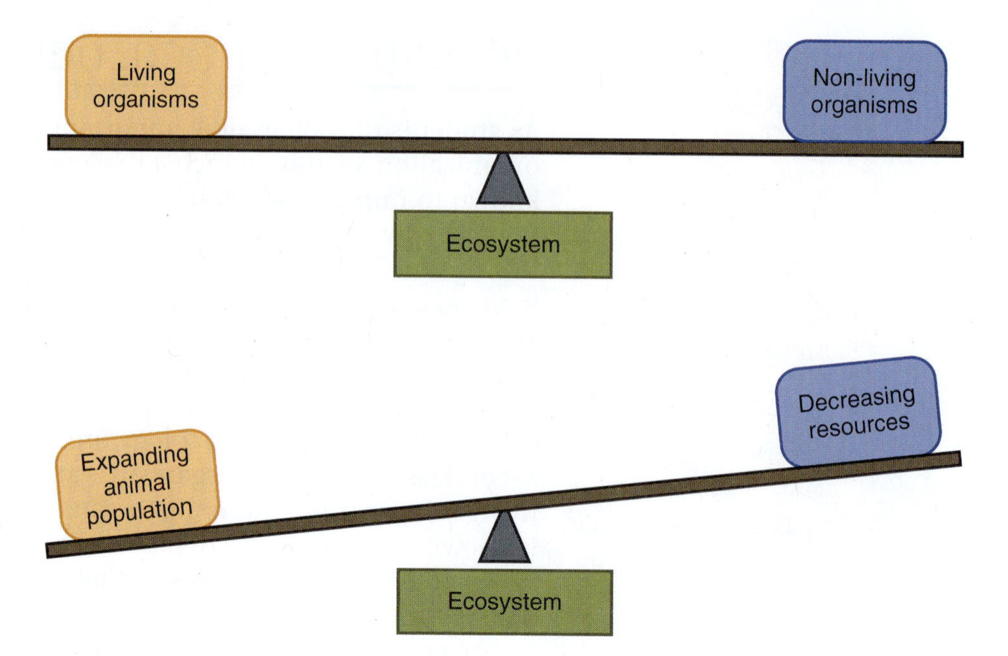

FIGURE 1–3 In theory, a balanced ecosystem would exist when living organisms and nonliving resources are maintained at constant levels. In practice, absolute balance seldom occurs.

FIGURE 1–4 An unbalanced ecosystem exists when the community of living things uses up nonliving resources faster than they can be replaced.

When an ecosystem supports all of the living community within it, and nonliving resources are maintained at constant levels, it is said to be balanced (Figure 1–3). A balanced ecosystem is rare, however, because the slightest change in any component of the ecosystem will affect all of the other components. If a population of organisms increases more rapidly than a resource is replenished in the system, it will eventually deplete the resource and cause the death of the other organisms. Such an ecosystem is unbalanced (Figure 1–4).

The term **terrestrial** applies to organisms that live on land; **aquatic** applies to organisms that live in water environments. Almost all organisms that inhabit Earth's terrestrial and aquatic environments have some common needs: sunlight, water, and nutrients from the soil (Figure 1–5). Plants depend on nutrients and moisture from the soil along with sunlight to supply their energy needs. All animals and insects derive their energy from the food they eat. Some eat plants as all or part of their diets, but even those animals and insects that eat diets of meat derive their energy from the plants that were eaten by their prey (Figure 1–6).

ecosystem, food chain
surface water, silt, erosion
water, soil, pollution

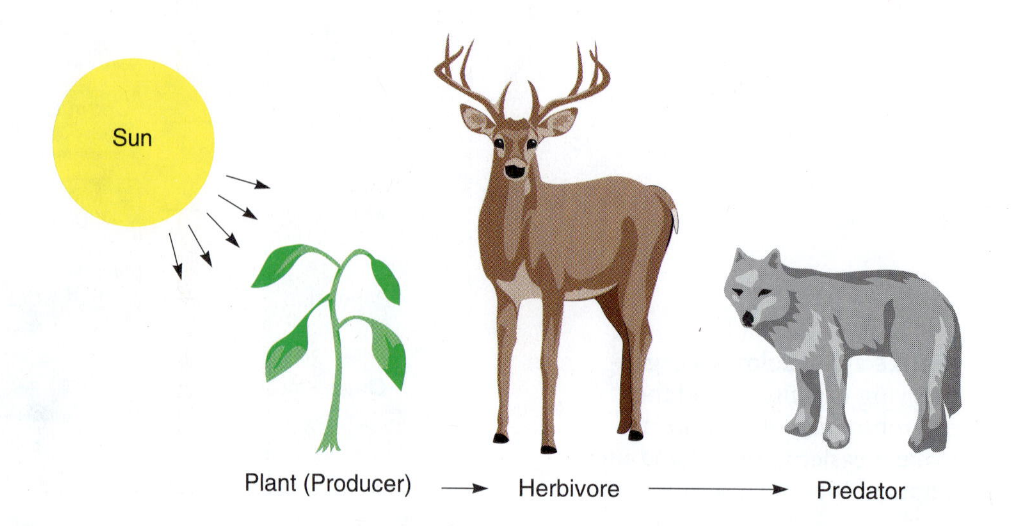

FIGURE 1–5 The energy flow in an ecosystem proceeds from sunlight to plants to herbivores and finally to carnivores.

science connection profile

BIOSPHERE II

Biosphere II is an experimental human-made environment in the Arizona desert. Its original purpose was to see whether humans could create a controlled, balanced, and sustainable living environment. It was sealed to prevent the movement of air or water into or out of the facility. As an experimental prototype of an environment to sustain human life on the moon or another planet, Biosphere II contained many scientific features to produce food and oxygen for people and carbon dioxide for plants. Water was recycled by plants and through evaporation.

As the first real attempt to create an artificial ecosphere, the original plan was for the people who lived in Biosphere II to remain sealed inside the structure for two years. Shortly after the structure was sealed in 1991, however, one person was removed because of an injury. Controversy followed the return of this person to the system following medical care. It has been alleged that additional supplies were taken into the system at that time. Several scientists contend that this invalidated the original scientific study.

After several months of operation, it became evident that the environment was unbalanced. Oxygen levels in the closed environment became too low, and the people living in the biosphere began to suffer from oxygen deficiency. Many of the birds, animals, and insects that were supposed to live in the environment died, except for cockroaches and ants. Eventually, oxygen was added to the environment to raise the oxygen back up to a level that was considered safe. During the remainder of the experiment, the people raised food for themselves and their animals using the resources that were available within the biosphere. This experiment was a good learning experience that illustrated the difficulty of creating a stable environment.

Biosphere II (Courtesy of Shutterstock.)

Columbia University managed the facility from 1995 to 2003 as part of the university's Earth Institute. Since mid-2007, the University of Arizona has operated Biosphere II as a research facility.

FIGURE 1–6 The polar bear depends on the ocean to provide seals, its major source of food. (Courtesy of the U.S. Fish and Wildlife Service. Photo by Dave Olson.)

FIGURE 1–7 Most of the soil erosion in a forest occurs immediately after intensive logging operations or forest fires. It is important that ground cover is quickly reestablished to protect the soil.

POLLUTION

Another serious problem that affects most and perhaps all of the environments in the world is pollution. **Pollution** is a condition in which a harmful material comes in contact with a life-sustaining resource such as water or soil. In many instances, the organisms that depend on the resource to sustain their lives are injured or damaged. Among the restrictions that humans introduce into natural environments are the substances that we make. Some of these materials are not easily degraded when they are no longer useful. On the other hand, waste materials generated by plants and animals seldom become serious problems in nature because they cycle back to the soil to be used again.

One of the great tragedies of our time has been the loss of significant amounts of topsoil (Figure 1–7). Most of this soil has been carried to lakes and oceans where it lies in great deposits of silt beneath the water. Water that is found in springs, streams, rivers, reservoirs, ponds, lakes, and oceans is known as **surface water**. Silting is the leading cause of surface water contamination, and serious damage is also sustained by the soil. For many years, it was a common practice to clear forests and native vegetation from the land and to produce crops until the soil was no longer fertile or until erosion of topsoil made farming unprofitable. Many farms were then abandoned, particularly in the Atlantic coast and southern regions of North America. Eventually, some of these lands have been reclaimed by native plants, but the damaged soils remain. The practice continues today throughout the rain forests of the Amazon river basin.

INTERNET KEY WORDS

industrial waste, pollution
landfills, water pollution

Every segment of society must become accountable for the waste materials that it creates. The agriculture industry must assume full responsibility for the chemicals and other toxic materials that are by-products of production as well as the waste materials it produces while finding ways to produce adequate supplies of food and fiber. **Industrial waste** also has been a serious environmental problem for many years (Figure 1–8). Waste materials include a variety of harmful chemicals, poisonous metallic compounds, acids, and other caustic materials that are by-products of mining or industrial processes and manufacturing activities. In addition, cities and towns have become major sources of pollution. Excess chemicals from lawns and gardens seep into the drainage systems along with used oil, antifreeze, paint, and many other hazardous substances. Much of this material flows back to rivers and streams, frequently bypassing waste-treatment facilities.

FIGURE 1–8 Mine tailings and industrial wastes are responsible for some serious pollution problems that harm natural resources.

Many waste materials can be changed to reduce or eliminate pollution. **Solid wastes** include most of the materials that are gathered by municipal trash and garbage collectors for disposal. Much of this trash is buried in landfills. In recent years, we have learned that highly toxic liquids often ooze out of the landfills and pollute local water and soil (Figure 1–9). Materials buried in landfills tend to remain for long periods of time and do not break down as quickly as many people expected. Much of it will still be there centuries from now in much the same form as when it was buried.

Some communities burn the combustible portion of their solid waste as a source of energy after separating out metals, glass, and plastic materials for recycling (Figures 1–10A and B). Although this approach is

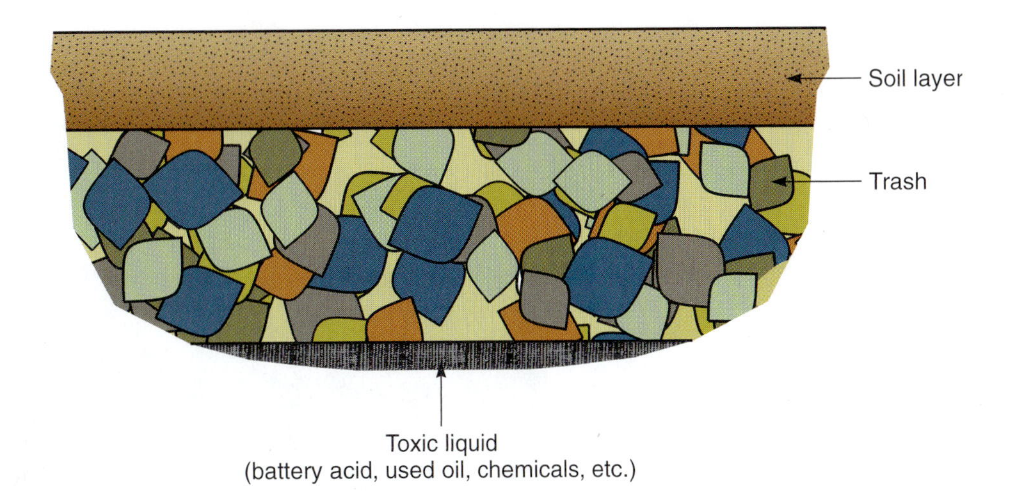

FIGURE 1–9 The layers of a landfill.

considered to be a good alternative to burying garbage, better solutions for handling solid wastes in all communities are still desirable.

Care must be taken to avoid creating air pollution from the burning process, which releases large amounts of carbon monoxide and sulfur dioxide into the air along with ash and other gases. Many of these pollutants are harmful to people, animals, and plants in the surrounding area. New technologies have been developed that are effective in trapping pollutants, but many of these processes are expensive and their use is limited.

pesticides, types
pesticides, harm
pesticides, benefits

Pesticides are chemicals that are used to control pests such as insects, rodents, or weeds. When a pesticide is used only to control insects, it is called an **insecticide** (Figure 1–11). A pesticide that is used to control plants is called an **herbicide**. A **rodenticide** is used to poison

FIGURE 1–10A Plastic is one of the many products that we currently recycle. (Courtesy of Getty Images.)

FIGURE 1–10B Metals can be recycled again and again. Old cars, cans, and other items made of metal can be melted down more efficiently than new ore can be refined. (Courtesy of Getty Images.)

FIGURE 1–11 Chemicals that control weeds and pests can be dangerous to the environment. They must be used and stored properly. (Courtesy of the U.S. Department of Agriculture.)

rodents such at mice, gophers, and rats. Large amounts of pesticides are used each year on lawns, gardens, golf courses, and farms to control unwanted species of plants, insects, and rodents. All of these materials can be dangerous to the environment when they are improperly used.

Empty pesticide containers are dangerous when proper disposal methods are not used. Most pesticides are sold in metal or plastic containers. Those who use these materials should carefully rinse empty containers with water and dispose of the container and the rinse water according to the directions on the label. Most empty pesticide containers are considered to be **hazardous materials**. Such materials are treated as threats to the environment, and their disposal is controlled by law.

INTERNET KEY WORDS

petroleum, environment
Exxon Valdez oil spill
underground fuel tanks, leakage

Petroleum is an oily, flammable liquid that occurs naturally in large underground deposits. In its basic form, we call it *crude oil,* and from this basic material a large variety of products are manufactured. Gasoline, diesel fuels, and heating oil are the best-known and most widely used products obtained from petroleum. When petroleum or petroleum products are spilled or leaked into the environment, they are often hazardous to the organisms that live there (Figure 1–12). Thousands of aquatic animals, particularly fish and waterfowl, die each year from spills of crude oil in surface waters.

FIGURE 1–12 A live, oil-covered goldeneye duck injured by a petroleum spill. (Courtesy of the U.S. Fish and Wildlife Service.)

Petroleum products are also damaging to the environment when they leak or spill on land. Leaking underground fuel tanks have polluted groundwater, contaminating drinking water and destroying plant and animal life. Poisonous fumes from petroleum spills are hazardous to the health of humans as well as other animals and also create potential fire hazards.

Environmental laws require the inspection of underground tanks to detect leaks. In some cases, leaks have been found after many years of environmental pollution (Figure 1–13).

Spills of this type are often difficult to find and expensive to clean up. Some states have even required that underground fuel tanks be removed to locations above the ground where they can be inspected more thoroughly for leakage.

interest profile

EXXON VALDEZ OIL SPILL

One of the most damaging environmental accidents in history occurred on March 24, 1989, off the coast of Alaska. The *Exxon Valdez*, a large oil tanker owned and operated by the Exxon Oil Corporation, ran aground, damaging the ship and spilling 11 million gallons of crude oil into the ocean at Prince William Sound. Despite a desperate effort to contain the spill, the oil soon became widely dispersed, covering a large area along the coast and killing massive numbers of birds, marine mammals, and fish. The ship's captain and owners were prosecuted for negligence that led to the spill, but received relatively little punishment. Even nearly 20 years later, we do not know the full extent of the damage to this ecologically sensitive area.

A spill of oil or any other chemicals in an environmentally sensitive area requires a massive cleanup effort. (Courtesy of the U.S. Fish and Wildlife Service.)

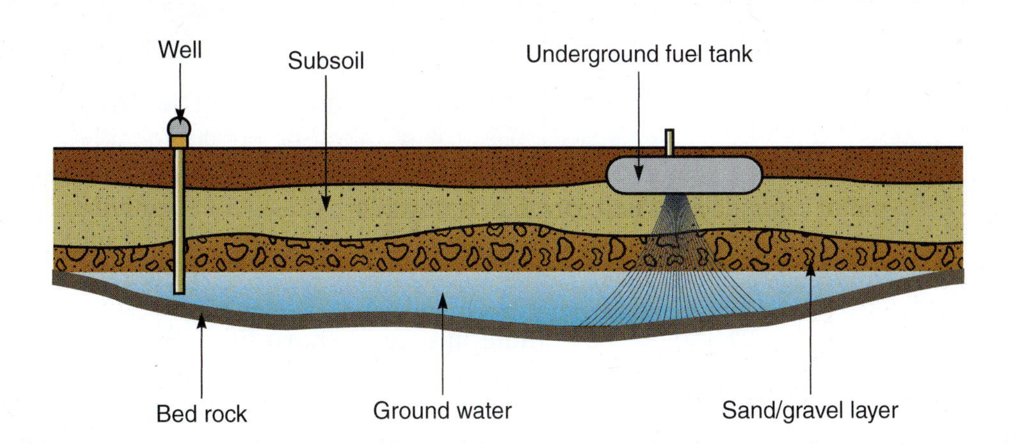

FIGURE 1–13 Underground fuel tank leakage.

INTERNET KEY WORDS

forest, erosion, clear cut
acid rain, cause, effect
law of conservation of matter

Much of our surface water is polluted by industrial wastes. Most of the pollutants that affect surface water also create problems in **groundwater**, or the water that is located beneath the soil surface. It is stored in large naturally-occurring underground reservoirs where it occupies the spaces between soil particles such as sand, gravel, and rocks. A large underground reservoir is also called an **aquifer**. Once an aquifer becomes polluted, it becomes nearly impossible to clean it up. It is of utmost importance that we prevent pollution in the first place.

In addition to human-made pollutants, a serious pollutant in surface water is the silt that comes from soil erosion following a fire. Inadequate planning of logging activities such as massive unprotected clear-cuts and poorly constructed roads also contribute to soil erosion (Figure 1–14). Other pollutants that cause serious damage to forest environments include **acid precipitation** or acid rain and chemicals from mine tailings. Much of the air pollution that is found in North America comes from automobile exhaust and the by-products of burning. The burning process releases large amounts of carbon monoxide and sulfur dioxide into the air along with other gases and ash. Each of these

FIGURE 1–14 Poor construction or improper use of forest roads can cause serious silt pollution in streams and rivers during heavy rainfall.

Law of Conservation of Matter

Nothing is wasted in nature. Matter may change from one form to another, but it cannot be created or destroyed by natural physical or chemical processes.

FIGURE 1–15 The law of conservation of matter.

materials can be harmful to the health of plants, animals, and humans. Entire forests are at risk because of acid rain and snow, and life no longer exists in some rivers and lakes that have been polluted by waste materials from mining or industry.

New technologies have been developed to prevent pollution and clean up polluted resources, but there is still much that must be done. A basic law of physics is the **law of conservation of matter**. Nothing is wasted in nature. Matter may change from one form to another, but it cannot be created or destroyed by natural physical or chemical processes (Figure 1–15). This law of nature applies to all living and nonliving resources. The law of conservation of matter applies as much to waste materials as it does to anything else. We may change the form of waste materials to make them more compatible with the environment, but we cannot destroy them. With this in mind, we must properly dispose of all waste materials to prevent serious environmental problems.

Conservation of Natural Resources

INTERNET KEY WORDS

preservationist, resources

The use or nonuse of natural resources for the purpose of protecting them has been the topic of much debate. Some policy makers—**preservationists**—are of the opinion that natural resources should be preserved from use by humans. To some degree, they have prevailed by

Interest Profile

USE OF SNOWMOBILES IN YELLOWSTONE PARK

One of the most hotly contested issues with regard to protected lands is being played out in the court system. For many years, a tourism industry has developed in Wyoming, Montana, and Idaho wherein winter tours have been conducted in Yellowstone Park using snowmobiles to access the area. During President Bill Clinton's administration, a court order was handed down that was intended to control the noise level of snow machines and gradually reduce the number that are allowed to enter the park. The court order was successfully appealed, allowing winter visitors to continue their visits to the park. The case remains open, pitting environmentalists against business owners in the region.

having helped set aside lands for national parks, national monuments, and wilderness areas in which timber harvests and some other human activities are no longer allowed.

Of course, even preservationists have trouble agreeing on which resources should be put aside and preserved from human use. There is debate among them over how much of our natural resource base should be protected from human uses other than appreciating these areas for their natural beauty or protecting them as critical habitat for endangered or threatened species of plants and animals. There are also many opinions concerning which human activities should be allowed on preserved lands.

A more moderate view of natural resource management is practiced by **conservationists**. These individuals subscribe to a plan that makes use of natural resources while taking steps to protect them from damage resulting from overuse and abuse. With this intent, soil and water conservation districts have been established throughout the United States and Canada. Each district is governed by a local board of directors. One purpose of soil and water conservation districts is to acquire funding for sharing the costs of implementing projects that conserve land, improve water quality, promote conservation through education, and provide technical natural resource services.

Interest Profile

MISSION STATEMENT OF THE SOIL AND WATER CONSERVATION DISTRICT, GASTON COUNTY, NORTH CAROLINA

To provide leadership and conservation assistance to the people of Gaston County to improve and sustain their soil, water, air, plant, and wildlife resources.

Looking Back

Ecology is the branch of science that describes relationships between living organisms and their environments. It includes all of the activities that affect the environment and the organisms living within it. An ecologist is a scientist who works in the science of ecology. An environmentalist is a political activist with a special interest in some aspect of the environment. Organisms are organized into populations, communities, and ecosystems that overlap and react with each other. Pollution of the environment by waste materials is one of our greatest problems as a society. Pesticides must be applied according to instructions on the container label because they can be dangerous to the environment when proper use directions are not followed. When they are thrown away, they become pollutants.

The law of conservation of matter states that the form of matter can be changed, but matter cannot be created or destroyed by ordinary physical or chemical processes. Humans are capable of changing matter into a variety of materials that are not found in nature. Environmental scientists spend their careers solving problems such as pollution and studying ways to maintain or restore the environment.

Self-Evaluation

Discussion and Essay

1. Define environmental science and list some issues with which it is concerned.
2. How is a scientist whose specialty is ecology different from an environmentalist?
3. How do scientists organize living organisms to aid their study of ecology?
4. How is a biosphere or ecosphere different from an ecosystem?
5. What factor determines whether an organism is terrestrial or aquatic?
6. Why is the pollution of natural resources considered to be a serious problem?
7. To maintain or improve environmental quality, what are some ways in which waste materials might be disposed?
8. What benefits and risks are associated with pesticide use on farms, lawns, roadsides, and gardens?
9. In what ways are industrial wastes dangerous to the environment?
10. What is the law of conservation of matter?
11. How is the pollution of the environment related to the law of conservation of matter?
12. In what ways might the science of ecology be of benefit to the world in the twenty-first century?

Multiple-Choice Questions

1. The branch of science that deals with relationships between organisms and the environments in which they live is
   a. zoology.
   b. ecology.
   c. biology.
   d. physics.
2. A scientist who studies relationships between living organisms and their environments is known as a(an)
   a. taxonomist.
   b. ecologist.
   c. zoologist.
   d. limnologist.

3. A group of similar organisms that is found in a defined area is known as a
   a. population.
   b. community.
   c. family.
   d. herd.
4. All of the ecosystems of the Earth when they are considered together are known as the
   a. ecosphere.
   b. community.
   c. population.
   d. atmosphere.
5. Which of the following organisms is aquatic?
   a. wolf
   b. moose
   c. hawk
   d. fish
6. Which of the following is not classified as surface water?
   a. a river or stream
   b. a pond
   c. a lake
   d. an aquifer
7. Which pesticide is used to control weeds?
   a. insecticide
   b. herbicide
   c. rodenticide
   d. fungicide
8. Which of the following is not a petroleum product?
   a. silt
   b. diesel
   c. heating oil
   d. gasoline
9. What is the name of the basic law of physics that states "Matter can be changed from one form to another, but it cannot be created or destroyed by ordinary physical or chemical processes"?
   a. the first law of energy
   b. the second law of energy
   c. the law of conservation of matter
   d. the law of kinetic energy

Learning Activities

1. Conduct a public awareness campaign to encourage all users of agricultural and industrial chemicals to dispose of waste materials in a legal and safe manner.
2. Develop an environmental research project to study the waste-disposal system that is used in your area. Identify actual and potential pollution problems that are associated with each of the ways the different waste materials are disposed. Invite local sanitation authorities to visit your class to discuss local pollution problems. Develop possible solutions for each local waste disposal problem that is identified.

TERMS TO KNOW

biotic
abiotic
producer
transformer
decomposer
biological synthesis
photosynthesis
respiration
decomposition
energy
first law of energy
radiant energy
chemical energy
kinetic energy
thermal energy
electrical energy
second law of energy
elemental cycle
carbon cycle
fossil fuels
nitrogen cycle
nitrogen fixation
nitrogen-fixing bacteria
denitrification
water cycle
transpiration
food chain
herbivore
primary consumer
secondary consumer
carnivore
food web
food pyramid
biological succession
primary succession
secondary succession
pioneer
climax community
niche
competitive advantage
competitive exclusionary principle
range of tolerance

CHAPTER 2

The Science Behind Environmental Science

Objectives

After completing this chapter, you should be able to

- distinguish between biotic and abiotic subsystems
- define the role of energy in the science of ecology
- explain the first law of energy
- explain the second law of energy
- describe the major events that occur in natural cycles such as the carbon, nitrogen, and water cycles
- investigate the importance of water to living organisms
- explain how a food chain is organized
- distinguish among food chains, food webs, and food pyramids
- define the terms ***producer***, ***herbivore***, and ***carnivore***
- identify the characteristics of primary and secondary consumers
- describe the characteristics of primary and secondary biological succession
- list differences between plants that are pioneers and those that make up a climax community
- explain the competitive exclusion principle
- predict the biological succession of organisms in a particular environment.

INTERNET KEY WORDS

law of conservation of matter
human waste environment
toxic waste disposal

Environmental science draws on many other sciences to gain an understanding of how the world works. It is an applied science that seeks solutions to problems in the environment by applying principles of science and mathematics. Basic science principles are addressed within this chapter that must be kept in mind as we explore ways to solve problems in the environment.

Conservation of Matter

As noted in Chapter 1, a basic law of physics is the law of conservation of matter: Matter can be changed from one form to another, but it cannot be created or destroyed by ordinary physical or chemical processes. This law holds true in the study of ecology.

The law of conservation of matter applies to everything that exists. Most organisms use only those materials that make up their food supply, and little waste material is generated. Any generated waste is capable of being recycled through natural processes. The human population, however, generates waste materials that are not so easily disposed of or recycled. Some of the materials that humans create persist in the environment almost indefinitely, and some of these waste materials are harmful or toxic to other organisms.

Waste materials have become a major problem in many of the world's industrialized nations: petroleum leaks and spills, by-products of industrial processes, solid wastes from our population centers, and pesticide residues from farms, gardens, and yards, to name just a few (Figure 2–1). These materials do not just go away when we are finished with them.

The law of conservation of matter applies also where waste materials are concerned. We may change the form of waste materials to make them more compatible with the environment, but we cannot destroy them. With this in mind, we must properly dispose of all waste materials, use fewer materials that are disposable, and recycle waste materials whenever possible (Figure 2–2).

FIGURE 2–1 Surface water pollution caused by oil or chemical spills is a major problem and causes serious damage to the environment.

FIGURE 2–2 Waste materials must be properly disposed of to prevent serious environmental damage.

The Science of Ecosystems

INTERNET KEY WORDS

ecosystem
biotic
abiotic
ecosystem
producer
transformer
decomposer

As discussed in Chapter 1, an ecosystem is a given set of organisms, organic residues, physical and chemical components, and conditions (e.g., light and temperature) that interact and transfer energy and matter in form and location. Ecosystems consist of **biotic** (living) subsystems as well as **abiotic** (nonliving) subsystems. An example of a biotic subsystem is the relationship among the plant and animal members of a food web or food chain. An example of an abiotic system is the water of a lake and the chemicals that dissolve from the atmosphere and land and affect the water's acidity level.

In its most basic sense, an ecosystem is an energy system. All the processes in an ecosystem depend on energy. In fact, we can say accurately that nothing happens in an ecosystem without the flow of energy. Every part of an ecosystem interacts with the other parts of the system and depends on them. Fish in a lake use oxygen from the water that is dissolved from the atmosphere. The plants in the lake use light from the sun and minerals from the lake bottom to grow.

To have a complete ecosystem, three component groups must be present: producers, transformers, and decomposers. **Producers** are basically green plants that produce new food (sugar) by means of photosynthesis. **Transformers** can take that primary source of food, incorporate other chemicals and energy forms, and change it into more complex organic compounds, foods, and tissue. **Decomposers** break the organic materials back down into their constituents for reuse in the ecosystem. One organism can be a member of more than one component group. For instance, a green plant can be both producer and transformer. A fungus can be both transformer and decomposer, as are animals.

INTERNET KEY WORDS

ecosystem, biological synthesis

Four fundamental biotic processes go on in ecosystems: synthesis, photosynthesis, respiration, and decomposition. The basic process is that of **biological synthesis**, or any change in the composition, shape, size, or structure of the plants or animals in the ecosystem. Examples are

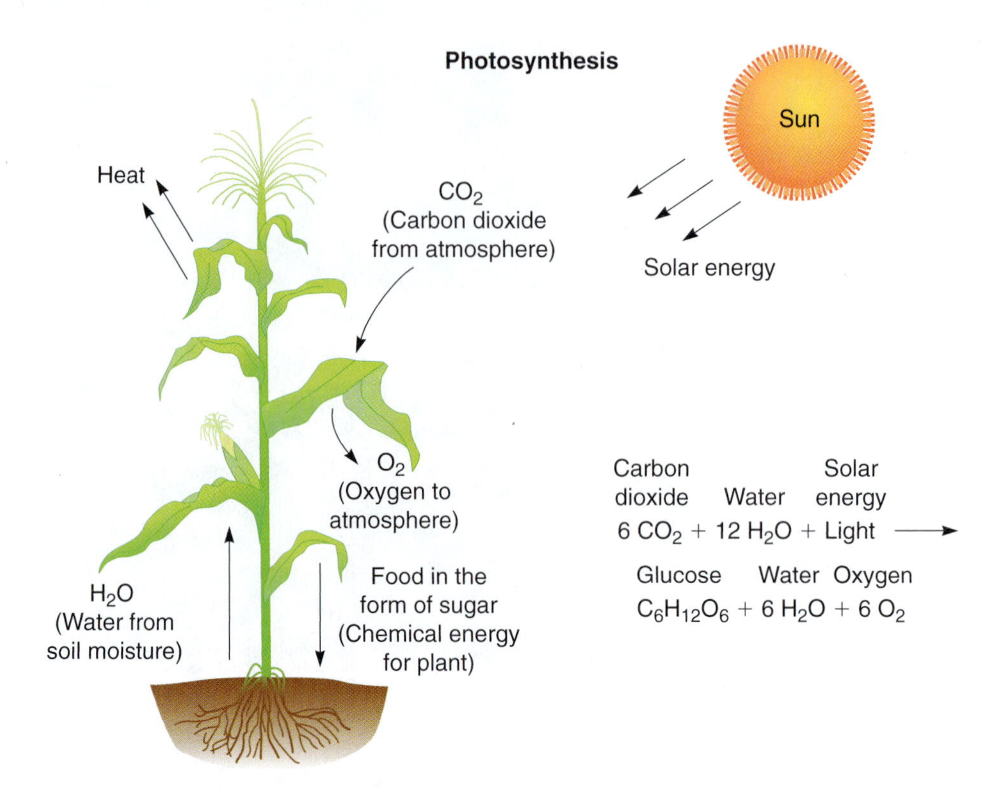

FIGURE 2–3 Photosynthesis is the process by which carbon dioxide is combined with water in order to store energy obtained from sunlight. Chlorophyll supports this reaction in which sugar and oxygen are produced.

photosynthesis
ecosystem respiration
decomposition process

the use of nutrients, minerals, and water to produce growth and reproduction in plants and animals.

In **photosynthesis**, plants convert water and carbon dioxide (CO_2) into sugar (Figure 2–3). The process requires the presence of a catalyst called *chlorophyll*, which generally gives the green color to healthy plants. It also requires energy from the sun energy that is incorporated into the sugar molecules. We have all heard that sugar is a "high energy" food. That is true because energy from the sun is stored in it. In fact, photosynthesis is the original source of almost all foods in the ecosystem. Photosynthesis also produces oxygen, which benefits all animals, including humans.

Respiration is a process that takes place within the individual cells of plants and animals. It involves the breakdown of foods into their components along with the release of energy (Figure 2–4). An example is the digestion of sugar into water and carbon dioxide, with the release of the stored energy being made available for use by plant or animal cells.

Decomposition is the process by which organic matter (plant or animal tissue) is reduced to organic compounds. Only by means of decomposition can the chemicals in plant and animal bodies be ready to recycle into the biotic subsystems in the ecosystem.

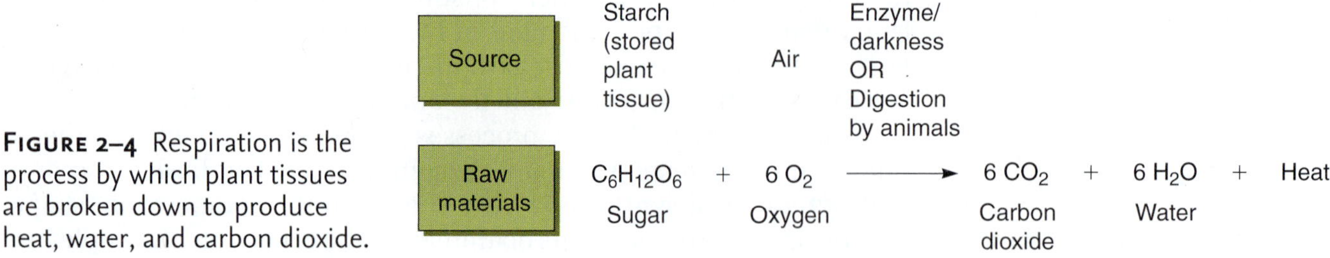

FIGURE 2–4 Respiration is the process by which plant tissues are broken down to produce heat, water, and carbon dioxide.

The Ultimate Concept in Ecology

An ecosystem can be defined in many different ways. In one sense, a terrarium in your classroom is an ecosystem. In another sense, the classroom itself makes up an ecosystem and the terrarium is simply a part of that ecosystem. In yet another sense, your whole school is part of an ecosystem that could be defined in geographic terms.

That brings us to what could be called the "ultimate concept in ecology." Everything on Earth is part of one or more ecosystems. In each system, if something happens to one part, it affects some or all the other parts of the system. The effects are often unpredictable and may be extreme.

Laws of Energy

Energy is the ability to do work or to cause changes to occur. It is the power or force that enables animals to move or the tides to flow. Energy flows through systems from areas where it is concentrated to areas where it becomes dispersed or unorganized. For example, food is the source of energy for animals. As it is digested, it gives off heat to warm the body of an animal, and it provides power to the muscles, making movement possible. Even the capacity of the brain to think depends on a supply of energy.

Energy also flows through entire ecosystems (Figure 2–5). Approximately two-thirds of the solar energy that passes from the sun to the Earth is trapped by land, water, plants, and atmosphere.

Solar energy in the form of sunlight is captured by plants and is stored as molecules of sugar and starch. When a plant is eaten by an animal, the energy is released from the plant cells, allowing the animal to do work, or it is stored in the animal's body in the form of fats, proteins, or carbohydrates until it is needed.

Energy cannot be recycled, but it can be stored for later use. When plants or animals die, the energy may be stored in the form of fossil fuels (coal or oil deposits) that store energy for long periods of time. It may also be released to the environment in the form of heat as the remains decompose. Sometimes energy is transferred to other animals when the remains of dead plants and animals are eaten and digested for food.

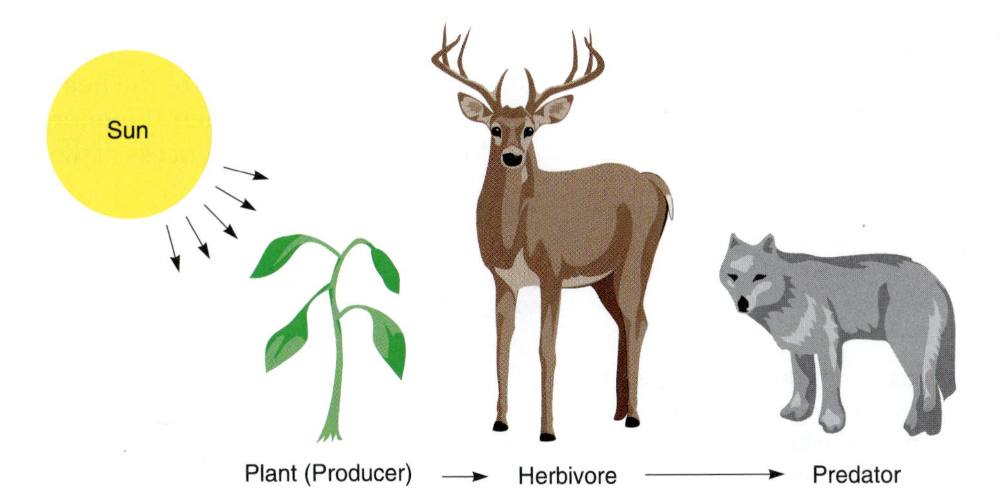

FIGURE 2–5 Energy flow in an ecosystem.

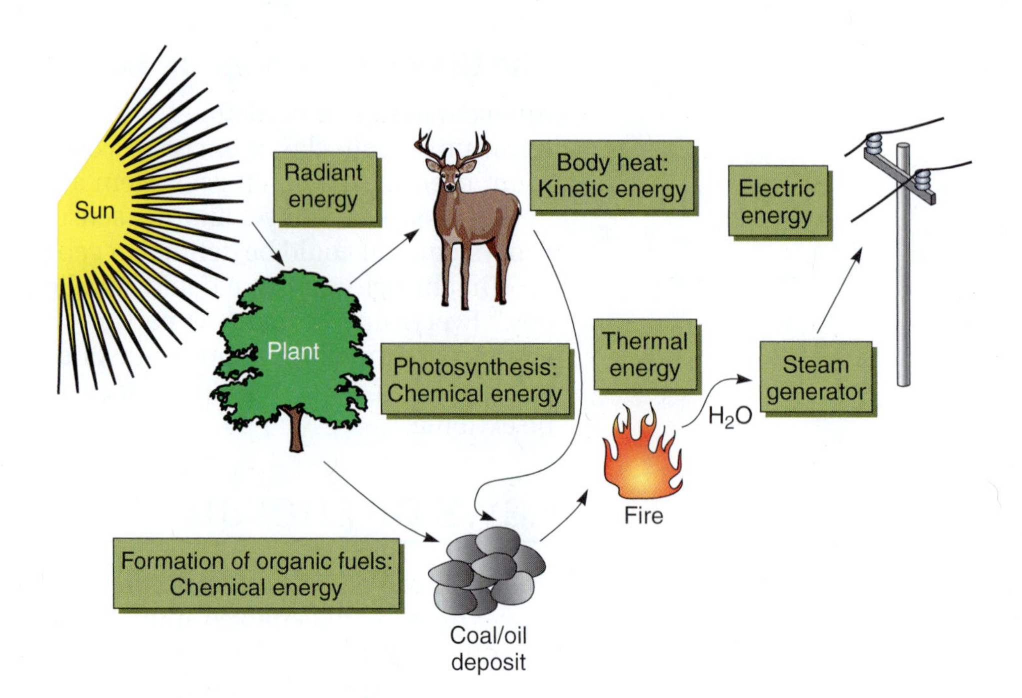

FIGURE 2–6 The first law of energy.

first law of energy
radiant energy
chemical energy
kinetic energy
thermal energy
electrical energy

The First Law of Energy

The **first law of energy** states that energy cannot be created or destroyed, only converted from one form of energy to another. For example, **radiant energy**, which comes from the sun, is converted to **chemical energy** (sugars and starches) by plant leaves during the process of photosynthesis. Chemical energy from a plant that is eaten by an animal is converted into body heat and **kinetic energy**, the energy associated with motion and movement.

Large deposits of coal or crude oil have been formed from decayed plant materials. During the formation of these materials, energy is stored as chemical energy. **Thermal energy** is created in the form of heat when these fuels are burned. Thermal energy can be converted to **electrical energy** by heating water to operate a steam engine that uses kinetic energy to drive a generator (Figure 2–6).

The Second Law of Energy

The **second law of energy** states that every time energy is converted from one form to another, some energy is lost in the form of heat (Figure 2–7). The heat that is lost is not destroyed—it simply becomes unavailable for later use. Every single change in the form of energy results in the loss of heat. It is possible for some of the heat that is lost from living plants and animals to be trapped when the energy in the food is converted from one form to another. The process of digestion releases heat into the body

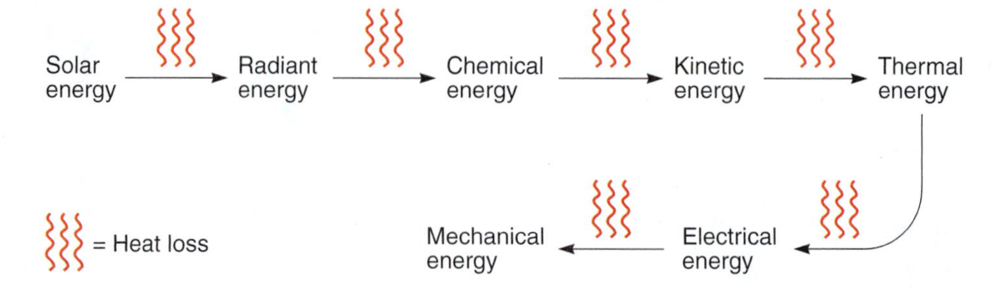

FIGURE 2–7 The second law of energy states that heat is lost each time energy is converted from one form to another.

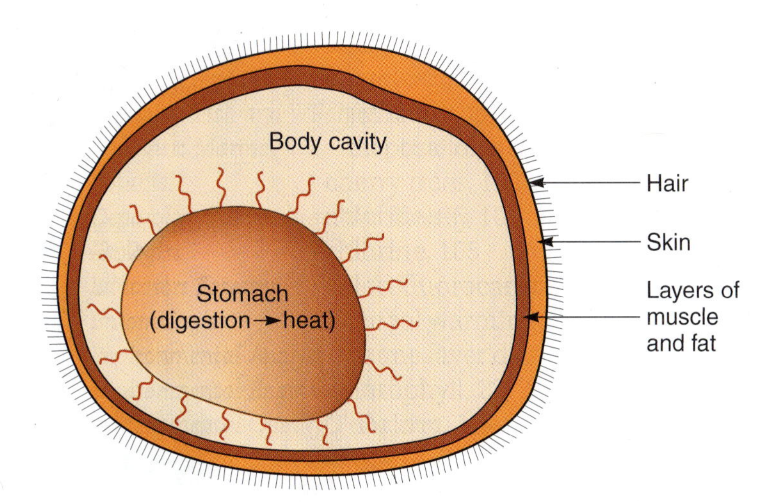

FIGURE 2–8 Maintaining body heat.

cavity where it is used to maintain body temperature in warm-blooded birds and mammals (Figure 2–8).

All living things respond to the laws of energy and all ecosystems on Earth depend on energy sources. Much of the controversy surrounding the energy conservation issue deals with this important principle: We cannot continue to use more energy than the Earth can replace without incurring serious consequences.

Natural Cycles

Only a few of the known elements that are found in the upper crust of the Earth, in water, or in the atmosphere are abundant in the tissues of living organisms. The most plentiful of these elements are carbon, hydrogen, oxygen, and nitrogen. These four elements account for 96 percent of the material that is found in living organisms. More than 30 other elements are known to make up the remaining 4 percent of living tissue.

These elements, which are so important in the formation of living organisms, are used over and over again. An atom of carbon may exist as part of a sugar molecule in a plant that is later eaten by an animal. It may then become part of the muscle of the animal. When the animal dies, the same atom of carbon may be passed into the soil or into the atmosphere when the tissue of the animal decomposes. Finally, it may be taken up by another plant to form new plant tissue. In this manner, elements cycle from living organisms to nonliving materials and back again. This circular flow of elements from living organisms to nonliving matter is known as an **elemental cycle**.

elemental cycle, carbon cycle

Cycles exist for all of the elements that make up living tissue. We now consider only the carbon and nitrogen cycles.

The Carbon Cycle

Carbon is the most abundant element found in living organisms. It makes up the framework of the molecules that are found in living tissue. In the absence of water, nearly half of the dry matter found in the bodies of animals or humans consists of carbon. The **carbon cycle** occurs as carbon atoms flow continuously from living organisms to the atmosphere, the oceans, and the soil (Figure 2–9). The respiration process of both plants and animals releases CO_2 into the atmosphere.

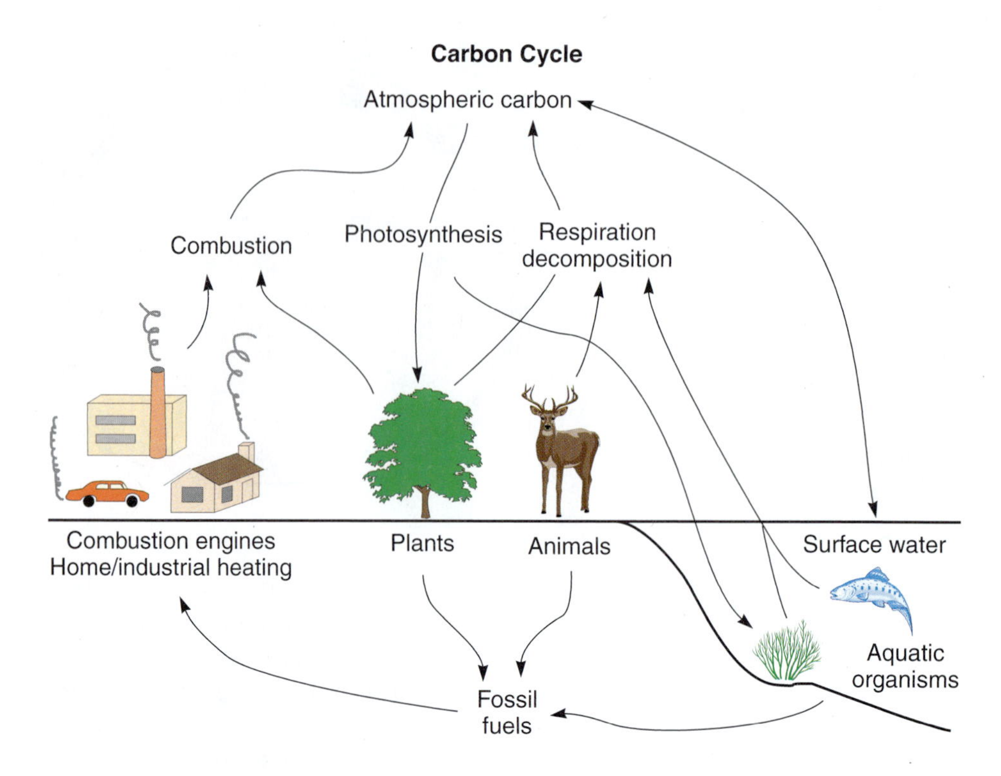

FIGURE 2–9 The carbon cycle.

fossil fuels,
carbon dioxide, ocean,
global warming

Using light during photosynthesis, plants take carbon dioxide from the atmosphere and turn it into sugars that are used to make new tissue such as roots, stems, and leaves. When plant tissue decays, carbon dioxide is released back to the atmosphere as a gas or it is converted over a long period of time to **fossil fuels** such as natural gas, crude oil, or coal. Sometimes, plant materials are eaten by animals. Similarly, when animals die, their bodies decompose, releasing carbon dioxide into the atmosphere as a gas, or the carbon from their bodies may be converted over long periods of time to fossil fuels. People mine or extract fossil fuels from the surface of the Earth for use as fuels and for other purposes. When these materials are burned, the combustion process also releases carbon dioxide to the atmosphere.

When the carbon content of the atmosphere is high, the oceans absorb large amounts of carbon dioxide and then release it to the atmosphere when atmospheric levels of carbon dioxide decrease. Until the last few decades, the atmosphere's carbon content has remained nearly the same because of this action by the oceans. Over the last 100 years, the burning of large amounts of fossil fuels has increased the levels of carbon dioxide in the atmosphere faster than the oceans can absorb the extra atmospheric carbon.

global warming, iron, algae

CO_2 and Global Warming

A heated debate is raging among scientists regarding global warming. One group warns that massive amounts of CO_2 being released into the atmosphere are resulting in global warming. Another group of scientists advise us that global warming is a concept that has not been proven using scientific evidence. They believe that temperature change is the result of natural cycles in the climate that are not well understood. It is certain, however, that long-term climatic change could have drastic consequences for natural and agricultural ecosystems.

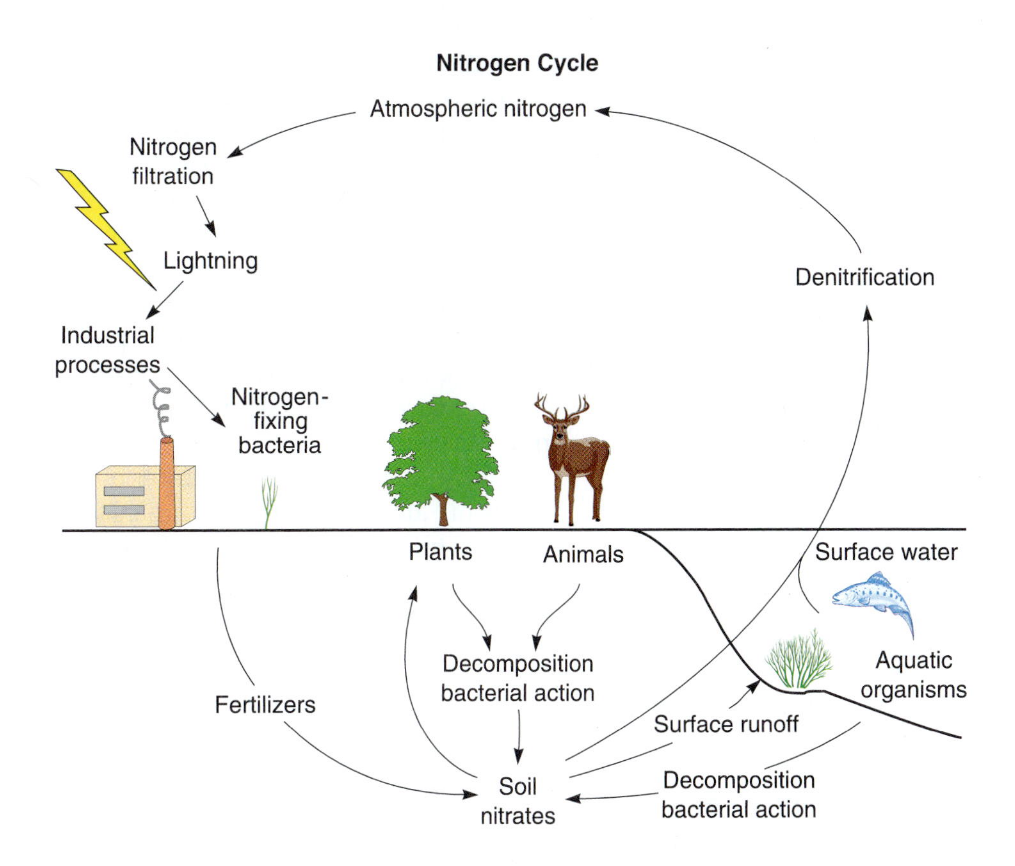

Figure 2–10 The nitrogen cycle.

If Earth's naturally occurring carbon cycle cannot accommodate massive additions of carbon to the atmosphere and oceans, can humans do anything about it? Research has demonstrated that seawater algae are capable of removing large amounts of CO_2 from the atmosphere. Adding iron as a nutrient to the water surface can stimulate algae growth. Scientists are divided, however, over what happens to the carbon that is absorbed by the algae. Some believe that it sinks to the ocean floor when the algae die. Others believe that it is lost back to the atmosphere. Some scientists worry that promoting algae growth on a massive scale could upset the delicate balance of the ecosystems of the oceans.

Figure 2–11 Nitrogen-fixing nodules on roots. (Courtesy of the U.S. Department of Agriculture's Agricultural Research Service.)

The Nitrogen Cycle

The circular flow of nitrogen from free gas in the atmosphere to nitrogen compounds such as nitrates in the soil and back to atmospheric nitrogen is known as the **nitrogen cycle** (Figure 2–10).

As the most abundant element in the atmosphere, nitrogen makes up approximately 80 percent of the air supply. In its elemental form (N_2), it is a colorless, odorless gas that cannot be used by plants or animals. It must be combined with oxygen or other elements before it becomes available as a nutrient for living organisms. Plants and animals use nitrogen compounds to form DNA and other important molecules such as proteins and vitamins.

Nitrogen fixation is a process by which nitrogen gas is converted to nitrates. This can occur in several different ways. **Nitrogen-fixing bacteria** are able to convert nitrogen gas to nitrates. Some forms of these bacteria live in the soil. Others live in nodules on the roots of clover, beans, peas, and other legumes (Figure 2–11). These types of plants are able to make their own nitrogen fertilizer. Some types of blue-green algae and fungi also are capable of nitrogen fixation.

Several industrial processes also convert nitrogen gas to nitrates. One process converts the gas to ammonia as a by-product of steel production; ammonia can also be obtained directly from natural gas. The ammonia is then converted to a form of nitrate that can be used for fertilizer, or it can be added directly to soil as a fertilizer.

Nitrogen fixation also occurs naturally in the atmosphere when lightning strikes. An electrical current passing through atmospheric nitrogen converts some of it to nitrogen compounds that can be used by plants. Nitrates also are released from animal wastes and dead and decaying plants and animals.

nitrogen cycle
denitrification

At the same time that nitrates are being produced from nitrogen gas, other nitrates break down to release nitrogen gas back into the atmosphere. This process of **denitrification** occurs when some forms of bacteria come into contact with nitrates. A similar process occurs when nitrates are carried by runoff water into surface water, which constantly exchanges nitrogen with the atmosphere.

The Water Cycle

nitrogen-fixing bacteria

Water is one of the most important resources in the environment (Figure 2–12). It provides a living environment for many species of organisms and supports the growth of plant life.

Water is a required nutrient for living things and makes up approximately 70 percent of the weight of living plants and animals (Figure 2–13). Living things require large amounts of water to survive. It controls the temperature of organisms, is a solvent for nutrients, and performs many other functions that maintain life. It dissolves nutrients and carries them to the tissues that need them. It stores heat and helps maintain a more constant temperature in the environment. Water action helps to create soil by breaking down rock into small particles. It also cleans the environment by diluting contaminants and flushing

Importance of water to the environment

- Soil formation.
- Living environment for plants and animals.
- Supports growth of plants.
- Dissolves and transports nutrients.
- Stores heat.
- Stabilizes temperature.
- Cools living organisms.
- Provides protection from natural enemies.

FIGURE 2–12 The importance of water in the environment.

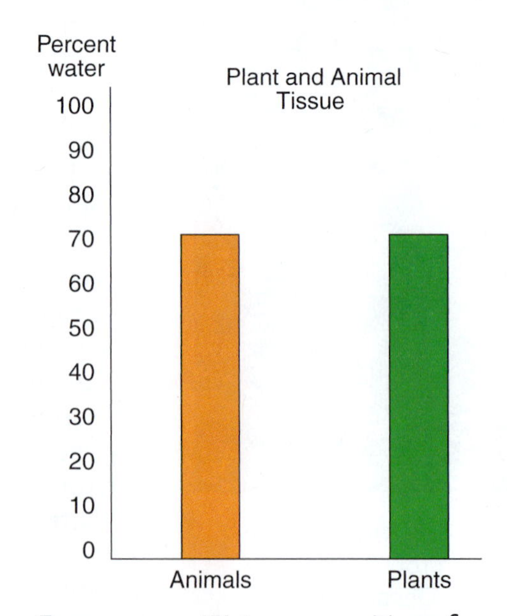

FIGURE 2–13 Water composition of living things.

FIGURE 2–14 Beaver pond, auxiliary dam, main dam, and lodge.

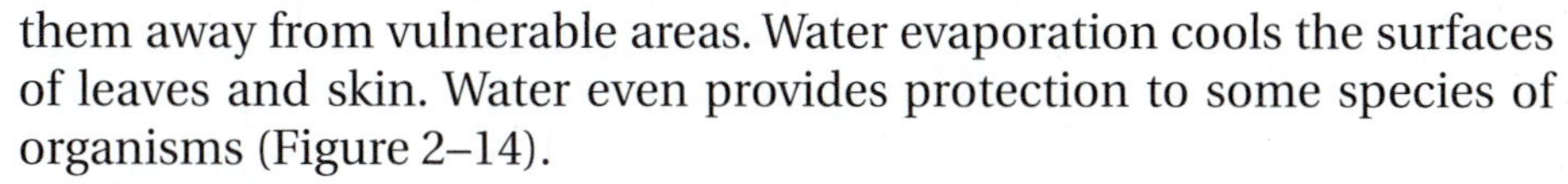

them away from vulnerable areas. Water evaporation cools the surfaces of leaves and skin. Water even provides protection to some species of organisms (Figure 2–14).

INTERNET KEY WORDS

water cycle

The **water cycle** occurs as water moves in the form of rain or snow from the oceans to the atmosphere, to the soil, to streams and rivers, and then back to the oceans (Figures 2–15 and 2–16). The energy that drives this cycle comes from two sources: solar energy and the force of gravity.

Water is constantly recycled. A molecule of water can be used over and over again as it moves through the water cycle. Solar energy is the source of heat that causes water to evaporate from the ocean. Additional water enters the atmosphere by evaporating from soil and plant surfaces, especially in areas of hot temperatures and high precipitation.

FIGURE 2–15 A watershed is an area in which rainwater and melting snow are absorbed. Some of the water will emerge from springs or from artesian wells at lower elevations.

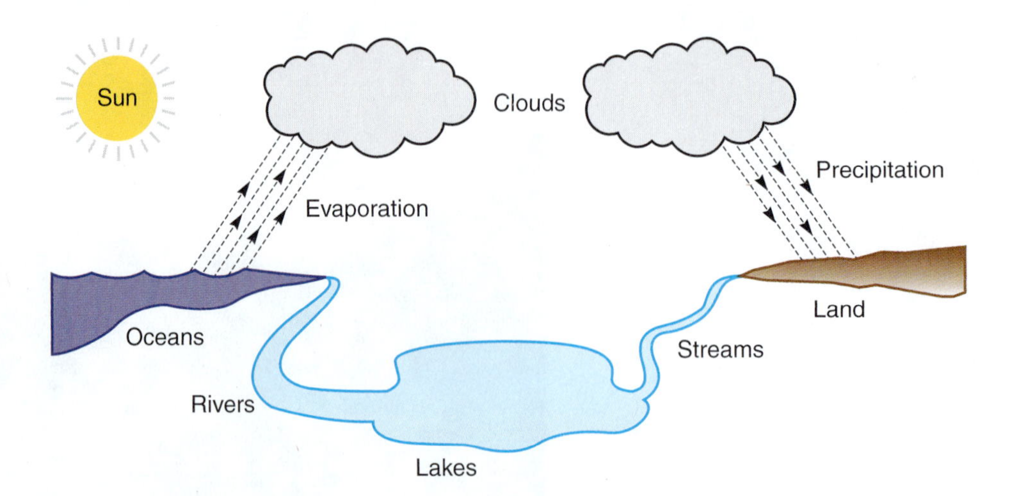

FIGURE 2–16 The water cycle.

water transpiration

Plants give up large amounts of water to the atmosphere through a process called **transpiration**. In this controlled evaporation process, plants lose water through pores in their leaf surfaces. These pores open wide when leaf surfaces are hot and close when leaf surfaces are cool. Transpiration also creates a negative pressure that moves water and nutrients into the plant from the soil (Figure 2–17).

All animals are also part of the water cycle. For example, mammals control body temperature by releasing water through their skin pores in the form of perspiration or sweat. Cooling takes place when the sweat evaporates from the skin's surface. Sweating in mammals occurs in much the same way that leaves transpire. Additional moisture is released to the atmosphere from the moist inner surfaces of the lungs (Figure 2–18).

Moisture from all of these sources builds up in the atmosphere and forms clouds, which release stored water down to the Earth's surface as rain or snow. Gravity causes the water to flow from high to low elevations. Large amounts of the Earth's water supply are stored for long periods of time in aquifers beneath the surface, in glaciers and in polar ice caps, in the atmosphere, and in deep lakes and oceans. Sometimes this water is stored for thousands of years before it completes a single cycle.

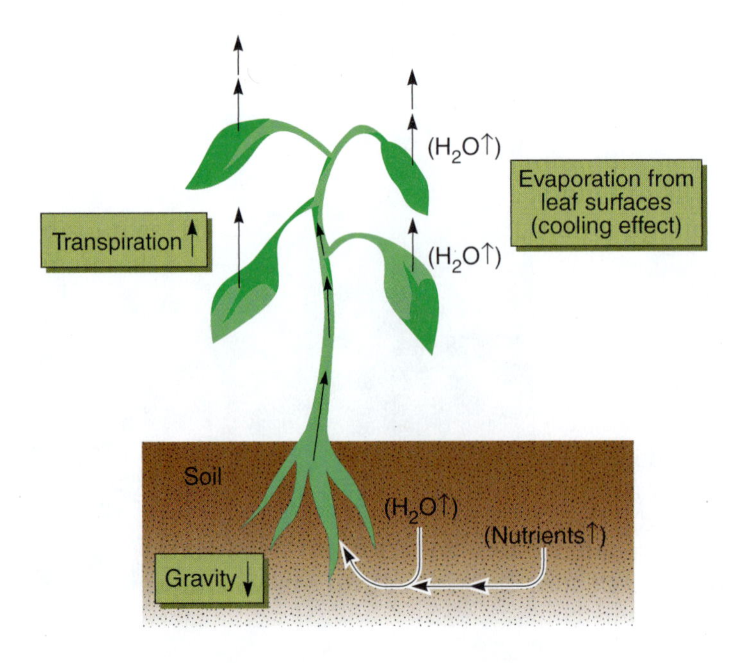

FIGURE 2–17 Effects of transpiration.

FIGURE 2–18 Cooling by evaporation occurs as water evaporates from the inner surface of the lungs during respiration.

Food chains

A **food chain** is made up of a sequence of living organisms that eat and are eaten by other organisms living in the community. Each member of the chain feeds on lower-ranking members of that chain. The general organization of a food chain moves from producer organisms (usually considered to be food plants) to **herbivores** (plant-eating organisms). These plant-eating organisms are also called **primary consumers**. **Secondary consumers**, or **carnivores**, are meat-eating organisms that eat primary consumers.

primary consumer, secondary consumer, carnivore

A typical food chain begins with a plant as a food source and ends with a large predator. For example, field mice eat the roots and seeds of meadow plants and are prey for raptors such as hawks and eagles and other predators such as foxes and coyotes (Figure 2–19).

FIGURE 2–19 An example of a food chain.

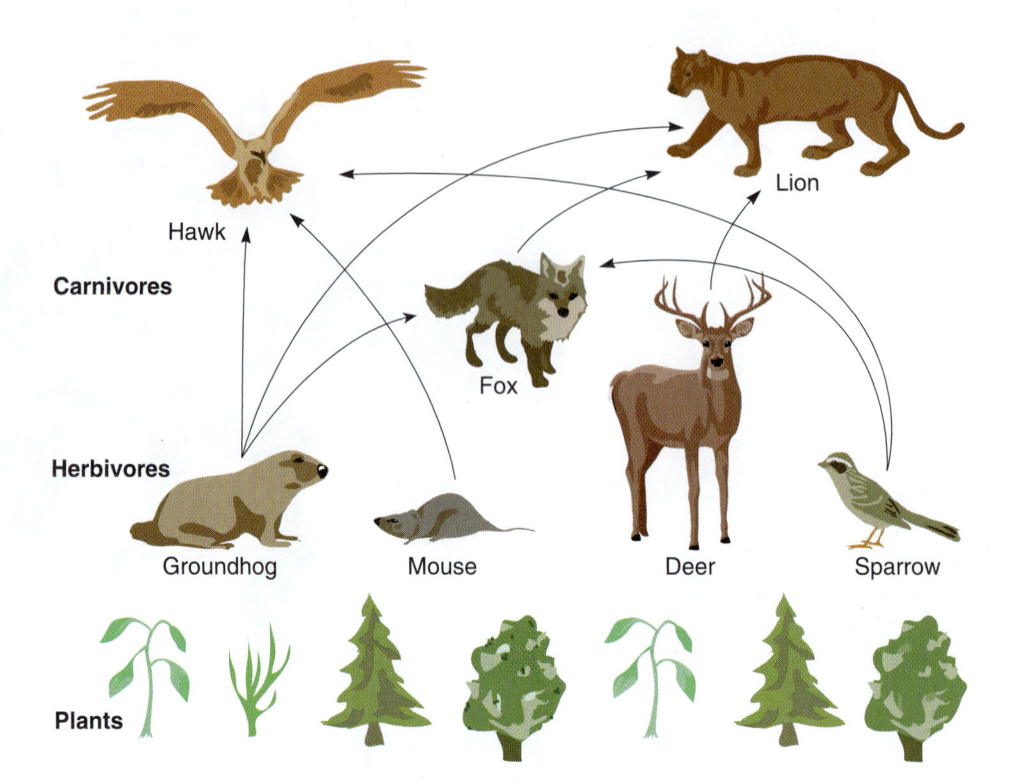

FIGURE 2–20 An example of a food web.

INTERNET KEY WORDS

food chain,
food web,
food pyramid

The final life form that participates in a food chain consists of bacteria. These bacteria are known as decomposers because they break down organic tissues from plants and animals that have died.

Most food chains become quite complicated because many predators will eat nearly any animal they can catch and kill. Each food chain is interwoven with other food chains to create a **food web** (Figure 2–20). A **food pyramid** arranges organisms in a ranking order according to their dominance in a food web.

The most versatile predators usually occupy the highest rank in a food web (Figure 2–21). These mammals or birds usually have few natural enemies, and they are capable of preying on a large variety of other species. They maintain their positions at the top of the food pyramid unless a stronger predator migrates into the area that is capable of competing more favorably for the existing food supply. Sometimes a new predatory species even preys on the species that previously occupied the highest rank in the pyramid.

When changes occur in the kinds of organisms that occupy an ecosystem, they affect nearly every other species in the ecosystem. The movement of humans into a new area has the effect of displacing the predators that occupy the top ranks in a food web. Humans assume these positions by preying on the herbivores in competition with the predators. They also control the size of the predator population by killing these animals or driving them out of the area.

FIGURE 2–21 A black bear is capable of preying on many other animals. (Courtesy of Getty Images.)

Humans and the Food Pyramid

Human dominance of the food pyramid has contributed to the controversy about natural environments and the movement to maintain and restore them (Figure 2–22). Extreme positions have been taken on both sides of the issue. Some people contend that human dominance over other species of organisms is a natural process that has evolved since the

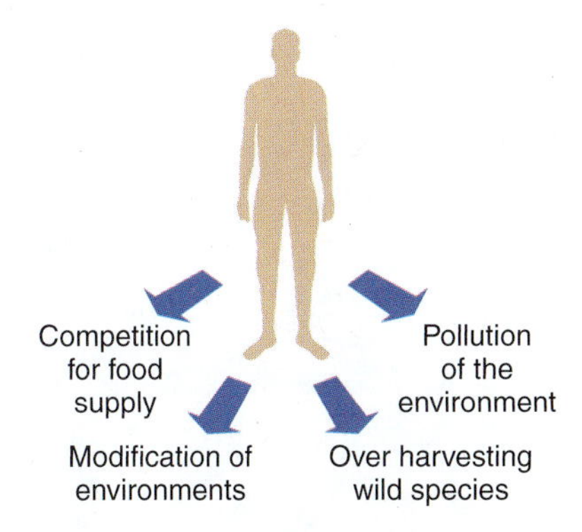

FIGURE 2–22 Human dominance.

beginning of human existence. Others recognize that the human species is the only species capable of changing its habits to preserve other species of organisms. They contend that humans have the moral obligation to protect natural resources and the organisms that depend on them (Figure 2–23).

FIGURE 2–23 We should not only appreciate but also care for our natural resources.

Biological succession

Animals and plants that were unable to adapt to the changes that occurred in their habitats as humans cleared land and established farms are present in fewer numbers or, in some cases, no longer exist (Figure 2–24). These native species have been replaced by different species that are capable of surviving in the new environments created by cultivated land (Figure 2–25). The change that occurs as one kind of

FIGURE 2–24 The American Wolf is an endangered species. (Courtesy of the U.S. Fish and Wildlife Service.)

FIGURE 2–25 The coyote is a highly adaptable animal that has learned to live with human neighbors. (Courtesy of Shutterstock.)

FIGURE 2–26 Lava flow that has cooled and hardened.

INTERNET KEY WORDS

biological succession, plants,
primary biological succession, lava flow
climax community, plants
competitive advantage, survival

living organism replaces another in an environment is called **biological succession** or ecological succession. Two forms of biological succession are known to exist. **Primary succession** occurs where organisms did not exist before, such as on a newly formed volcanic island where a recent lava flow has cooled and hardened (Figures 2–26 and 2–27).

Secondary succession occurs when an ecosystem is damaged or partly destroyed and remnants of the former community still exist. The changed environment will support only those organisms that are naturally found in an earlier stage of biological succession. Examples of secondary succession can be found in areas such as Yellowstone Park where large tracts of old-growth forests were destroyed by fire (Figure 2–28).

Biological succession is a continuous process. An environment is seldom stable and unchanging. Every time a living environment changes in some way, the creatures that seek shelter or food from it are also affected. Some organisms find it difficult to adjust to changes in their environments,

Ecology Profile

PRIMARY BIOLOGICAL SUCCESSION OVER A LAVA FLOW

The material that remains after hot, molten lava has cooled is sterile and without life. The surface of such a flow consists of hardened lava rock and cinders. Over time, simple plants such as lichens and fungi begin to grow on the hard surface. Tiny soil particles carried by wind become trapped in the lichen growth, and a soil base begins to develop on the hard lava surface. This makes it possible for simple plants to germinate and grow; these are called **pioneers**. The lichens begin to decay when they are acted on by small fungi that feed off the plants.

Hundreds or even thousands of years may pass before sufficient soil exists to enable complex plants to survive. Hardy plants, such as thistles and weeds, are the first complex species to grow in this environment. They are then followed by shrubs, trees, and other plants that are adapted to the conditions found in the area. The plants that occupy an environment when the succession of species is complete and plant populations become stable are called **climax communities.**

FIGURE 2–27 This plant is an example of primary succession on a lava flow.

FIGURE 2–28 Secondary succession occurs when remnants of the previous community still exist. (Courtesy of the U.S. Department of Agriculture.)

which may reduce their ability to survive. Other organisms may benefit from a change in their environment—for example, by making it easier to compete with other organisms in that particular environment.

A **niche** is the role that an organism fulfills in an environment. This role includes the organism's position in the food web, where it lives, and when it is active. Several animals appear to be competitors in a single habitat, but if they occupy different niches, competition between them may be minimal.

A **competitive advantage** exists when one organism is more able to survive in an environment than another. A species that enjoys such an advantage will increase in numbers as time passes, and the populations of organisms without a competitive advantage will decrease. Eventually, the weakest competitor is lost from the environment.

competitive exclusion principle, forest fire, secondary succession

The loss of an organism from a specific niche or habitat is an example of the **competitive exclusion principle**, which is the hypothesis that

science connection profile

SECONDARY SUCCESSION FOLLOWING A FOREST FIRE

Forest fires are sometimes responsible for the destruction of old-growth forests and the habitat they provide to animals, birds, and other organisms. The heat generated by a forest fire often sterilizes the soil, killing many of the soil organisms, and it frequently moves the stage of succession backward. The 1988 fires that burned large tracts of the pine forest in Yellowstone Park were destructive because there was an abundance of dead plant material on the forest floor. This created extreme heat that caused the forest canopy to burn.

After a destructive fire occurs, hardy grasses, thistles, and other pioneer plants are some of the first forms of vegetation to be found in the area. In some instances, living trees that have been protected by their thick bark also remain. In other instances, trees are reseeded because of fire. Some kinds of pinecones, for example, only release their seeds under intense heat.

Courtesy of the U.S. Department of Agriculture.

Secondary succession after a forest fire often begins with cheatgrass and crabgrass growth followed by tall grasses and food plants. Pine trees eventually invade the grasslands, establishing pine forests. Hardwood trees are usually the last species to come into an area. They grow up through the pine forest, and eventually become the dominant species.

INTERNET KEY WORDS

secondary succession

two or more species cannot coexist on a single scarce resource. Evidence exists, however, that species can coexist when neither species exercises dominance over the other. The Bighorn sheep is a good example of an organism that declined because it could not adjust to a changing environment (Figure 2–29). In the Grand Canyon, the wild burro, which is not a native species, has established itself in habitats formerly occupied by Bighorn sheep. The burros are more competitive grazers than the

FIGURE 2–29 Bighorn sheep. (Courtesy of the U.S. Fish and Wildlife Service.)

FIGURE 2–30 The comfort zone for the flamingo is extreme warmth.

sheep, and they have a competitive advantage when food supplies are inadequate for both populations.

The ability of an organism to survive changes in its environment depends on its **range of tolerance** to change. All organisms have comfort zones in which living conditions are matched with their survival needs (Figure 2–30). As the range of conditions to which an organism can adjust becomes greater, its ability to survive increases.

INTERNET KEY WORDS

pollution, fish survival

The range of tolerance of an organism for its environment is largely determined by its inherited ability to adjust to new environmental conditions. A bird that is able to gather or digest only one kind of food will be unable to survive if that food supply is lost from the environment. Some organisms depend totally on a single kind of shelter; others cannot tolerate pollution in the air and water (Figure 2–31). In addition,

FIGURE 2–31 Pollution in rivers, streams, and ponds can jeopardize populations of many fish species. By concerted efforts to clean up waters and by using restocking programs, we can avoid this problem. (Courtesy of the U.S. Department of Agriculture.)

the temperature range in the environment often determines whether an organism is capable of surviving. A narrow tolerance range makes it extremely difficult for an organism to survive environmental changes.

Looking Back

The law of conservation of matter states that the form of matter can be changed, but it cannot be created or destroyed by ordinary physical or chemical processes. Energy is the force that drives all systems in nature. It originates from the sun and is lost to outer space in the form of heat. The first law of energy states that energy cannot be created or destroyed, only converted from one form to another. The second law of energy states that each time energy changes from one form to another, some of it is lost in the form of heat.

The circular flow of an element from living organisms to nonliving matter is an elemental cycle. The carbon and nitrogen cycles are two of the most important elemental cycles. The water cycle is the circular flow of water from the oceans to the atmosphere in the form of water vapor, to the land in the form of dew, rain, or snow, to the rivers and streams, and then back to the oceans.

A food chain consists of a sequence of living organisms in which each member of the chain feeds on lower-ranking members of that chain. Overlapping food chains form food webs. The organization of a food pyramid progresses from organisms that produce food (plants) to animals that eat plants (herbivores) and then to animals that eat other animals (carnivores).

Self-Evaluation

Essay Questions

1. Define what energy is and discuss the role it plays in nature.
2. Define and explain the first and second laws of energy.
3. What are some examples of the different forms of energy?
4. What is an elemental cycle?
5. Illustrate the carbon and nitrogen cycles and explain what is happening in each.
6. Illustrate and explain the water cycle.
7. List some factors that make water important to all living organisms.
8. Describe the organization of a food chain.
9. How is a food chain different from a food web?
10. How is a food pyramid organized?
11. How is primary biological succession different than secondary biological succession? Give examples of each type of succession.
12. What is an example of the competitive exclusionary principle?

Multiple-Choice Questions

1. The conversion of energy from one form to another is always accompanied by the loss of
   a. water.
   b. matter.
   c. carbon.
   d. heat.

2. Which of the following forms of energy occurs as a result of photosynthesis?
   a. thermal
   b. electrical
   c. kinetic
   d. chemical
3. The most abundant element found in living organisms is
   a. carbon.
   b. hydrogen.
   c. oxygen.
   d. nitrogen.
4. Which of the following is the elemental cycle that is responsible for the formation of fossil fuels?
   a. the water cycle
   b. the nitrogen cycle
   c. the energy cycle
   d. the carbon cycle
5. The process of transpiration is important to which natural cycle?
   a. the water cycle
   b. the nitrogen cycle
   c. the conservation of matter
   d. the carbon cycle
6. Which of the following is a secondary consumer?
   a. producer
   b. carnivore
   c. herbivore
   d. food plant
7. The most simple arrangement of organisms in an environment in a ranking order that connects all of the producers to the primary and secondary consumers is called a
   a. food chain.
   b. food web.
   c. food network.
   d. food pyramid.
8. Primary succession occurs when
   a. organisms live in an area where they did not live before.
   b. an ecosystem is damaged or partly destroyed.
   c. remnants of a former community still exist.
   d. plants displace animals from an environment.
9. The ability of a particular organism to survive more easily in a shared environment than another is an example of
   a. the competitive exclusion principle.
   b. adaptive superiority.
   c. secondary succession.
   d. a competitive advantage.
10. The ability of an organism to survive changes in an environment is a demonstration of its
   a. range of tolerance.
   b. comfort zone.
   c. competitive exclusion principle.
   d. niche.

Learning Activities

1. Conduct a writing exercise by assigning students in teams of two to prepare the script for a two-minute television spot that explains one of the environmental science principles studied in this chapter. The script should name the principle, describe it, explain how it works in the environment, and explain why it is important. Each team should present its work to the class.
2. Take a field trip to a natural environment such as a park or nature center near your school. Have class members prepare and give reports on the habits of the organisms that are found there. Investigate and illustrate the ranking of each organism in the food pyramid. Suggest possible ways that the food chain might be organized and illustrate how the food chains overlap to create a food web.

TERMS TO KNOW

- environment
- nonexhaustible resource
- renewable resource
- nonrenewable resource
- exhaustible resource
- population level
- carrying capacity
- conservation
- soil conservation
- silt
- preservation
- multiple use
- biotechnology
- genetic engineering
- smog

CHAPTER 3

Ecosystem Management

Objectives

After completing this chapter, you should be able to

- identify the key elements of a living environment
- recognize differences among nonexhaustible resources, renewable resources, and exhaustible or nonrenewable resources
- explain how carrying capacity is related to population level
- distinguish between the concepts of conservation and preservation of natural resources
- evaluate the importance of conservation of natural resources such as soil, wildlife, minerals, and wildlife habitat
- discuss the concept of balance in natural ecosystems
- describe the impacts that modern industry has had on the ecosystems of North America
- discuss the role of ecology in human efforts to manage natural resources
- suggest ways that polluted environments might be cleaned up to make them safe for wild animals, plants, and humans

FIGURE 3–1 A freshwater ecosystem may have many different kinds of plants and animals living within its boundaries.

People have always lived in an **environment**. An environment—soil, water, plants, animals, energy, minerals, and atmosphere—provides everything for our survival (Figure 3–1). If any of our requirements had not been met, then we would have ceased to exist. At one time, our natural resources could be used without fear. There were fewer people, and our technology did not allow for rapid and massive use of natural resources: Trees were cut by saws and axes, and coal was harvested by hand. Those days are gone now.

People still live in an environment, but now we have the ability to use its resources on an unbelievably large scale. We can take water from a water table hundreds of feet underground, and with the aid of modern technology, we can level mountains, change the flow of rivers, and take energy from the atom. Sounds impressive, doesn't it?

INTERNET KEY WORDS

how to conserve natural resources

Unfortunately, technology and our growing population present a problem. In the future, we must live in the same environment—our "spaceship Earth"—and it is not getting appreciably larger. We have discovered that many of nature's resources are not going to last forever.

Resource Management

Wise use of our natural resources is not a new idea, but today more than ever it must become a common goal. Managing our natural resources for the future as well as the present must become a priority for all of us. This chapter explains several important ideas in natural resources conservation and management.

Managing natural resources implies that we must take an active role in caring for the environment (Figure 3–2). To do so, we must understand how natural systems work. For example, how does the water supply

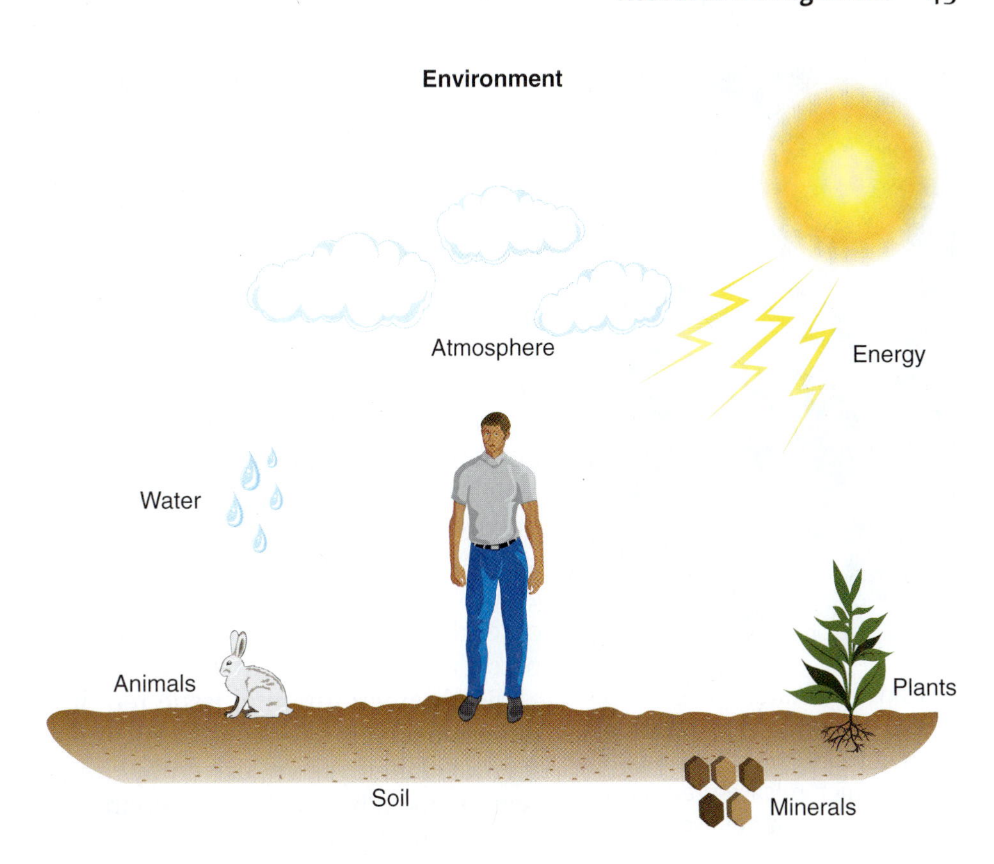

FIGURE 3–2 For humans, the environment includes all of the resources that surround us.

renew itself? How will we know when a healthy environment exists? What conditions must be present for healthy forests and other natural environments to continue into the future? What natural processes are at work to control destructive diseases and insects within an ecosystem? These and many other questions must be answered if we are to have a positive impact in any management effort.

In addition to understanding the natural processes that occur in an ecosystem, we must gather data about the health of the system. Data are needed concerning the population trends and growth of the organisms that occupy the environment. With this information, a manager can identify environmental issues at a much earlier stage. Accurate data also provide clues that are helpful in correcting problems or identifying events or conditions that lead to positive effects within the environment.

In some instances, natural resource managers simply seek to avoid harming a resource. An example of this kind of management is the use of hunting regulations for game animals. If a deer population is not reduced in the fall season, there will be an increase in the number of deaths during the winter season because of starvation. Once we know approximately how many deer can be sustained in an area during the winter months, hunting seasons and bag limits can be established. By reducing the number of deer to a level that can be sustained during seasons when their food supply is limited, a closer balance can be attained. The end result of this kind of management will be that no harm has been done to either the plant population on which the deer depend for food or the number of deer that survive winter conditions.

FIGURE 3–3 Automobiles, trucks, homes, and factories burn gasoline, oil, coal, and wood, which release products of combustion that pollute the air. (Courtesy of Michael Dzaman.)

A different kind of management must be implemented in an ecosystem that has become severely damaged from human activities. Steps must first be taken to stop the abuse of the resource. One example of this kind of management strategy is the passage of legislation that required catalytic converters on new cars. This device removes pollutants from exhaust gases (Figure 3–3). By removing nitrogen compounds, acid precipitation caused by vehicle exhaust gases was reduced. This management strategy has affected all of the ecosystems of the United States, especially those around large population centers.

Sometimes ecosystems are damaged so badly that large-scale cleanup operations are required. This kind of management has been made possible by the legislation that created the federal superfund for the cleanup of severely damaged environments. This fund has paid for cleaning up such sites as nuclear waste storage sites, mine tailings containing dangerous amounts of lead, polluted rivers, and many other sites. In each case, the environment has become poisoned and is no longer a safe place for living organisms.

Each environmental issue requires its own solution because many different factors influence environmental conditions and few environmental challenges are driven by the same factors. For this reason, each issue is usually considered separately. Many environmental problems are so complex that trained professional consultants are needed to find solutions that work.

The Nature of Resources

In a very real sense, everything in our environment could be considered a natural resource. Rocks may be used as gravel for our roads, facings for our buildings, or material for our statues. Wind, falling water, still air, resting water, minerals, insects—virtually everything around us can be considered a resource. When someone takes an object and uses it to perform work or change other parts of the environment, that object has become a resource. Even before its use, the potential for use makes the object a natural resource.

INTERNET KEY WORDS

example of exhaustible natural resources
example of renewable natural resources, trees

Those things that have become or show promise of becoming important to us are the natural resources we are concerned about in this book. Those natural resources may take many forms. More important, they may be capable of going on forever. On the other hand, they may be extremely limited. They may be usable over and over, or they may be gone forever with a single use. Let us look at these a little closer.

Nonexhaustible Resources

Natural resources that can last forever regardless of human activities are **nonexhaustible resources**. They renew themselves continuously. This does not mean that such resources are not limited or that human misuse cannot damage such resources—it certainly can.

A good example is surface water. If a gallon of water is taken from a river, another gallon will replace it. If a stream is dammed, the water will simply go elsewhere. If a watershed is damaged so that its rainfall does not soak into the ground, then the rainfall will simply go elsewhere.

FIGURE 3–4 The water cycle renews the supply of fresh water.

Little that we do will affect the total amount of water that comes to Earth in the form of precipitation.

Water supplies may be limited. As noted in Chapter 2, the water cycle makes water a nonexhaustible resource (Figure 3–4). We face many problems, however, as a result of the damage done to our water supply by pollution. Nevertheless, our water supply remains nonexhaustible for all practical purposes.

Another example is air. We use air to breathe, to grow plants, to fly airplanes, to power windmills, and to dry food and clothes. We can damage the air with pollution. We could even make it unusable, as many environmentalists would argue, but we cannot use it up. It also is nonexhaustible for all practical purposes.

Renewable Resources

Natural resources that can be replaced by human efforts are considered to be **renewable resources**. On the one hand, simply because a resource is renewable does not mean it will never be used up. On the other hand, it is possible to use such resources and yet have as much left afterward as before that use. Conservation practices for renewable resources should restrict their use to ensure that they are used no faster than they are regenerated. This rate of use is expected to sustain a constant resource supply for as long as they may be needed.

A forest is one example of a renewable resource. In this country, we use more wood today than ever before, yet we produce more wood each year than we use. The types of wood, however, have changed. We no longer harvest as many large hardwoods as we once did, but we have no foreseeable shortage of wood or wood products. This is true because of the advances made in forestry in both woodland management and genetics.

Another example is our fish and wildlife population. In our nation's past, there have been times of great waste. Huge droves of passenger pigeons were destroyed, and the popular food bird became extinct. Great herds of bison, or buffalo, were killed for their hides and meat, and this great American natural resource neared extinction. With techniques of game management, however, the numbers of bison have rebounded (Figure 3–5). Fish populations, too, respond readily to fisheries management techniques.

Exhaustible Resources

Many of our natural resources exist in finite quantities. Those limited resources that cannot be replaced or reproduced are known as **nonrenewable resources** or **exhaustible resources**. In the case of exhaustible resources, we cannot manage them for renewal. They do not renew themselves; once they are gone, they are gone forever. We can conserve our exhaustible resources. We can learn how to use less. We may try to find more of them. We may even be able to recycle some of them; but once these resources are gone, we simply have to do without them.

Conservation of nonrenewable resources is accomplished by reducing the rate at which these resources are used so that they last longer. This can involve recycling metals or even plastics (derived from petroleum)

FIGURE 3–5 The bison of Yellowstone National Park. (Courtesy of U.S. Fish and Wildlife Service.)

to reduce the necessity for extracting new materials. Conservation of these resources must ensure that they are not used up before we learn to replace them with other resources.

INTERNET KEY WORDS

example of nonrenewable resources, exhaustible
oil shortage
soil erosion

Even though an exhaustible resource exists in a finite (limited) supply, that does not mean it is necessarily a limited resource. Many exhaustible resources exist in such huge amounts that they are practically nonexhaustible. For instance, there is so much coal on the planet that, even though it is exhaustible, there is no practical limit. In addition, there is so much iron ore that, although iron is an exhaustible resource, it has no practical limit.

One especially important exhaustible resource is oil. We constantly hear of the "energy crisis." There is only so much oil in the ground, and when we have removed all we can find, it is gone, so we must develop other sources of energy. Another example is our mineral resources. We use lead, cobalt, zinc, and other minerals to make our goods. We depend on these mineral resources for our way of life. We must manage those resources to make them last as long as possible.

Soil also probably fits into this category. Soil is constantly being formed by nature. We can improve existing soil, make it more fertile, move water to it, and supply missing minerals. We can even make soil substitutes in small quantities, but we cannot really make soil. Only nature can do that. Why, then, is soil not a nonexhaustible resource? It is exhaustible because nature makes soil so slowly. A soil destroyed by improper use will probably be replaced—in 500,000 years—but that is not renewal that is useful to us. Thus, soil is a nonrenewable or exhaustible resource (Figure 3–6).

Balance of Nature

We have all read or heard about the balance of nature, but is nature really balanced? If nature were perfectly balanced, there would be little change. As a gallon of water flowed to the ocean, another gallon

FIGURE 3–6 Soil is a nonrenewable or exhaustible resource. (Courtesy of the U.S. Department of Agriculture, Natural Resources Conservation Service.)

would evaporate and enter the hydrologic cycle. As one rabbit died, another would be born. Rivers would not change their course. Ice ages would never occur. Clearly, this is not the case. There is, however, a kind of balance.

In reality, the forces of nature constantly counteract each other. The result is a constant change of nature. Change is different from balance. Balance implies no change. The reality is that change in the environment is both continuous and natural. What is important is that the changes in nature be gradual to allow time for organisms such as humans to adapt to them. Gradual change is what we are really talking about when we refer to a balance of nature. Keep that idea in mind whenever the term balance of nature is used.

Some plant and animal species develop. Other ones become extinct. Fires, insects, or diseases destroy a forest. Grasses and brush replace the forest, only to be replaced eventually by a new forest. People also change things. We clear forests, plow fields, drain swamps, and build cities and highways. Such massive changes in the ecosystems affect every living thing, including humans. Managing our natural resources wisely means many things. For one, it means controlling nature so that we can use its resources without destroying or at least permanently upsetting its balance.

Carrying Capacity

A **population level** can be defined as the number of a given species of plant or animal in a given area at a particular point in time. **Carrying capacity** refers to the ability of an ecosystem to provide food and shelter for a given population level. Population levels cannot exceed the carrying capacity of the ecosystem for long.

Carrying capacities are affected by the food chain of the species in question. For example, quail eat insects and plant seeds. A covey

carrying capacity, nature farming, soil depletion

of quail must have an adequate supply of food to thrive, reproduce, and even survive. The population level of quail in a given area cannot exceed the area's carrying capacity. Population levels are also affected by water availability, shelter, and predators. All these factors, as well as diseases and parasites, help keep species populations at acceptable levels. When a population exceeds its ecosystem's carrying capacity, then diseases, predators, or starvation reduce the population level. That is the way of nature.

The events in one part of an ecosystem affect other parts of the system. The most profound human effect on the ecosystem has been agricultural production of food and fiber. Farming has drastically increased the carrying capacity of the world for humans by increasing the amount of food that can be produced on the land. Intensive crop and livestock production have enhanced that aspect of the ecosystem. Drastically altering one aspect of the ecosystem, however, also affects other parts of the system. When we harvest crops, for example, we remove much of the organic matter that would normally be returned to the soil by decomposition. This has profound long-term effects on an ecosystem. Modern conservation practices such as incorporating crop residues and green manure crops into the soil directly apply ecological principles that seek to counterbalance unsustainable practices.

Human Population and Resource Use

Around the time Christ was born, the world's human population was approximately 300 million. That figure was a great increase over the estimated 10 million at the beginning of the new stone age (around 6000 B.C.), and the rate of increase has jumped in the past few centuries to unbelievable levels (Figure 3–7). As late as 1800, there were fewer than 1 billion living humans. The number grew to 1.6 billion by 1900, skyrocketed to 4.3 billion in 1979, and was more than 6 billion in 1999 (Figure 3–8).

The issue of human population levels and the world's carrying capacity is an important one. If we are to continue to feed, shelter, and clothe ourselves into the next century, we must plan now and also begin to better manage our natural resources using improved natural resource management skills and tools. There is no immediate danger of worldwide starvation or death from disease. We are not running out of natural resources, but the challenge is before us. Managing our natural resources carefully is more important than ever before.

Balancing Population Needs with Resources

The abuse or misuse of natural resources—soil, minerals, water, and air—can no longer be allowed. Even though our society needs food in large quantities and thus requires modern, large-scale farming, farmers and agricultural concerns must protect our soil and not allow it to be destroyed by erosion. Even though our society—homes, farms, and factories—produces vast amounts of waste, it can no longer be simply

Year	Population (in billions)	Year	Population (in billions)
6000 B.C.	0.01	1950	2.52
A.D. 1	0.30	1960	3.02
1000	0.31	1970	3.70
1250	0.40	1980	4.44
1500	0.50	1990	5.27
1750	0.79	2000	6.06
1800	0.98	2010	6.79
1850	1.26	2020	7.50
1900	1.65	2030	8.11
1910	1.75	2040	8.58
1920	1.86	2050	8.91
1930	2.07	2100	11.20
1940	2.30	2200	stabilized at or near 11.60 billion

United Nations, *World Population Prospects: The 1998 Revision* (New York: Author) and the World Resources Institute.

FIGURE 3–7 Estimated and Projected World Population Over 8,000 Years.

water pollution

discharged into streams, rivers, lakes, and oceans. Although water pollution has become a serious world problem, solutions can and should be found and implemented. Even though our society—consumers, industry, and technology—requires metals and minerals, they are not inexhaustible and can and should be recycled and used more than once and substituted where possible with renewable or nonexhaustible resources. In the end, we have no choice but to use our natural resources as fully and wisely as we can. In working for today, we must keep an eye on the future. One way to do that is to follow wise conservation practices.

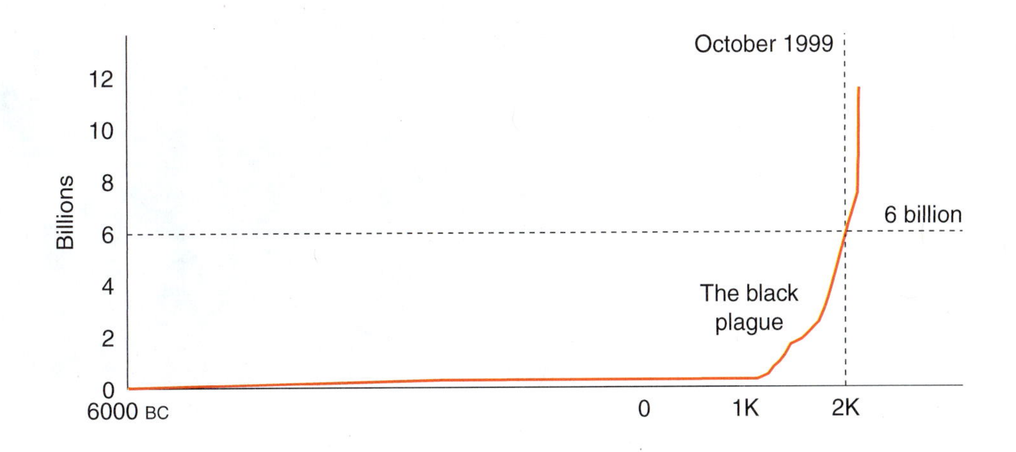

FIGURE 3–8 World population over 8,000 years (forecast through 2250).

conservation

Conservation is the practice of protecting natural resources against harm and waste. It involves using less of a resource than is available so that future generations may also benefit from its use. Conservation of resources does not necessarily mean that the resources are not used, only that they are not wastefully or carelessly used. Although conservation may bring to mind thoughts of nature and beauty, its true test is whether society benefits from its undertakings.

Conservationists believe in using natural resources to produce long-range benefits for people while taking good care of those resources. Conservation does not set aside resources simply to have them. A forest is not something to be prized simply for its own sake. It is a natural resource to be managed for wood production, as a guard against erosion, and as a sponge to soak up rainfall to meet the world's water needs. It is also a place to hunt, hike, camp, picnic, and study. To the conservationist, for example, a deer is a natural resource (Figure 3–9). It provides pleasure for the naturalist. It reproduces and provides food and recreation for the hunter. Just as the forest is managed partially to harvest trees, a deer population is managed partially to harvest deer.

soil conservation

One of the most important conservation goals is conserving topsoil. To maintain wildlife habitats, soil must be protected from damage. Without good soil, it is impossible to grow the plants that are required by people and wild animals for food and shelter. **Soil conservation** is the practice of protecting soil from the destructive forces of wind and water. Preventing soil erosion is important for two primary reasons: preventing water pollution and maintaining viable farmland. Soil erosion is the greatest source of water pollution known to humans. Many tons of soil are lost each year by wind and flowing water (Figure 3–10). Farmland

FIGURE 3–9 To a conservation-minded person, a deer is a natural resource. (Courtesy of U.S. Fish and Wildlife Service.)

Figure 3–10 When the wind is particularly strong at the ground level, it can cause serious damage to the soil. (Courtesy of U.S. Department of Agriculture, Natural Resources Conservation Service.)

soils likewise must be maintained to support the production of crops, because abundant crop yields on existing farmland slows down the conversion of wildlife habitat to new farmland.

Both wind and water carry soil particles to new locations. Massive amounts of **silt**—tiny soil particles—become suspended in water as it flows over exposed surfaces. When silt-laden streams enter lakes and ponds, the rate of flow is reduced and the particles settle to the bottom, forming large deposits of silt that eventually fill the lake or pond. Some ancient civilizations such as the Egyptians benefited from upstream erosion because the rivers flooded their farms each spring, depositing new silt and keeping their soils fertile and productive. One problem today is that some land-management practices result in harmful amounts of silt being carried and deposited downstream beyond the natural floodplains.

Preservation

Some things are worth having and guarding just for their own sake. The Liberty Bell, for example, is a national treasure. Yosemite National Park, the Grand Canyon, and the American bald eagle are other examples. Such heirlooms of our heritage should be preserved. Why? Simply because we, as a people, value them. No other reason is needed. To help guarantee that these national treasures received legal protection and were administered in the public trust, Congress established the National Park Service in 1916. By 1970, the service administered some 30 million acres around the United States. An additional 7.4 million acres were included in state parks.

INTERNET KEY WORDS

natural resource preservation versus conservation

We must also balance our desire for **preservation** with our needs as a people. Preservation is a part of conservation, but only a small part. In seeking to preserve a natural area, such as millions of acres of forest in Alaska, we must ask several questions. Can we use such resources without destroying them, as we can with forests? Can we afford to set aside such resources? Which is more important—economic growth or the preservation of nature?

With this last question, the growth of the human population surely must be considered. Is it important to preserve a tiny fish from extinction? After all, once it is gone, it is gone forever from the Earth. Or is it more important to keep workers employed such as would happen if a dam were built? Such questions are being asked in this country each year. Conservationists and preservationists would probably answer from different perspectives. We all live in the same ecosystem. We must learn to work together. We need a new national attitude toward managing our natural resources.

MULTIPLE USE

A concept that originated with foresters seems to be gaining popularity among all conservationists: **multiple use** of resources. A forest can be used for recreation as well as for the production of wood. A lake can be used as a water reservoir, for fishing, and as a flood-control measure. Windbreaks can be seeded with plants that produce food for wildlife resulting in improved hunting. Winter cover crops can be grains that are useful to migrating birds.

The concept of multiple use is highly productive. It encourages us to plan natural resource management activities to produce more than one benefit. It allows us to combine soil and water management, forest management, mineral and energy management, and recreation—all at the same time (Figure 3–11).

FIGURE 3–11 This multiple-use area serves as wildlife habitat, conserves soil and water, and is used for human recreation and education.

interest profile

A NATIONAL ATTITUDE

Stewart Udall, who was Secretary of the Interior during the mid-1960s, had this to say:

> The conservation goal of America's third century as a nation must be the development and protection of a quality environment which serves both the demands of nature for ecological balance and the demands of man for social and psychological balance. The landscapes and cityscapes that comprise the face of our continent present a partial statement about the state of our civilization, in much the same manner as the cut of a man's clothes tells something of the man himself. As the republic nears its 200th birthday, the cut of the countryside bespeaks ambivalence. The need for a new national attitude toward our environment has grown until today it is an absolute necessity for human survival.
>
> Technology has stretched and magnified our natural resource potential in many areas. It has also supplied a harassed people with an infinite number of painkillers and tranquilizers. But it cannot provide us with one square inch of additional planetary surface, nor do more than gloss over the mounting environmental insults to humanity. It becomes increasingly apparent that runaway population, noise, and psychological pressures of too-close living will eventually run us out of space and nervous energy even if food and minerals and fuels were never to flag. Any new national attitude toward natural resources must take into consideration at least two factors—the quality of human life we seek to establish, and the specific meld of environmental ingredients that make up that quality.

Reclaiming Damaged or Polluted Resources

Pollution of a resource is usually quite difficult to overcome. It requires finding the origin of the pollutants and reducing their release at the source. It also may be necessary to remove the pollutants from the contaminated area. If this cannot be done, then ways must be found to dilute or break them down into nontoxic substances.

INTERNET KEY WORDS

biotechnology, genetics
genetic engineering

The science of **biotechnology** is relatively new, but it has important environmental applications. Scientists in this field are altering the genetic makeup of living organisms to help researchers in many fields, including medicine, agriculture, and environmental sciences. Through a biotechnology practice known as **genetic engineering**, the genes of bacteria, for example, have been modified to enable them to ingest and break down pesticides and other chemical pollutants. This is an important scientific advancement that may help us to reclaim damaged resources.

INTERNET KEY WORDS

oil spill cleanup

Most of the pollution to our water resources comes from untreated sewage and industrial waste. Since the Clean Water Act was first passed by Congress in 1972, billions of dollars have been spent to build treatment plants to reduce waste discharges into waterways. This law, which was strengthened in 1987, is administered by the Environmental Protection Agency. The goal of the Clean Water Act is to clean wastewater well enough that we will be able to swim and fish in it. It is now illegal for cities to dump sewage that has not been treated, and industries are required to eliminate discharged pollution by using the best practicable technology that is available (Figure 3–12).

FIGURE 3–12 The goal of the Clean Water Act is to clean wastewater well enough that humans will be able to use it for swimming and fishing. (Courtesy of Getty Images.)

INTERNET KEY WORDS

sewage dumping
Clean Water Act

Water that contains nitrates and phosphates can be naturally cleaned by bacteria or plants before it is released back into streams, a process that occurs as water passes through marshes and swampy areas (Figure 3–13). It is also possible for humans to create marshes that perform the same function.

Human-created lagoons, for example, are ponds where wastewater is stored for three or four weeks. During this period, bacteria and algae metabolize dissolved nutrients and many of the solids. Several lagoons that are linked together in a series are capable of sufficiently cleaning water so that it can be released into streams and rivers.

FIGURE 3–13 Marshes and swamps act as natural water purifiers.

FIGURE 3–14 Exhaust gases from cars, buses, and trucks are the greatest sources of air pollution. (Courtesy of Getty Images.)

Preserving Air Quality

The greatest hazard to wild animals and fish comes from the effects of acid rain on surface waters and on plants that provide food and shelter. The weak acids form when rain combines with pollutants in the atmosphere; they are capable of killing forests and destroying living organisms in streams and lakes. These problems are evident in North America and in many other industrialized nations.

Car and truck exhaust gases are the greatest sources of atmospheric pollution (Figure 3–14), but factories and electrical power plants that burn coal and petroleum products also emit large amounts of polluted gases into the atmosphere. Ultraviolet light from the sun reacts with atmospheric pollutants and adds to the atmospheric haze. The result of this atmospheric pollution is a great cloud of polluted air called **smog**. Although the problem has lessened since the development of vehicles that burn unleaded gasoline, exhaust gases continue to pose threats to environments near large population centers because more cars and longer distances are driven every year.

INTERNET KEY WORDS

preserving air quality
atmospheric pollution

Looking Back

All living things live in an environment. It provides everything we need for our survival. An environment's ability to provide food and shelter determines the carrying capacity for each species. Once the carrying capacity for a species is exceeded, part of the population will die. The growth of the human population has been possible because natural environments have been modified to support our needs. Some of the most pressing concerns of humanity are to maintain the living environment and manage our natural resources appropriately while protecting other organisms that also draw their needs from the environment.

Natural resources can be classified according to availability. Nonexhaustible resources include such things as sunlight or water; these are continuously renewed by nature. Renewable resources such as wildlife or forests can be replenished through human efforts. Nonrenewable or exhaustible resources, such as coal or oil, cannot be replaced. Once used up, they are gone forever. Soil conservation is vital to preserve a resource that, like air and water, is essential to all living things. Damage to natural resources is a problem that all of us must work to overcome. Sometimes that requires massive cleanup efforts to remove pollutants from the environment.

Self-Evaluation

Essay Questions

1. List the key elements that make up the environment.
2. Define and give examples of nonexhaustible natural resources.
3. Define and give examples of renewable natural resources.
4. Define and give examples of exhaustible natural resources.
5. Is there an accurate balance in nature? What would such a balance in nature mean?

6. How does the carrying capacity of an ecosystem affect the population level for any organism that is found there?
7. In what ways is the conservation of a natural resource different from its preservation?
8. Identify the major destructive forces that contribute to soil erosion. Describe how these forces erode soil.
9. Why is soil conservation important?
10. Explain the multiple-use concept of resource management. Why is it important?
11. How is conservation related to wildlife and natural resources?
12. What are some conservation practices that are used to preserve and restore wildlife populations and improve habitats?
13. List some ways in which genetic engineering is used to reclaim damaged and polluted resources.

Multiple-Choice Questions

1. Which of the following is not a key element in the human environment?
   - a. legislation
   - b. animals
   - c. minerals
   - d. water
2. A term describing a resource that is capable of replacing itself through reproduction or new growth is
   - a. recycling.
   - b. conservation.
   - c. renewable resource.
   - d. nonrenewable resource.
3. Which of the following is a renewable resource?
   - a. coal
   - b. water
   - c. soil
   - d. oil
4. Which of the following is not a renewable resource?
   - a. wildlife
   - b. soil
   - c. forest
   - d. plants
5. The number of a particular species of plant or animal in a given area at a specific point in time is an example of
   - a. carrying capacity.
   - b. multiple use.
   - c. population level.
   - d. biologistics.
6. The practice of using natural resources while protecting against harm and waste is called
   - a. conservation.
   - b. preservation.
   - c. resource renewal.
   - d. biotechnology.
7. The practice of maintaining an environment and the resources within it in their natural state simply because we value them is called
   - a. conservation.
   - b. preservation.
   - c. resource renewal.
   - d. biotechnology.
8. Which of the following is not considered to be a destructive element contributing to soil erosion?
   - a. solar heat
   - b. wind
   - c. flowing water
   - d. steep slopes

9. Damage to a water supply from contamination with sewage or other waste material is called
   a. flooding.
   b. remediation.
   c. pollution.
   d. reclamation.
10. A serious form of air contamination that is caused by car exhaust fumes and industrial gases is
   a. smog.
   b. fog.
   c. inversions.
   d. meteors.

Learning Activities

1. Make a list of some of the nonexhaustible, renewable, and exhaustible natural resources you have used today.
2. Make a bulletin board to show the differences between nonexhaustible, renewable, and exhaustible resources.
3. Organize a conservationist-versus-preservationist debate about the best future for a national forest or park near your community.

section TWO

Ecosystems and Natural Resources

CHAPTER 4
Air

CHAPTER 5
Water

CHAPTER 6
Land and Soil

CHAPTER 7
Forests

CHAPTER 8
Grasslands and Wetlands

CHAPTER 9
Wildlife Biology and Management

TERMS TO KNOW

- air
- water
- soil
- habitat
- sulfur
- hydrocarbons
- nitrous oxides
- tetraethyl lead
- carbon monoxide
- radon
- radioactive material
- chlorofluorocarbons (CFCs)
- ozone
- particulates
- pest
- toxic
- asbestos
- greenhouse effect

CHAPTER 4

Air

Objectives

After completing this chapter, you should be able to

- define the term *air* and identify its major components
- determine major sources of air pollution and identify procedures for maintaining and improving air quality
- analyze the importance of air to humans and other living organisms
- determine the characteristics of clean air
- describe common threats to air quality
- describe important relationships between plant life and air quality
- discuss the greenhouse effect and global warming
- list practices that lead to improved air quality

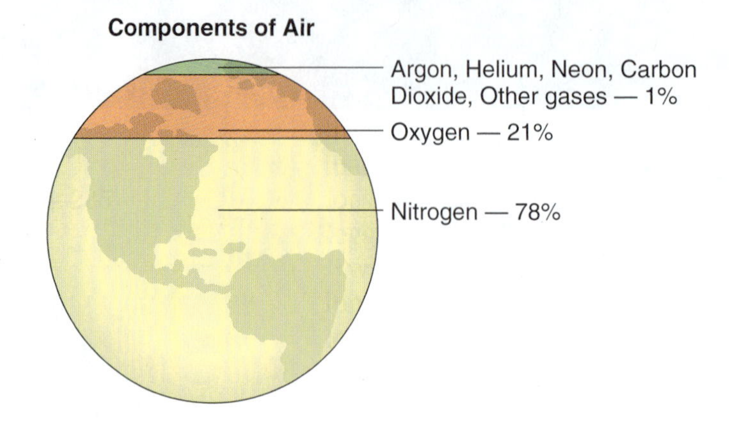

FIGURE 4–1 The atmosphere of the Earth is composed mostly of nitrogen and oxygen and small amounts of a few other gases.

air quality standards, pollutants, air pollution organizations

Life as we know it on our planet requires a certain balance of unpolluted air, water, and soil. **Air** is a colorless, odorless, tasteless mixture of gases. It occurs in the atmosphere around the Earth and is composed of approximately 78 percent nitrogen, 21 percent oxygen, and a one percent mixture of argon, carbon dioxide, neon, helium, and other gases (Figure 4–1). **Water** is a clear, colorless, tasteless, and nearly odorless liquid. Its chemical makeup is two parts hydrogen to one part oxygen. **Soil** is the top layer of the Earth's surface that is suitable for the growth of plant life.

Air Quality

Without a reasonable balance of air, water, and soil, most organisms would perish. Slight changes in the composition of air or water may favor some organisms and cause others to diminish in number or in health. Unfavorable soil conditions usually mean inadequate food, water, shelter, and other factors related to **habitat**, which is the area or type of environment in which an organism or biological population normally lives (Figure 4–2).

Threats to Air Quality

The mixture of gases we call air is absolutely essential for life. The air we breathe should be healthful and life supporting. Air must contain approximately 21 percent oxygen for human survival. If a human stops breathing and no life-supporting equipment or procedures are used, the brain will die in approximately four to six minutes. Air may contain poisonous materials or organisms that can decrease the body's efficiency, cause disease, or cause death through poisoning.

FIGURE 4–2 Good habitats for plants, animals, and humans need clean air, clean water, and productive soil. (Courtesy of Michael Dzaman.)

FIGURE 4–3 Automobiles, trucks, homes, and factories burn gasoline, oil, coal, natural gas, and wood, all of which release the by-products of combustion that pollute the air. (Courtesy of Getty Images.)

Even though the Earth's circumference at the equator is 24,902 miles, the abuse of the atmosphere in one area frequently damages the environment in distant parts of the world. Air currents flow in somewhat constant patterns, and air pollution will move with them. However, when warm and cold air meet, the exact air movement will be determined by the differences in temperature, terrain, and other factors. Because we cannot control the wind, humans have an obligation to keep the air clean for their own benefit as well as that of society at large.

There are major worldwide threats to air quality, including sulfur compounds, hydrocarbons, nitrous oxides, lead, carbon monoxide, radon gas, radioactive dust, industrial chemicals, and pesticide spray materials, among others (Figure 4–3). Most of these products not only are poisonous to breathe but also have other damaging effects.

INTERNET KEY WORDS

acid rain, sulfuric acid

Sulfur

Sulfur is a pale yellow element that commonly occurs in nature. It is present in coal and crude oil. When it combines with these and other fuels in the presence of oxygen, it forms harmful gases such as sulfur dioxide. Most smoke and exhaust from homes, factories, or motor vehicles contains some of these harmful sulfur compounds unless special equipment is used to remove them. Once these invisible gases are in the air, they combine with moisture to form sulfuric acid, which falls as *acid rain.* Acid rain damages and kills trees and other plants and also has a corrosive effect on metals.

Hydrocarbons

As the twentieth century proceeded, **hydrocarbons** became serious problems as the numbers of factories and motor vehicles increased. In the United States, hydrocarbon emissions (by-products of combustion or burning) are held in check by (1) special emission-control equipment on automobiles and (2) special equipment called *stack scrubbers* in large industrial plants. Hydrocarbon output is controlled on automobiles by crankcase ventilation, exhaust gas recirculation, air injection, and such engine refinements as four valves per cylinder. Without this equipment, air pollution would be much more intense in our major cities and in heavy stop-and-go traffic on major highways.

INTERNET KEY WORDS

air quality, hydrocarbons, nitrous oxides

Nitrous Oxides and Lead

Nitrous oxides are compounds that contain nitrogen and oxygen. They constitute approximately five percent of the pollutants in automobile exhaust. Although this seems like a small amount, they are damaging to the atmosphere and must be removed from exhaust gases. This group of chemicals is the most difficult and perhaps the most costly pollutant to remove from emissions. Scientists and engineers have partly solved the problem by developing and installing catalytic converters in automotive exhaust systems.

Hot exhaust gases from the combustion engine flow through a honeycomb of platinum metal in the catalytic converter. The reaction converts the nitrous oxides into harmless gases. Before 1986, all gasoline contained **tetraethyl lead**, a colorless, poisonous, oily liquid that improved the burning qualities of gasoline and helped control engine knocking. However, tetraethyl lead ruined catalytic converters, so they could not be used until a substitute for tetraethyl lead was found in the 1970s.

INTERNET KEY WORDS

tetraethyl lead gasoline

science connection profile

NATURAL SELECTION AS A RESULT OF POLLUTION

Air pollution affects many aspects of life on Earth. One example was observed in England during the Industrial Revolution. In the early 1800s, coal was largely used as fuel to power factories. Burning coal polluted the air with dark soot, some of which would settle out and fall to the Earth.

The phenomenon of natural selection was demonstrated dramatically with one form of insect life because of coal pollution. Small-winged moths spend daylight hours hiding on the bark of trees from predators, primarily birds. Their wings are camouflaged to match the trees they hide on. In time, the industrial soot darkened the bark of the trees. Most of the moths had lighter colored wings, so they became easy targets for predators. The few moths that had darker wings were able to survive and reproduce because the tree bark was now darker. Later, in the 1900s, this pollution was reduced. The tree bark returned to its original color, as did the wing color of the moths.

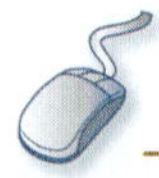

INTERNET KEY WORDS

pollution, radon pollution, carbon monoxide, radioactive dust pollution, chlorofluorocarbon pollution, ozone pollution

Tetraethyl lead is still used in third world countries, and lead and nitrous oxides are still major pollutants of the atmosphere. The levels of pollution, however, have been reduced as the use of lead in gasoline has declined. Today in Western nations, only fuel for small airplanes contains tetraethyl lead.

Carbon Monoxide

Carbon monoxide is one automotive gas that cannot be removed with current technologies. This colorless, odorless, and poisonous gas kills people in automobiles with leaking exhaust systems or when engines are run in closed areas without adequate ventilation. Victims fall asleep and die. Carbon monoxide emissions may be reduced by keeping engines in good repair and properly tuned.

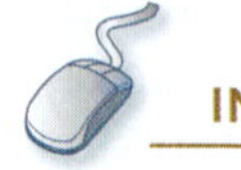

INTERNET KEY WORDS

air quality, carbon monoxide, radon, radioactive air pollution

Radon

Radon has become a hazard in homes in many parts of the United States. This colorless, radioactive gas is formed by the decay of radium. It moves up through soil and flows into the atmosphere at low and usually harmless rates. However, a hazardous condition can develop if a house or other building is constructed over an area where radon gas is being emitted. The gas can accumulate in buildings that have cracks in basement floor or walls, or it can enter through sump holes. The problem can be prevented by tightly sealing all cracks or providing continuous ventilation either below the basement floor or throughout the building (Figure 4–4).

Radioactive Dust and Materials

Radioactive material is matter that emits radiation. Of growing environmental concern are dust from an atomic explosion or other nuclear reaction and materials contaminated by atomic accidents or wastes. The damage from radioactivity ranges from skin burns to sickness to hereditary damage to death. Controversy over the possibility of worldwide contamination and other hazards from serious nuclear accidents has led to a reduction in the construction of atomic-powered electric-generation plants.

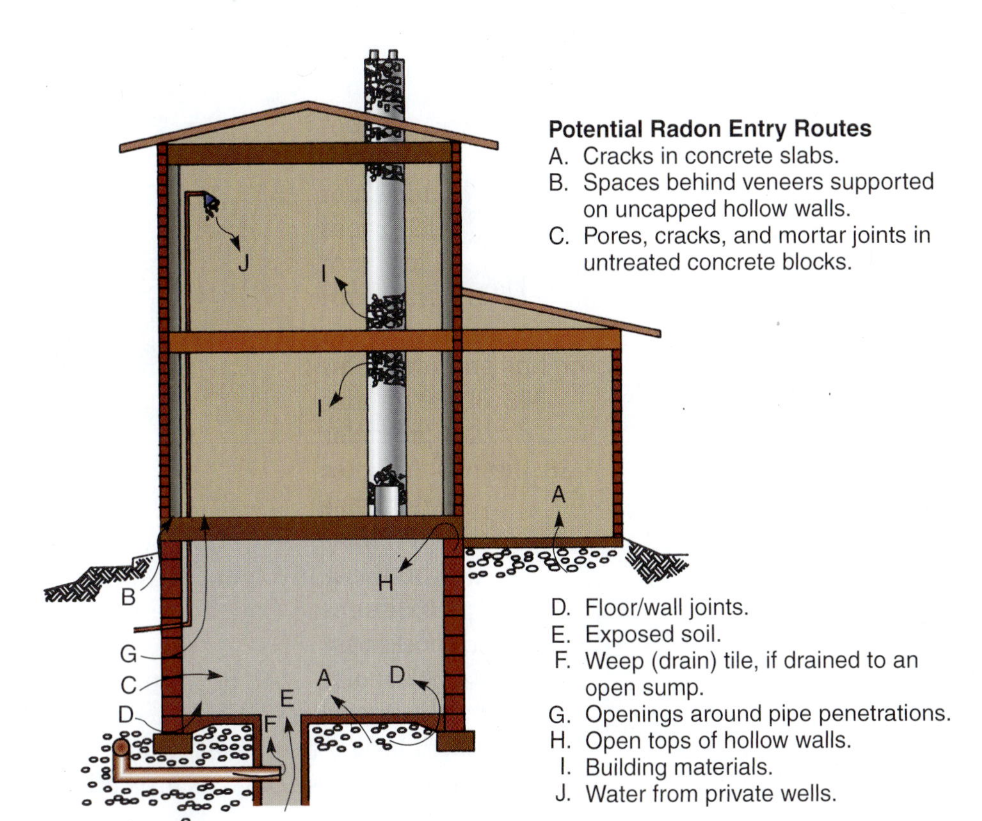

FIGURE 4–4 Ventilation systems must be correctly designed and carefully maintained to keep interior areas free of radon gas pollution.

Chlorofluorocarbons

Chlorofluorocarbons (CFCs) are a group of molecular compounds consisting of chlorine, fluorine, carbon, and hydrogen. They are used as aerosol propellants and refrigeration gas. These materials are highly stable. Once released from an aerosol can or cooling system, they bounce around in the air and eventually float upward into the high atmosphere. It is believed that CFCs will survive in the upper atmosphere for about 100 years. Meanwhile, their chlorine atoms destroy ozone molecules without themselves being destroyed. In newer equipment and consumer products, CFCs are being replaced by less-polluting agents.

Ozone (O_3) is a molecule that exists in relatively low quantities in the lower atmosphere but in relatively greater quantities in a protective layer approximately 15 miles above the Earth's surface. It filters out harmful ultraviolet rays from the sun. There is evidence that the ozone layer is being damaged (see "Hole in the Ozone Layer" later in the chapter). Most living organisms will be exposed to the damaging effects of ultraviolet rays, which include skin cancer and damage to the body's immune system.

In 1987, at an historic international conference, at least 37 nations agreed to schedule cutbacks in the production of CFCs. However, given the possible damage by this pollutant, it may make sense to stop all production immediately.

particulate pollution, global warming, greenhouse effect, climatic change

Particulates

Small particles that become suspended in air are known as **particulates**. These tiny particles appear as smoke or dust clouds. Particulates eventually will settle out of the air because of gravity, but they are so light in weight that the slightest breeze keeps them suspended. They are

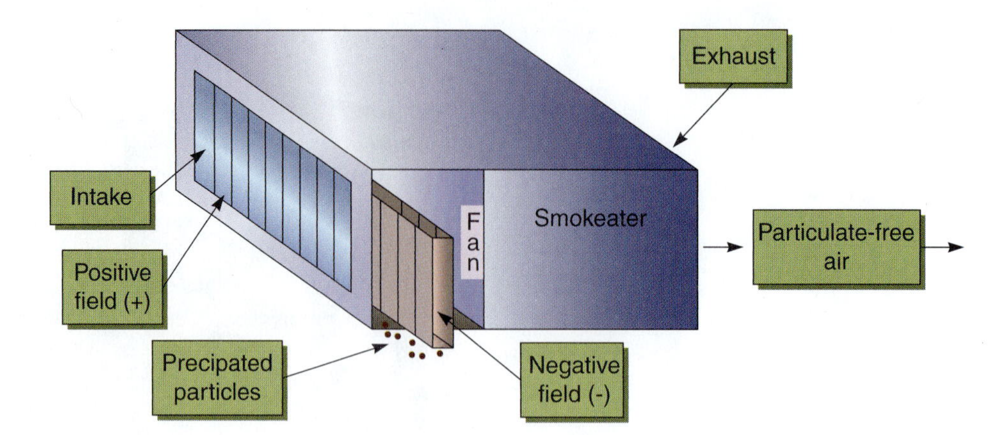

FIGURE 4–5 Electrostatic precipitators remove particles from the air.

especially harmful to people who suffer from respiratory diseases such as asthma or emphysema. Most particulate matter coming from industrial processes can now be removed from gas emissions by a process called *scrubbing*.

Particles also can be removed from gases by passing the gases over two oppositely-charged electrical fields. The first field charges the particles in the gases, and the second attracts and holds them (Figure 4–5). These devices are called *electrostatic precipitators.*

Pesticides

A **pest** is a living organism that acts as a nuisance or spreads disease. Examples include house flies, cockroaches, fleas, and mosquitoes. A pesticide is a material used to control pests. Many pesticides are chemicals mixed with water so they can be sprayed on plants, animals, soil, or water to kill or otherwise control insects, weeds, rodents, and other pests. Spray materials are pollutants if they carry **toxic** (poisonous) materials or are harmful to more than the target organism. Such sprayed materials are generally harmful to the air if they are not used exactly as specified by the government and the manufacturer. Poisons may be thinned out or diluted by air movement, but excessive toxic materials can overburden the ability of the atmosphere to cleanse itself. Abuse of chemicals to control pests is an area of growing concern in maintaining air quality.

pesticide spray, air pollution, asbestos removal codes, asbestos, health

Asbestos

Asbestos is a heat and friction-resistant material. In the past, it was used extensively in vehicle brakes and clutch linings, shingles for house siding, steam and hot-water pipe insulation, ceiling panels, and other products. Unfortunately, asbestos fibers are damaging to the lungs and cause disease and death. There are now state and federal laws and codes requiring the removal of asbestos from public buildings, industrial settings, and general use.

One serious aspect of air pollution is that pollutants are carried by the wind to other areas. This damages the environments of wild animals and fish, particularly those that live near large cities. Some regions along the eastern and western coasts of North America have sustained considerable amounts of damage from acid rain.

The solution to these problems is to remove as much of these pollutants as possible before they are released to the environment. We also would be wise to cut back on the amount of emissions that are produced.

The best way to do it is to create effective mass transit systems in our cities that reduce dependence on personal cars. In addition, we should research new industrial processes that require less energy.

The Greenhouse Effect and Global Warming

One factor that makes life possible on Earth is the planet's proximity to the sun. If the distance between the sun and Earth were any less, the planet's surface would become too hot to sustain life. If the distance were any greater, the temperature would be too cold to support many life forms.

Besides being life-sustaining, sunlight also has other effects on living organisms. For example, the excessive heat and drought of summer is frequently accompanied by parched and withered crops. Medical science also has found a definite link between extensive exposure to the sun and the occurrence of melanoma, a usually deadly form of skin cancer. Furthermore, some scientific evidence suggests that skin-damaging and life-threatening ultraviolet rays from the sun now reach the Earth's surface with greater intensity than in the past.

The Greenhouse Effect

When sunlight passes through a clear object, such as a glass window, it heats the air on the opposite side of that object. If the warmed air is not cooled or flushed out, heat builds up under the glass—for example, under skylights, auto windshields, and in greenhouses. The glass also absorbs some of the energy from the light and gives it off or radiates it as heat to the interior area. The overall result is a buildup of heat that causes the interior to become warmer than the outside. This heat buildup from the rays passing through the clear object and the resulting heat being trapped inside is known as the **greenhouse effect**.

The sun's rays include many different colors of light and types of rays. Some of the more familiar rays are ultraviolet and infrared. Ultraviolet rays are known for their extensive skin damage and other life-threatening effects from overexposure. Infrared rays are emitted from any warm object, such as a hot stove, glass warmed by the passage of ultraviolet rays, or an open fire.

The gases in the Earth's atmosphere serve as a relatively clear object through which sunlight passes. The crust of the Earth absorbs, radiates, and reflects heat back into the air above it. The atmosphere encircles the Earth and creates the greenhouse effect (Figure 4–6).

Scientists do not agree on the concept of global warming, but they do report that the buildup of heat within the global "greenhouse" is increasing and changing average temperatures. This increased warming trend is believed by many people to be a serious threat to the environment and to life itself. Scientists contend that the greenhouse effect must be stabilized or reduced if we are to protect air quality (Figure 4–7).

ozone, layer, hole

Hole in the Ozone Layer

Beginning in the 1970s, the ozone layer over the South Pole was found to be less dense, or thinner, than in the past. A thinner ozone layer lets more

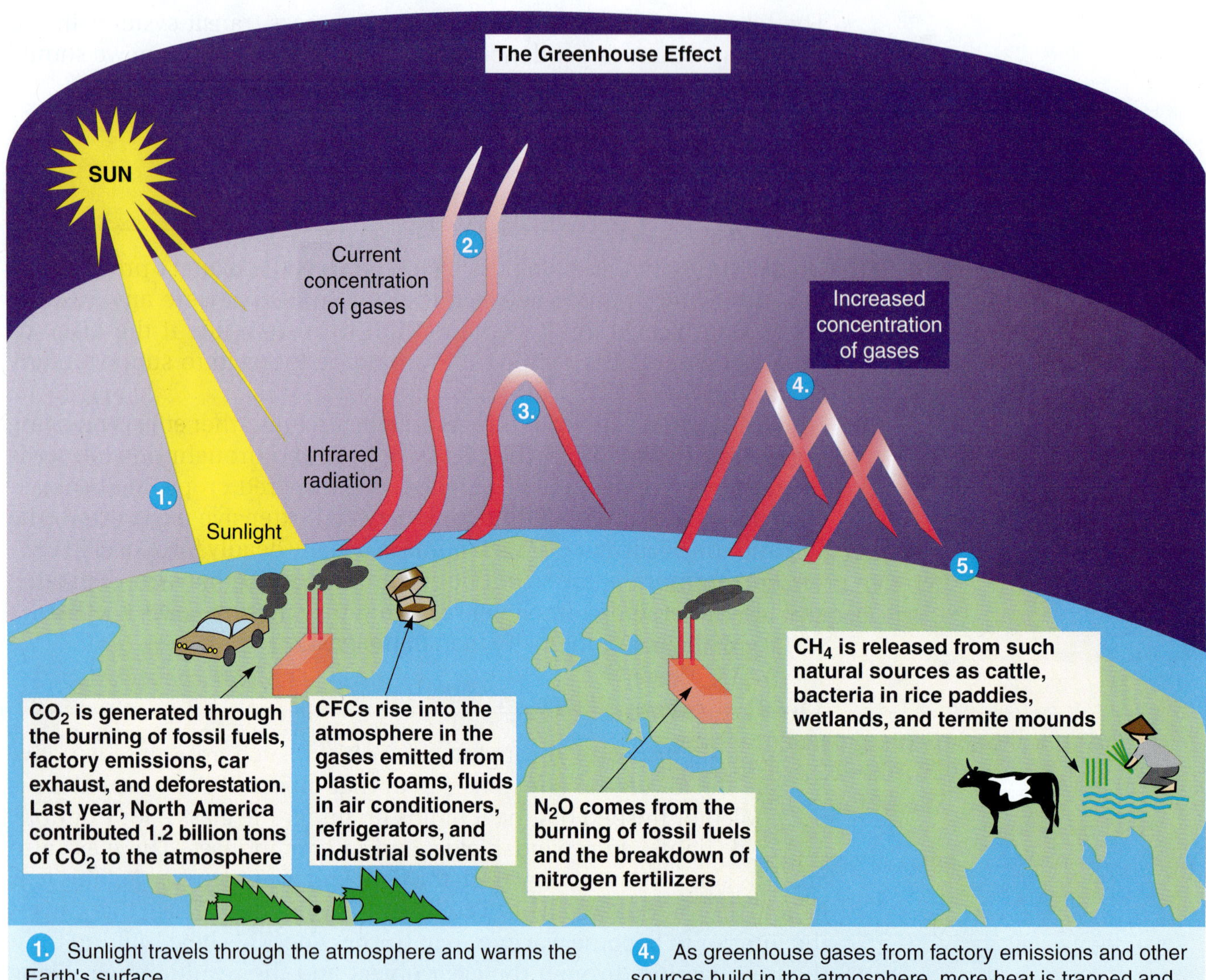

FIGURE 4–6 The greenhouse effect on land, sea, and air. (Adapted from material provided by the Electric Power Research Institute.)

of the damaging ultraviolet rays through. One specific thinner area over Antarctica—referred to as a "hole in the ozone layer"—was enlarging at an alarming rate. Scientists identified the buildup of CFCs as the cause of the thinning and the hole. Of equal concern is the buildup of air pollutants that could contribute to global warming even where the ozone layer has been damaged less than it was over the South Pole. Scientists and governments of the world are trying to find ways to make continued improvements in living conditions without experiencing additional reductions in air quality and other environmental factors.

Global Warming: A Topic for Scientific Debate

Scientists around the world are divided on the issue of global warming. Some of them insist that a recent upward trend in the temperature

Predicted Change	Global Average	Regional Average
Temperature	+4° to +9° F	−5 to +18° F**
Sea level	+4 to +40 in	
Precipitation	+7% to +15%	−20 to +20%
Direct solar radiation	−10 to +10%	−30 to +30%
Evaporation/Transpiration	+5 to +10%	−10 to +10%
Soil moisture	?	−50 to +50%
Runoff	Increase	−50 to +50%

Source: Schieder, S. (1990). Prudent planning for a warmer planet. *New Scientist,* 128(1743).

**To interpret, if the greenhouse effect produces the results that some scientists predict, these kinds of changes could occur. The average global temperature could increase between 4° and 9° F. Regional changes could range from a drop of 5° to an increase of 18° F in different parts of the world. Other categories of possible changes are given.

FIGURE 4–7 Global warming is believed to be caused by the greenhouse effect.

Framework Convention on Climate Change

of Earth's surface is evidence of dangerous warming of the global environment if current trends continue (Figure 4–8). They cite human activities as factors that contribute to the intensity of the greenhouse effect and to global warming. Other scientists cite the lack of scientific data to support the global warming theory. They believe that the trend toward a rise in global temperatures may be partly the result of long-term climatic cycles and weather patterns. Daily temperatures have

Carbon Dioxide (CO_2) The relative contribution of the greenhouse gas CO_2 to the global warming trend is expected to be about 50 percent by 2020

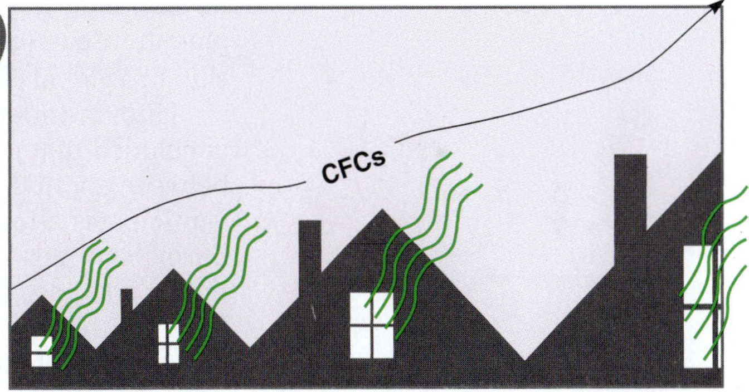

Chlorofluorocarbons (CFCs) About 25 percent of greenhouse effect by 2020

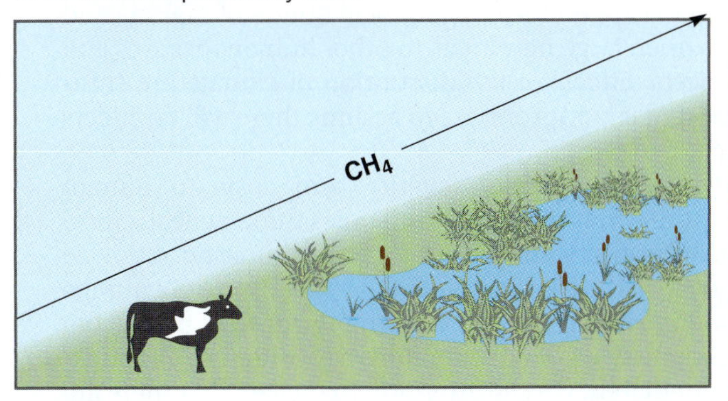

Methane (CH_4) About 15 percent of greenhouse effect by 2020

Nitrous Oxide (N_2O) About 10 percent of greenhouse effect by 2020

FIGURE 4–8 Increases in the concentrations of four major pollutants and their predicted effects on global warming by 2020.

been recorded for a little more than 100 years, and we do not have actual historical records from which to draw conclusions on climatic cycles and weather patterns of longer duration. An example of this line of thinking is evident in the following statement by Dr. Roy Spencer, senior scientist for climate studies at NASA's Marshall Space Flight Center:

> The adjusted satellite trends are still not near the expected value of global warming predicted by computer climate models. The Intergovernmental Panel on Climate Change's (IPCC) 1995 estimate of average global warming at the surface until the year 2100 is +0.18 deg. C/decade. Climate models suggest that the deep layer measured by the satellite and weather balloons should be warming about 30% faster than the surface (+0.23 deg. C/decade). None of the satellite or weather balloon estimates are near this value.

In contrast with this scientific view is one group of scientists' statement on global climatic disruption issued in June 1997:

> We are scientists who are familiar with the causes and effects of climatic change as summarized recently by the Intergovernmental Panel on Climate Change (IPCC). We endorse those reports and observe that the further accumulation of greenhouse gases commits the Earth irreversibly to further global climatic change and consequent ecological, economic, and social disruption. The risks associated with such changes justify preventive action through reductions in emissions of greenhouse gases. In ratifying the Framework Convention on Climate Change, the United States agreed in principle to reduce its emissions. It is time for the United States, as the largest emitter of greenhouse gases, to fulfill this commitment and demonstrate leadership in a global effort.
>
> Human-induced global climatic change is under way. The IPCC concluded that global mean surface air temperature has increased by between about 0.5 and 1.1 degrees Fahrenheit in the last 100 years and vanticipates a further continuing rise of 1.8 to 6.3 degrees Fahrenheit during the next century. Sea-level has risen on average 4–10 inches during the past 100 years and is expected to rise another 6 inches to 3 feet by 2100. Global warming from the increase in heat-trapping gases in the atmosphere causes an amplified hydrological cycle resulting in increased precipitation and flooding in some regions and more severe aridity in other areas. The IPCC concluded that "The balance of evidence suggests a discernible human influence on global climate." The warming is expected to expand the geographical ranges of malaria and dengue fever and to open large new areas to other human diseases and plant and animal pests. Effects of the disruption of climate are sufficiently complicated that it is appropriate to assume there will be effects not now anticipated.
>
> Our familiarity with the scale, severity, and costs to human welfare of the disruptions that the climatic changes threaten leads us to introduce this note of urgency and to call for early domestic action to reduce U.S. emissions via the most cost-effective means. We encourage other nations to join in similar actions with the purpose of producing a substantial and progressive global reduction in greenhouse gas emissions beginning immediately. We call attention to the fact that there are financial as well as environmental advantages to reducing emissions. More than 2000 economists recently observed that there are many potential policies to reduce greenhouse-gas emissions for which total benefits outweigh the total costs. . . .

The United Nations Framework Convention on Climate Change met in Milan, Italy, in 2003. There were 188 parties to the convention, and the following statement is part of an issued press release: "Ministers agreed that climate change remains the most important global challenge to humanity and that its adverse affects are already a reality in all parts of the world."

Precautions Against Global Warming

Reversing the perceived trend toward continued global warming and the problems it could bring may require real changes in the way we do some things. With good research, wise government, and environmentally-sensitive business, we can slow down or stop the decline of our air quality brought about by human-caused pollution. Even small-engine lawn and garden equipment is now seen as a serious contribution to pollution, and changes are needed to address this problem. For instance, leaf blowers create pollution through engine exhaust and by the consumption of electricity while expelling dust and dirt into the air. Experts now observe that a gasoline lawn mower can cause 50 times more pollution per horsepower than a modern truck engine (Figure 4–9). Similarly, a lawn mower running for just one hour may create as much pollution as an automobile traveling 240 miles. A chain saw running for two hours emits as much pollution from hydrocarbons as a new car running from coast to coast across the United States (Figure 4–10).

Carbon Dioxide (CO_2)

Carbon dioxide is a major product of combustion. Our robust appetites for food, clothing, consumer goods, heated and air conditioned spaces, and transportation have led to a cultural lifestyle that requires huge

FIGURE 4–9 A lawn mower may create more pollution per horsepower than a modern truck engine.

FIGURE 4–10 A chain saw running 2-hours emits as much pollution as a new car driving coast to coast across the United States.

volumes of fuels to be burned to raise food, manufacture goods, and move vehicles. Our highways in and around large cities are clogged with automobiles. Interstate highways are loaded with trucks and rivers and other waterways carry heavy shipping traffic. Homes and commercial buildings use extensive volumes of electricity for light and temperature control and farms and factories have huge machines and heating devices—all consuming fuel that expels carbon dioxide and other pollutants into the air.

Green plants can use some carbon dioxide from the air and convert it into plant food, oxygen, and water. However, in most parts of the world, the mass of green plants is being reduced even as the expulsion of pollutants increases dramatically. The practice of slash-and-burn agriculture in the jungle regions of the world threatens to remove vast areas and volumes of plant growth with little hope of replacement. Increased carbon dioxide in the atmosphere is projected to account for 50 percent of the increase in global warming by 2020.

Chlorofluorocarbons

Since the discovery of the damage that CFCs do to the ozone layer, other propellants have been used in pressurized spray containers, and CFCs are now recaptured from refrigeration units before they are discarded. Still, the escape of CFCs into the air and their effect on the ozone layer is projected to account for 25 percent of the increase in global warming by 2020.

global warming, myth

Methane (CH_4)

Most methane gas comes from naturally decaying plant materials such as leaves and debris on the soil surfaces of forests and jungles. Some of it is a product of decaying organic matter. Methane rises from piles, pits, and other accumulations of decaying animal manure, peat bogs, and sewage. Some large farms are now using carefully engineered systems to capture the gas from large manure-holding areas. The methane is then used as fuel for engines driving generators to provide electricity for the farm and for sale. Methane is projected to account for 15 percent of the increase in global warming by 2020.

air pollution, crop yields
nitrous oxide, pollution

Nitrous Oxide (N_2O)

Nitrous oxide has long been a troublesome pollutant from gasoline engines. The ever-increasing number of automobiles, trucks, tractors, heavy equipment, aircraft, chain saws, lawn mowers, boats, and other engine-driven applications has offset the tremendous improvements made in emission reduction from individual engines. Because nitrous oxide emissions are still increasing, it is projected that they will account for 10 percent of the increase in global warming by 2020.

Air and Living Organisms

Oxygen in the air is consumed by plants and animals during a process called *respiration*. Animals, as well as humans, use oxygen to convert food into energy and nutrients for the body. Animals breathe in or inhale to obtain oxygen. They exhale (breathe out) carbon dioxide gas. Plants release oxygen during the day through the process of photosynthesis

science connection profile

OVERCOMING THE EFFECTS OF AIR POLLUTION

U.S. Department of Agriculture (USDA) scientists are tackling what could be the toughest conflict of the century: the battle to breathe. Whereas ozone depletion is a serious problem in the upper atmosphere, ozone as a product of combustion hovers just above the Earth's surface as a pollutant that decreases air quality.

Researchers have estimated that U.S. farmers experience at least 1 billion dollars per year in lost crop yields because of air pollution. Cutting the level of ozone in the air by 40 percent would mean an extra $2.78 billion for agricultural producers.

USDA and University of Maryland scientists discovered that treatment of plants with the growth hormone ethylenediura (EDU) can reduce damage by ozone. EDU alters enzyme and membrane activity within the leaf cells where photosynthesis takes place. A single drenching of soil with EDU effectively protected some plants from damage and reduced the sensitivity of others to excessive levels of ozone in the air. Similarly, injection of EDU in the stems of shade trees in highly polluted areas could protect them from damage.

In addition to ozone, other major air pollutants that are damaging to crops include peroxyacetetyl nitrate, oxides of nitrogen, sulfur dioxide, fluorides, agricultural chemicals, and ethylene. The task of reducing the amounts of pollutants in the air and finding ways to reduce the effects of pollutants on living organisms will continue to challenge future generations.

A biological aide uses a porometer to measure the impact of CO_2 enrichment of the air on the transpiration rate and stomata activity of a soybean leaf. (Courtesy of the U.S. Department of Agriculture.)

INTERNET KEY WORDS

improving air quality
improving photosynthesis

(chlorophyll in green plants enables them to use light, carbon dioxide, and water to make food and release oxygen).

Maintaining and Improving Air Quality

Air quality can be improved by reducing or avoiding the release of pollutants into the air and removing existing pollutants. Specific practices that can help reduce air pollution include:

- stopping the use of aerosol products that contain CFCs;
- providing adequate ventilation in tightly constructed and heavily insulated buildings;
- having buildings checked for the presence of radon gas;
- using exhaust fans to remove cooking oils, odors, solvents, and sprays from interior areas;

science connection profile

RESEARCH TO IMPROVE PHOTOSYNTHESIS

Improving the efficiency of photosynthesis is one area of current interest in plant research. Scientists have identified an enzyme that plays an important role in the process. When it is present in large amounts, plant growth increases. Conversely, small amounts of the enzyme cause plants to grow slowly. In addition, the enzyme is known to react with oxygen during hours of darkness to reverse photosynthesis. This process is called *respiration*.

Scientists hope to be able to stimulate plant growth through more efficient photosynthesis by interrupting respiration during darkness. Total plant yields will increase if science is successful in maintaining daylight gains in plant tissues.

- regularly cleaning and servicing furnaces, air conditioners, and ventilation systems;
- maintaining all systems that remove sawdust, wood chips, paint spray, welding fumes, and dust to ensure that they function most efficiently;
- keeping gasoline and diesel engines properly tuned and serviced;
- keeping all emissions systems in place and properly serviced on motor vehicles;
- observing all codes and laws regarding outdoor burning;
- reporting any suspicious toxic materials or conditions to the police or appropriate authorities;
- reducing the use of pesticide sprays as much as possible; and
- using pesticide spray materials strictly according to label directions.

Regulating Air Quality

The United States Congress has passed various laws to prevent the loss of air quality and clean up existing air quality problems. The first significant laws were the Clean Air Act of 1963 and the Air Quality Act of 1967. These required reductions in releases of industrial pollutants into the atmosphere. The laws have been expanded and updated several times since then.

The 1970 Clean Air Act has helped to reduce air pollution. Tall smoke stacks that disperse harmful gases over larger areas instead of eliminating their release altogether are no longer legal for pollution control. The current law requires each state to develop an implementation plan that describes how the state will meet the act's requirements.

The Clean Air Acts of 1970 and 1990 are the most far-reaching of the air quality laws. From these versions, a series of clean air amendments has been approved. New federal and state agencies have been created to interpret and enforce the laws. Among these are the Environmental Protection Agency (EPA), the Office of Air Quality Planning and Standards, the Alternative Fuels Data Center, and the Commission for Environmental Cooperation. Each government office has created new regulations and standards for air quality.

preserve wildlife

Among the federal standards that have been implemented are the Clean Air Act's National Ambient Air Quality Standards, the Clean Air Act's New Source Performance Standards, the Prevention of Significant Deterioration, Air Guidance Documents from the EPA's Office of Air and Radiation, and updated air quality standards for smog (ozone) and particulate matter. Each of these air quality standards is intended to reduce air pollution and improve air quality.

As people begin to experience the effects of pollution on the atmosphere in the form of lung and skin diseases, there will be greater motivation to solve the problems created by harmful atmospheric gases. Humans probably will not do much specifically to improve air quality for wild animals, but wild creatures will benefit when humans improve air quality for themselves.

Looking Back

Air becomes polluted by waste gases and particles from the combustion of wood, coal, and petroleum products. All of these resources can be protected, and attempts to reclaim damaged or polluted resources are being made using the technologies available through modern science.

Student Activities

1. Write the terms to know from the beginning of the chapter and their meanings in your notebook.
2. Make a pie chart that illustrates the components of air.
3. Stand in a safe location five feet or so behind and to the side of a parked truck or bus with its gasoline engine running. Notice the smell of the exhaust. Stand in the same relative position from an automobile that is less than three years old or has traveled less than 40,000 miles. What differences do you observe in the truck and automobile exhausts? Why? Which do you think causes more air pollution?
4. Examine three different aerosol cans. Which products use chlorofluorocarbons as the propellant? Why is it unwise to use such products?
5. Talk with an automobile tune-up specialist about the effect of engine adjustments on the content of exhaust gases.
6. Ask your teacher to invite an air pollution specialist to your class to discuss the problems of air pollution in your town, county, or state.
7. Do a research project on the greenhouse effect and its relationship to global warming.
8. Obtain a radon test kit and test for radon in your home.
9. Create four original drawings showing the four major pollutants. Under each drawing, write one sentence that explains where this pollution comes from and what can be done to improve air quality.

Self-Evaluation

Discussion and Essay

1. Identify the gases that make up earth's atmosphere and list the approximate percentage of each.
2. What are the major pollutants that damage the air and where do they come from?
3. Discuss the effects of acid precipitation on living organisms and the environments in which they live.
4. How is radon gas a problem in the air we breathe, and what can be done about it?
5. In what ways do Chlorofluorocarbons (CFCs) affect the ozone layer of the earth?
6. Why are particulates harmful, and how can they be removed from the air?
7. What has been done to reduce the hazard caused by asbestos particles that are suspended in the air?
8. What are some of the points of disagreement among scientists over the highly controversial topic of man-caused global warming?
9. Discuss the effects of acid precipitation on wild creatures and the environments in which they live.
10. Name the most significant sources of air pollution.
11. Suggest ways that air pollution can be reduced or eliminated.

Multiple-Choice Questions

1. Air is
   a. 78 percent argon.
   b. 21 percent nitrogen.
   c. 21 percent oxygen.
   d. 10 percent carbon dioxide.
2. Pure water is
   a. a mixture of gases.
   b. metallic tasting.
   c. one part hydrogen to two parts oxygen.
   d. odorless.
3. Without proper air to breathe, a human can survive only approximately
   a. 6 minutes.
   b. 12 minutes.
   c. 2 hours.
   d. 12 hours.
4. Radon gas is a widespread threat to air quality
   a. on the highway.
   b. in factories.
   c. in homes.
   d. in wooded areas.
5. Radioactive dust is probably caused by
   a. improperly adjusted furnaces.
   b. cracks in basement floors.
   c. a damaged ozone layer.
   d. nuclear reactions.
6. Chemicals used to kill insects are called
   a. pests.
   b. pesticides.
   c. pollutants.
   d. toxic materials.
7. One ingredient *not* associated with photosynthesis is
   a. carbon dioxide.
   b. oxygen.
   c. radon.
   d. water.

8. Chlorofluorocarbons have been found to damage
   a. aerosol sprays.
   b. the ozone layer.
   c. refrigeration units.
   d. water pumps and equipment.
9. Poisonous gas we cannot remove from auto exhaust is
   a. carbon monoxide.
   b. hydrocarbons.
   c. nitrous oxides.
   d. radon.
10. The most reliable source of information on the use of a pesticide is
   a. experienced applicators.
   b. an extension service.
   c. personal experience.
   d. the product label.
11. The buildup of heat resulting from sunlight passing through glass and heating trapped air in the interior area is called
   a. the greenhouse effect.
   b. infrared energy.
   c. radiation.
   d. ultraviolet.
12. A product of decaying plant or animal matter is
   a. chlorofluorocarbons.
   b. methane.
   c. nitrous oxide.
   d. ozone.

Matching

1. carbon monoxide
2. chlorofluorocarbons
3. hydrocarbons
4. lead
5. nitrous oxides
6. ozone
7. radon
8. pests
9. photosynthesis
10. sulfur

a. tetraethyl
b. pale yellow
c. five percent of auto exhaust
d. damages ozone layer
e. filters ultraviolet rays
f. diseases, insects, weeds
g. chlorophyll, light, carbon dioxide
h. causes death from auto exhaust
i. leaks into houses
j. pollutant from autos and factories

Learning Activities

1. Identify the pollutants that are most often responsible for reducing air quality. Create a bulletin board in the classroom that illustrates the sources of these pollutants.
2. Invite an official from a government agency such as the EPA or the Department of Environmental Quality (DEQ) to do a class presentation on air quality. After the presentation, conduct a visual inspection of your community to identify potential sources of air pollution. Be sure to note that visible emissions, such as steam, do not necessarily indicate serious pollution problems. Consider ways to control pollutants that are identified.
3. In groups of three or four students, identify one source of air pollution. Make a suggestion for a new law, in the form of a main motion, that would improve the air quality in your community.
4. Create four original drawings that show the four major pollutants. Under each drawing, write one sentence that explains where this pollution comes from and what can be done to improve air quality.

TERMS TO KNOW

potable
ion
cation
anion
salt
organic matter
freshwater
saltwater
tidewater
universal solvent
precipitation
evaporation
watershed
water table
fertility
saturated soil
free water
gravitational water
capillary water
hygroscopic water
purify
pH
acid
alkaline
thermal stress
heat capacity
solubility
salinity
buffer
water hardness
turbid
dissolved oxygen
saturation
aeration
anaerobic

CHAPTER 5

water

Objectives

After completing this chapter, you should be able to

- define water and related terms
- identify uses of water in the environment that qualify it as an essential nutrient
- describe the makeup of water in nature
- distinguish between anions and cations
- distinguish between freshwater and saltwater
- explain why water is described as the universal solvent
- illustrate a food chain that exists in a water environment
- define some relationships that exist between land and water
- clarify the function of a watershed
- describe natural processes that occur in wetlands and marshes to remove pollutants from surface water
- explain how land functions as a reservoir to store water
- list sources of pollution that threaten the purity of groundwater in aquifers
- distinguish the differences among the forms of soil moisture: free water, gravitational water, and hygroscopic water
- identify some ways to improve water quality
- list some of the characteristics of water that must be considered in evaluating water as a living environment
- list some tests that are used to evaluate the suitability of water as a living environment for plants and animals
- explain how aquatic plants and animals may be affected by trace elements and gases that are dissolved in water environments
- suggest several practices that each of us might implement in our homes and communities to conserve clean water

FIGURE 5–1 Seventy percent of the Earth's surface is covered with water. (Courtesy of Getty Images.)

We live on the water planet. Most of the Earth's surface is covered with water (Figure 5–1), and the oceans and lakes are vast. Most people around the world, including in the United States, live near an ocean, river, lake, or stream. Those who do not live near such bodies of water must have access to water from deep wells. Like the tissues of plants and other animals, our bodies are approximately 55 to 70 percent water, depending on maturity, fat content, and the individual's sex. We can survive only a few days if our supply of **potable** water—that is, drinkable and free from harmful chemicals and organisms—is cut off.

Flying over the vast continents of the Americas, Europe, Asia, and Africa, we might feel that the landmass of the Earth is an endless resource (Figure 5–2). The great oceans of the world, however, combine to provide an even larger area. Even so, both land and water resources have become limited, and there is genuine concern that we are rapidly depleting them. In developed countries, a safe and adequate water supply generally faces only temporary shortages and mild inconveniences such as restrictions on water use for nonessential tasks such as washing cars and watering lawns. In third world countries, however, a safe water supply is a luxury.

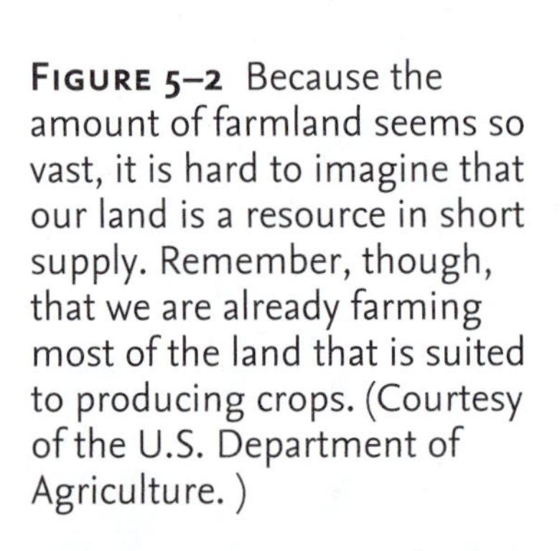

FIGURE 5–2 Because the amount of farmland seems so vast, it is hard to imagine that our land is a resource in short supply. Remember, though, that we are already farming most of the land that is suited to producing crops. (Courtesy of the U.S. Department of Agriculture.)

Although there seems to be a sufficient volume of water in most areas of the world, supplies of safe water are generally insufficient because of misuse, poor management, waste, and pollution.

The Nature of Water

Water is essential for all plant and animal life (Figure 5–3). It dissolves and transports nutrients to living cells and carries away waste products. It stores heat and helps maintain a more constant temperature in the environment. Evaporating water cools the surfaces of leaves and plants and the bodies of animals and humans when the temperature is uncomfortably high. Water serves so many useful functions that a sufficient supply of potable water is one of the first considerations for a healthy community.

The action of water helps to create soil by breaking down rock into small particles. It cleans the environment by diluting contaminants and flushing them away from vulnerable areas. Water also provides protection to some species of animals (Figures 5–4 and 5–5).

Figure 5–3 Clean, fresh water serves many life-supporting functions, including carrying nutrients, transporting waste, working as a coolant, providing homes for aquatic plants and animals, supplying oxygen for fish, and operating as a cleanser for humans and animals. (Courtesy of U.S. Department of Agriculture.)

Figure 5–4 Water environments are used by beavers and some other animals as protection from predators.

FIGURE 5–5 Pure water is one of the most valuable and important resources in the environment.

INTERNET KEY WORDS

ions, water
cations, anions, water

Water is formed by the association of two hydrogen atoms and one oxygen atom: H_2O or HOH. In nature, water is not a pure substance. To some degree, it consists of many dissolved and suspended substances. Unless the water is particularly muddy, these substances consist predominately of dissolved ions. **Ions** are elemental forms or groups that carry an electrical charge in solution. Positive ions are called **cations**, and examples are the elements calcium, potassium, magnesium, and sodium. Negative ions or **anions** are molecular substances such as carbonate, bicarbonate, sulfate, and chloride (Figure 5–6).

Substances that dissolve readily in water to form simple ions are called **salts.** Because during evaporation many of the dissolved ions in water would combine to form salt compounds, the content of simple ions in solution is frequently referred to as *salt content.* A sample of natural surface water also contains **organic matter**, which consists of living material, its excretions, and decomposing material. This is usually only a small percentage of a residue. Organic matter can affect water chemistry by requiring oxygen for living processes as the complex organic matter is built up from and torn down into simple compounds. Gases are also present in water, either dissolving at underground pressures in the case of groundwater or under normal pressure at the surface from the gas mixture of the air. Some gases enter the water as they are formed by living agents in water or by chemical release from mud deposits in the bottom.

Fresh Versus Salt Water

Most of the water on Earth is saltwater, not freshwater; except for transportation, it is not suitable for human use. **Freshwater** refers to water that flows from the land to oceans and contains little or no salt. The water in our oceans and some inland lakes contains heavy concentrations of salt. This water is called **saltwater**. Similarly, our bays and tidewater rivers contain too much salt for domestic or household use. **Tidewater** refers to the water that flows up the mouth of a river with rising or inflowing ocean tides. Saltwater is not fit for animal consumption or for plant irrigation.

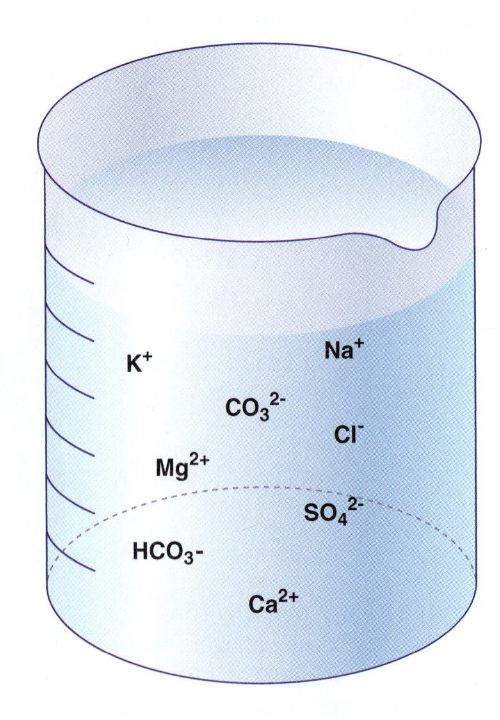

FIGURE 5–6 Cations (positive ions) and anions (negative ions) in water. (Courtesy of Rick Parker.)

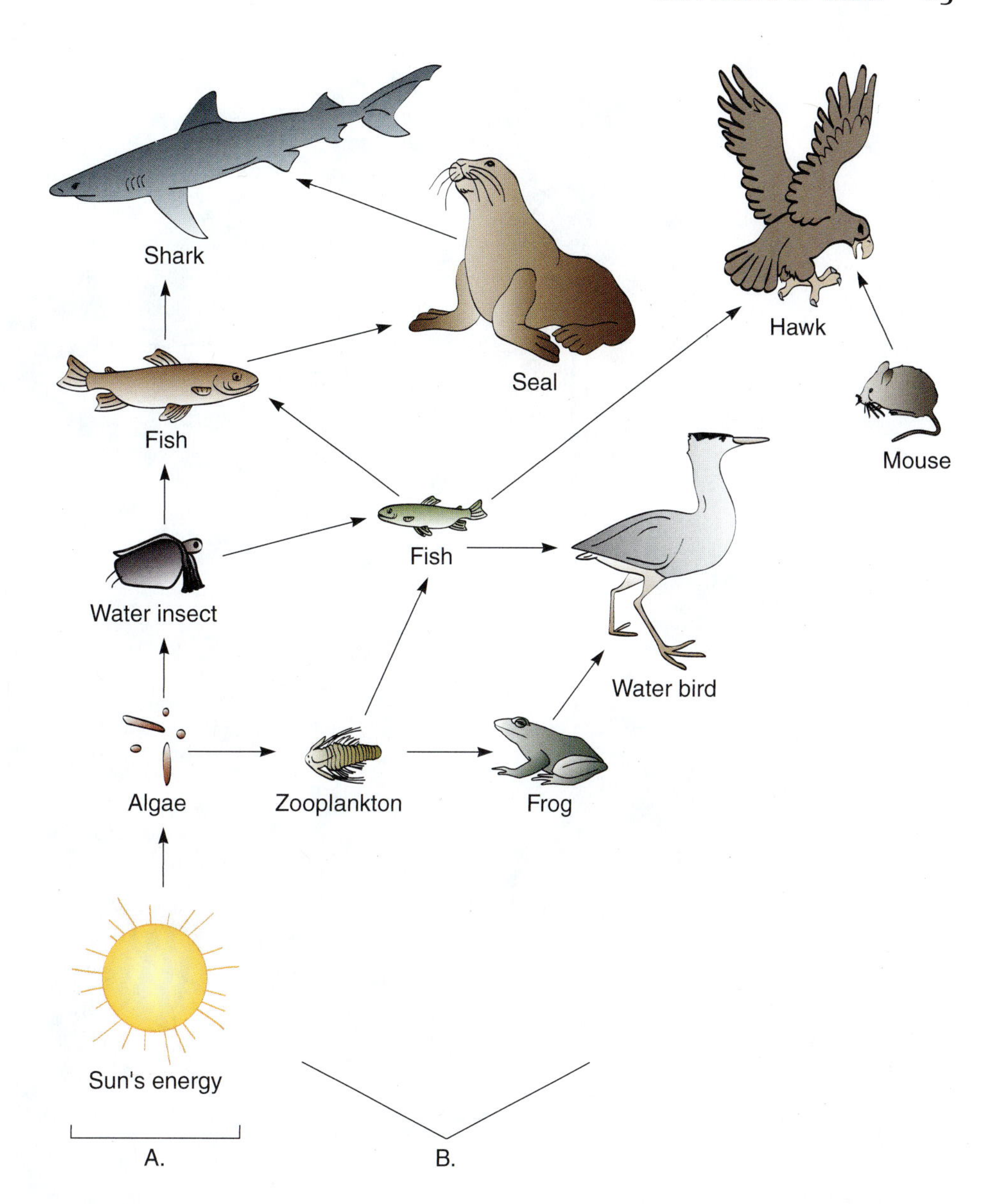

FIGURE 5–7 (A) A food chain model showing nutritional energy flowing in one direction (B) A food web model showing the nutritional energy flow throughout the ecosystem.

food chain web

water, universal solvent
water pollution, oceans, rivers

Aquatic Food Chains and Webs

A *food chain* is made up of a sequence of living organisms that eat and are eaten by other organisms living in the community (Figure 5–7). Each member of the chain feeds on lower-ranking members of the chain. The general organization of an aquatic food chain moves from organisms known as *producers* (food plants) to *herbivores* (plant-eating water insects). Herbivores are eaten, in turn, by *carnivores* (meat-eating animals).

The Universal Solvent

Water has been described as the **universal solvent** (a substance that dissolves or otherwise changes most other materials). Nearly every material will rust, corrode, decompose, dissolve, or otherwise yield to the presence of water. Therefore, water is seldom seen in its pure form. It generally has something in it.

Some minerals in water are healthful and give the water a desirable flavor. However, water sometimes carries toxic or undesirable chemicals or minerals. Water also may contain decayed plant or animal remains,

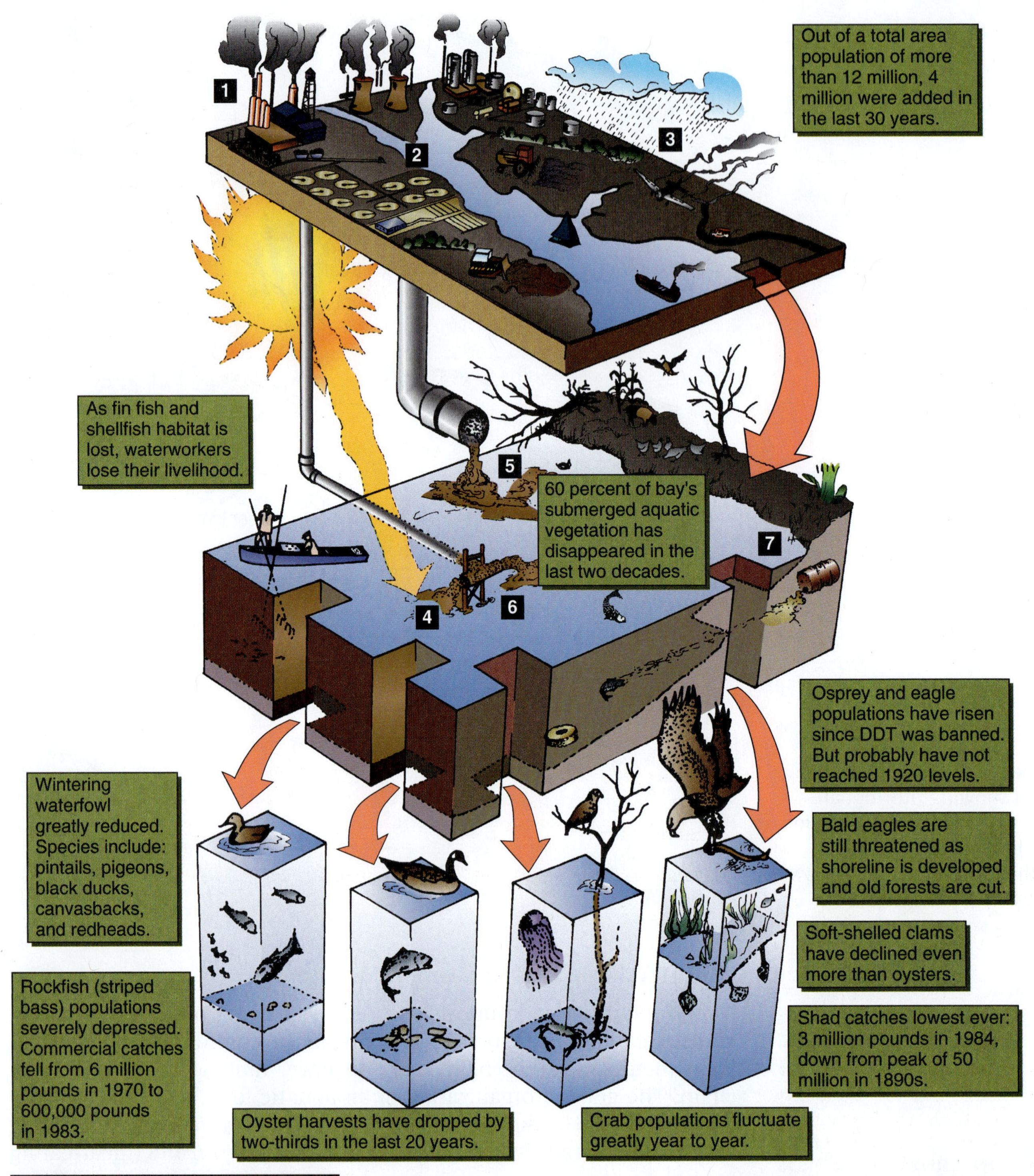

Bay Facts

Size: 195 miles long, 3.5 to 30 miles wide. Average depth: 24 feet.
Drainage basin: 64,000 square miles, 50 major tributaries feed it.
Water: Fresh water from tributaries mixes with salt water from Atlantic. Bay acts like sink, trapping pollutants. Only one percent are flushed out to sea.
Plants, animals: More than 2,000 species in bay shoreline.

1 Industrial wastes: Thousands of commercial, industrial facilities discharge water containing toxic chemicals, metals, nitrogen, phosphorous into bay. Also: cooling needs can lead to corrosion, chlorine contamination.

2 Municipal sewage: Water discharged from treatment plant contains nitrogen, phosphorous, toxic chemicals.

3 Rain runoff: Sediment from farms, forests, urban areas carries fertilizers, pesticides, herbicides. Over past 30 years, farmers have doubled the amount of fertilizers they use. Since 1960, herbicide use has tripled.

4 Nutrients: High nutrient levels are most severe on northern, middle bay areas and tributaries, leading to excessive algae growth and less light filtering down to allow grasses to grow below water surface. This gives waterfowl less food. Algae dies quickly, using up oxygen in the water and making it unsuitable for aquatic life. This leads to fewer fish, shellfish, oysters.

5 Toxic chemicals: Rain washes sediment into bay. This often includes chemicals, herbicides, pesticides, toxic wastes from industry.

6 Chlorine: It's used in bay for disinfecting drinking water, sewage, industrial processes. It's suspected of hindering spawning runs of migrating fish.

7 Heavy metals: Cadmium, mercury, copper from industrial wastewater, sewage treatment plants can be toxic. Lead from auto exhaust, iron and zinc from industrial discharge, shore erosion.

FIGURE 5–8 The water resources on which the world depends are becoming stressed because of pollution and the inability of aquatic species to maintain stable populations. (Adapted from and used with permission of *The Washington Post* and *MD Magazine*, Autumn 1988.)

disease-causing organisms, or poisons. Ocean water may be described as a thin soup. It is like our blood. It gathers and transports nutrients, and it is the habitat for microorganisms. Also like blood, it carries life-supporting oxygen and neutralizes and removes wastes.

Scientists tell us that the purity levels of our rivers, bays, and seas are in trouble. People are polluting air, water, and land faster than nature can cleanse and purify these resources (Figure 5–8). The sea contains all of the dissolved elements carried by the rivers through all of the Earth's history to low places on the planet's crust. The water itself may evaporate and be carried back to the land as moisture in the clouds. Eventually it may fall to the land again as rain or snow. However, the minerals remain behind in the ocean. Many of those substances may be toxic or otherwise threatening to organisms living in the sea. The water that falls as rain or snow would be pure if it were not contaminated by pollutants as it falls through the air.

The Water Cycle

INTERNET KEY WORDS

properties of water, hydrologic cycle

Moisture evaporates from the Earth, plant leaves, freshwater sources, and the seas to form clouds in the atmosphere. Clouds remain in the air until warm air masses meet cold air masses. This causes the water vapor to change to a liquid and fall to the Earth's surface in the form of rain, sleet, or snow. Large amounts of the Earth's water supply are stored for long periods of time. Storage occurs in aquifers beneath the surface of the Earth, in glaciers and polar ice caps, in the atmosphere, and in deep lakes and oceans. Sometimes this water is stored for thousands of years before it completes a single cycle. Gravity draws the water back into the ground and causes it to flow from high elevations to low elevations. The cycling of water among the water sources, atmosphere, and surface areas is called the *water cycle*. The water cycle was discussed in greater detail in Chapter 2.

science connection profile

FOOD CHAINS AND WEBS

Many species of animals and plants live in or near water environments. Bays and oceans provide excellent support for the growth of algae, which in turn become major food sources for water-dwelling insects. Insects provide nutritional energy for shellfish. Shellfish are eaten by larger fish, which are finally consumed by top-level predators such as sharks. In this example, energy is passed from one organism to another in the form of food. This model, referred to as a *food chain*, describes the interdependence of plants and animals for nutrition. All food chains begin with the sun, which is the primary energy source for most living organisms.

Producers at the base of the food chain capture solar energy and convert it into nutrients such as sugars and proteins. Plants and algae are good examples. The next step in the food chain takes place when a consumer eats a producer. Animals that depend on producers or other animals for food are called *consumers*. In a simple food chain, energy is passed from the sun to the producers and then to consumers in one straight line. In most ecosystems, however, feeding relationships are much more complex than a single food chain suggests. *Food webs*, which are made up of many different overlapping food chains, represent the sum of all feeding relationships in an ecosystem. Using the example on page 79, it is easy to see the involvement of other organisms. For example, birds will also eat fish, as do sharks. Water insects may eat different aquatic plants than algae. Both foods chains and food webs explain the feeding relationships in given ecosystems and can be observed on land and in water.

The energy that drives this cycle comes from two sources: solar energy and the force of gravity. Water is constantly recycled. A molecule of water can be used over again and again as it moves through the cycle. Solar energy trapped by the ocean is a source of heat that causes evaporation. Additional water enters the atmosphere by evaporating from soil and plant surfaces, especially in areas of hot temperatures and high precipitation.

Relationships of Land and Water

Precipitation

Land and water are related to each other in many ways. Land in cold regions or high altitudes retains moisture on its surface in the form of snow. This moisture is then released gradually to feed the streams and rivers after the **precipitation** (moisture from rain and snow) has stopped falling. Precipitation is caused by the change of water in the air from a gaseous state to a liquid state. It then falls to the land or bodies of water. Moisture-laden, warm-air clouds contact cold-air masses in the atmosphere and the result is precipitation. Clouds are formed by water changing from a liquid to a gas when it is evaporated by air movement over land and water. **Evaporation** means changing from a liquid to a vapor or gas.

Water and Watersheds

watershed, rain, snow

A **watershed** is a large land area in which water from rain and snow is absorbed into the soil to emerge as springs of water or artesian wells at lower elevations (Figure 5–9). The springs merge to form streams and rivers (Figure 5–10). Each watershed is separated from other watersheds by natural divisions or geological formations, and each is drained to a particular stream or body of water. Watersheds are valuable because they act like huge sponges, soaking up water from precipitation and melting snow and releasing it slowly.

Forested areas provide ideal conditions for controlling the flow of water as it enters and leaves a watershed. Vast expanses of forests and

FIGURE 5–9 A forested watershed is ideal because much of the seasonal moisture tends to be absorbed rather than running off the land.

FIGURE 5–10 A clean, pure water supply is available only when the source watershed is well managed.

other vegetation are needed to regulate a uniform flow of water in rivers and streams. A watershed that is forested is superior to other watersheds because trees tend to slow the melting of snow, allowing it to be absorbed into the soil instead of running off the soil surface. Slow-flowing water infiltrates into the soil and comes out of the earth purified from most contaminants. Trees aid in water absorption by reducing the depth to which soils become frozen. This allows water to seep into the soil sooner than it does in unprotected areas. Snow usually lasts several weeks longer in forested areas than it does in open areas, helping to maintain constant stream flows (Figure 5–11).

Precipitation on land surfaces can cause severe flooding and soil erosion unless plants are available to slow the flow of water over the land. Where forests have been cleared from watersheds, flooding of lowland areas becomes much more severe, occurring when melted snow or runoff water from rain exceeds the rate at which the water can be absorbed into the soil. Frozen soil prevents water from being absorbed down through the soil profile. Deep-rooted plants improve the infiltration of water into the soil by breaking up hard layers. Smaller, more fibrous roots tend to hold soil particles in place when water moves across the soil surface.

Streams that flow through forests are protected from the heat of the sun by the shade of trees and brush growing along their banks. This is important to the survival of some kinds of fish such as trout that require cool

FIGURE 5–11 A healthy watershed is important in stabilizing stream flows.

FIGURE 5–12 Logs and fallen trees that become submerged in streams provide cover and protection for fish and enhance their habitat.

FIGURE 5–13 Stream bank vegetation provides protective cover for fish.

water temperatures. Fish also need places in the water where they can go to escape predators (Figure 5–12). In this way, trees and brush contribute to fish survival by providing cover on and beneath the surface of the water (Figure 5–13). Cover plants also provide insect habitat, and the fish feed on insects that drop into the water from these plants.

Marshes and Wetlands

INTERNET KEY WORDS

natural water purification,
marshes
wetlands

Some of the most valuable resources in nature are marshes, swamps, and estuaries (Figure 5–14). These areas are natural water-treatment facilities that remove many contaminants from surface waters. They require no supplemental energy sources, and they function without human interference. The cleansing agents are plants, bacteria, and other aquatic organisms. Phosphates, nitrates, and some other contaminants of water

FIGURE 5–14 Marshes and swamps act as natural water-treatment facilities by removing a variety of contaminants from surface water. (Photo courtesy of the U.S. Department of Agriculture.)

are also nutrients to plants and other organisms found in slow-moving water environments that take up such contaminants from surface water. The water flowing out of marshlands is substantially cleaner than it was when it entered.

Water is also cleansed by plants as they take in contaminated water and release large amounts of clean water vapor into the atmosphere through the process called *transpiration,* a controlled evaporation process by which plants lose water through pores in their leaf surfaces (Figure 5–15). These pores open wide when leaf surfaces are hot and close when leaf surfaces are cool. Transpiration creates a negative pressure that draws water and nutrients into the plant from the soil.

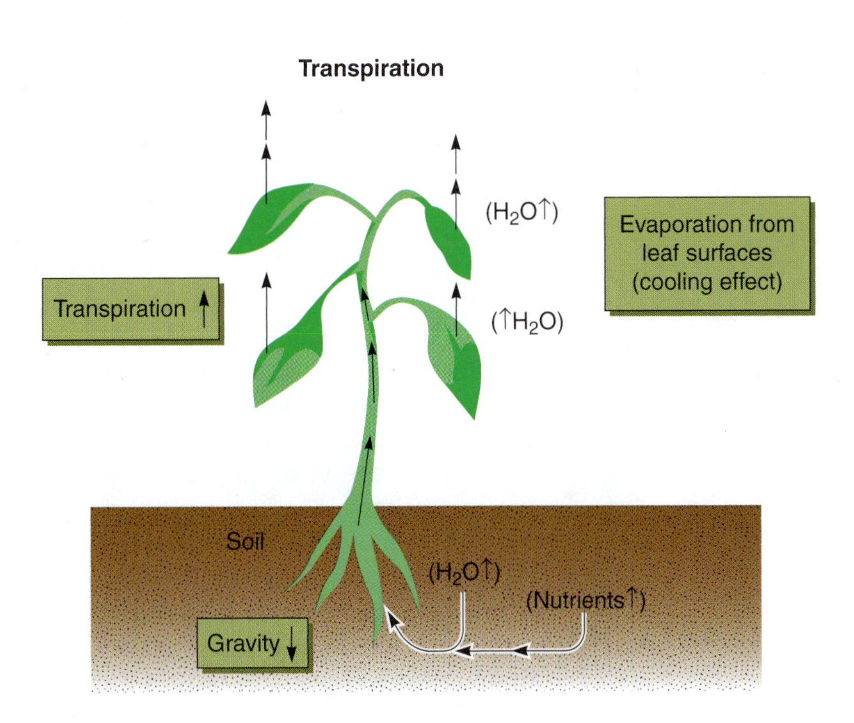

FIGURE 5–15 Transpiration releases water from the surfaces of plant leaves into the atmosphere.

Interest Profile

CONSTRUCTED AND NATURAL WETLANDS

Constructed wetlands are human-made marshlands that can be used to treat polluted water. They contain plants that tolerate standing water and provide habitat for bacteria that are capable of breaking down materials that cause water pollution. As water passes slowly through the marsh, these bacteria and other microbes cleanse the water by consuming pollutants and changing them to harmless compounds. Some water plants also take up pollutants.

A natural wetland performs the same function of removing impurities from the water. Many natural wetlands, however, have been drained to prepare land for farming. Currently, we are beginning to recognize the value of the wetlands.

Constructed wetlands usually consist of rectangular plots arranged in a series that are filled with gravel or porous soil and lined to prevent pollutants from leaching into the groundwater. Plants are added to the plots to simulate a natural marshland. Constructed marshlands do a good job of mimicking the advantages of natural marshlands. They also help to create an excellent wildlife habitat.

Land as a Reservoir

In more ways than one, land serves as a container or reservoir for water. Where water soaks down into the soil, it forms a **water table**; below that level, soil is saturated or filled with water. Water held below the water table may run out onto the Earth's surface at a lower elevation in the form of springs. Springs feed streams that in turn form rivers and flow into bays and oceans. Since ancient times, people have known to dig wells below the water table to extract water for human needs.

water table
human-made water reservoir
groundwater, terms

Even above the water table, excess water held in the soil is taken up by plant roots, from which water travels throughout the plant. Much of it evaporates from the leaves through transpiration to contribute to the moisture supply in the atmosphere.

Water also helps soil by improving its physical structure. It is also essential for microorganisms that live in the soil and contributes to the soil's **fertility** (the amount and type of nutrients in the soil).

Groundwater

Much of the potable water supply is groundwater. It is stored underground in huge natural aquifers that underlie much of the land surface. It fills in the spaces between fractured bedrock, stones, gravel, and sand. Some of the aquifers in the United States and Canada are massive. The largest is the Ogallala Aquifer that underlies the Great Plains states. Aquifer groundwater is the source of the water that flows in the rivers and streams that crisscross the continent. Aquifers also supply drinking water and irrigation water obtained through wells and pumps.

Despite the vastness of the groundwater supply, the water level in many aquifers is dropping. Efforts are underway in some regions to recharge the aquifers with excess water from rivers and streams. One thing is clear: We must follow fundamental rules when we take water from aquifers. First, we must not pump groundwater at a rate exceeding an aquifer's recharge rate. Second, we must not allow future priorities to divert recharge water to other uses. Third, we must not allow aquifers

science connection profile

SAFE-GUARDING GROUNDWATER

Authorities estimate that nearly 330,000 tons of pesticides are applied yearly on U.S. crops. Added to this are the fertilizers that are used on farms, yards, gardens, parks, and so on and the chemicals used to control fleas, flies, mosquitoes, ticks, termites, roaches, and other pests of livestock and humans. Although agricultural chemicals have enabled us to feed and clothe ourselves and many others in the world, pollution from these chemicals has become a threat to our air, water, and land.

Applying pesticides and fertilizers where they are intended and keeping them there until they change or biodegrade into harmless products is a major goal in protecting our water supplies. To help safeguard our surface water and groundwater, the Agricultural Research Service of the U.S. Department of Agriculture (USDA) has developed a specific strategy that has helped shape research priorities. The ultimate goals of the plan are to: (1) provide U.S. farmers with cost-effective best-management practices that will ensure ample supplies of food and fiber at a reasonable cost while reducing pesticide movement into the groundwater; (2) identify the factors that accelerate or retard pesticide movement; and (3) provide computer models that will quickly and accurately predict contamination.

USDA water research is being concentrated in three areas to protect against contamination of groundwater: (1) agricultural watershed management, (2) irrigation and drainage management, and (3) water quality protection and management.

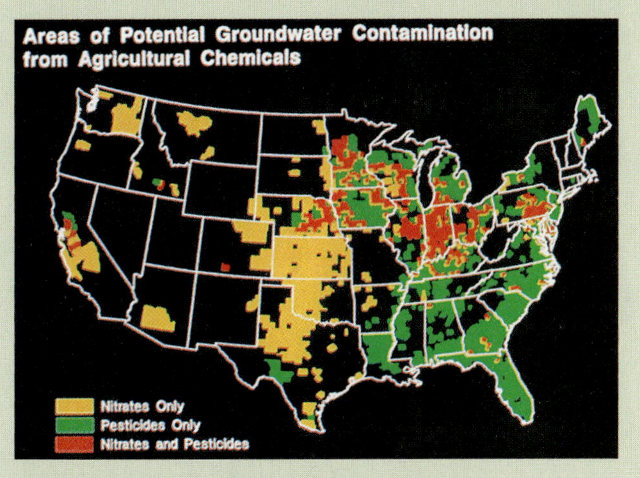

Groundwater contamination by agricultural chemicals is considered a risk in many states. (Courtesy of the U.S. Department of Agriculture, Agricultural Research Service, August 1989, p. 2—USDA/ERS, AER 576. Neilson & Lee, "The Magnitude and Cost of Groundwater Contamination from Agricultural Chemicals: A National Perspective," 1987.)

The computer models currently being used and developed are decision-enhancing tools that use a database and computer program to help select management practices such as pesticide processes related to movement in various soils, climates, and other environmental conditions. These models integrate data from many sources that are frequently not available to decision makers. New technologies, such as advances in slow-release formulations, improved pesticide application scheduling, and selective placement, are also under study.

to become polluted by sewer systems, chemicals, pesticides, or other natural or human-produced contaminants.

Improving Water Quality

agricultural watershed management examples

How can we reduce water pollution? How can soil erosion be reduced? What is the most productive use of water and soil that does not pollute or lose these essential resources? These important questions deserve answers now. Farmers have long appreciated the value of these resources and generally have used them wisely. Economic conditions, governmental policies, production costs, farm income, personal knowledge, and other factors, however, also influence the use of conservation practices by farmers and other land users. Every citizen, business, agency, and industry affects air and water quality. Similarly, all of us have some influence on how our land and water resources are used.

science connection profile

GROUNDWATER BASICS

A **saturated soil** occurs when water fills all of the spaces or pores in the soil. If soil remains saturated for too long, plants will die from a lack of air around their roots. The water that drains out of soil after it has been wetted is called **free water** or **gravitational water.** Gravitational water feeds wells and springs. When gravitational water leaves the soil, some moisture—**capillary water**—remains, and this can be taken up by plant roots. Water that is held too tightly for plant roots to absorb is called **hygroscopic water** (Figure 5–16).

Groundwater is easily polluted by chemicals and manure through abandoned wells and other depressions. In addition, constant irrigation can increase salt concentration. These and other problems are being addressed by soil conservation and water-management districts.

FIGURE 5–16 Although plants use only capillary water, hygroscopic water contributes to soil structure, and gravitational water is held in reserve for the future use of both plants and animals.

how to purify water

Improved water quality can be achieved by proper land management, careful water storage and handling, and appropriate water use. Once clean water is mixed with contaminants, it is not safe for use for other purposes until it is cleaned and purified. **Purify** means to remove all foreign material. Several general efforts can help to improve water quality: controlling water runoff from lawns, gardens, feedlots, and fields; keeping soil covered with plants; constructing livestock facilities so manure can be collected and spread on fields; and, where feasible, using no-till or minimum tillage practices to produce crops.

More specifically, each of us can follow these guidelines: Dispose of household products carefully. Many products under kitchen sinks, in basements, or in garages are threats to clean water (Figure 5–17). Never pour paints, wood preservatives, brush cleaners, or solvents down the sink. It is also unwise to wash such materials from equipment or pour left-over materials onto city streets because the contaminants can flow into storm drains and eventually enter the water supply, rivers, and oceans.

Use all products sparingly and completely. Put solvents into closed containers, allow the suspended materials to settle out, and then reuse

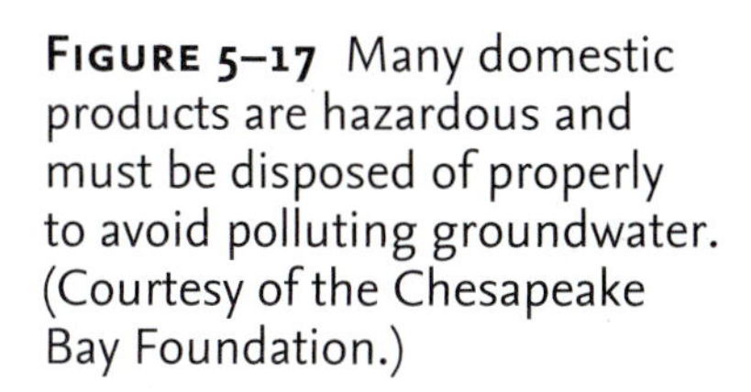

FIGURE 5–17 Many domestic products are hazardous and must be disposed of properly to avoid polluting groundwater. (Courtesy of the Chesapeake Bay Foundation.)

the solvent. Empty containers should be stuffed with newspaper to absorb all liquids before the containers are discarded. Avoid spilling or dumping gasoline, fuel, or oil on the ground or in storm drains. Turn over used petroleum products to truck and automobile service centers for recycling, and send empty containers to approved disposal sites. Keep chemical spills from running or seeping away. Do not flush the chemicals away; they will damage lawns, trees, gardens, fields, and groundwater. Instead, sprinkle spills with an absorbent material such as soil, kitty litter, or sawdust and then place the chemical-laden absorbent material in a strong plastic bag and discard it according to local recommendations.

INTERNET KEY WORDS

algae bloom

Practice sensible pest control. Many insecticides kill all insects—both harmful and beneficial ones. Insecticides also pollute water. Therefore, whenever possible, use agricultural practices such as crop rotations and resistant plant varieties instead of insecticides. Encourage beneficial

science connection profile

INDICATOR SPECIES

One way scientists can tell that water has been polluted is by studying *indicator species*, or sensitive organisms that show the state of an environment's overall health. The four categories of indicator species in water are fish, aquatic invertebrates (including worms, snails, and aquatic insects), algae, and aquatic plants. A pollutant in water affects these organisms in different ways. Depending on the type of pollution, the organisms can die off, have their numbers reduced, or increase their numbers. An algae bloom is an example of how an indicator species is affected by pollution.

From farmlands on which fertilizer has been too heavily applied, runoff water may contain excess fertilizer. As this fertilizer enters a lake or river, it can work on algae as it does on crops and encourage growth. The result is an overgrowth called an *algal bloom*. These are often seen late in the growing season and indicate to scientists that a fertilizer pollutant is in the water. Specific problems may arise from such a bloom. Algae can produce toxins that are harmful to fish, animals, humans, and other organisms. When the algae die and decompose, oxygen is consumed and oxygen levels in the water fall, often killing other organisms that depend on the oxygen.

Indicator species are organisms that are sensitive to pollution. They are used to determine the suitability of a living environment.

insects and insect-eating birds by improving habitat around lawns, gardens, and fields. Eliminate pools of stagnant water to prevent mosquitoes from laying eggs and developing larvae. Follow all pesticide label instructions exactly.

Properly maintain your septic system (if your home has one). Avoid using excessive water. Do not flush inappropriate materials down the toilet. Do not plant trees where their roots might interfere with sewer lines, septic tanks, or field drains. Do not run tractors and other heavy equipment over field drains.

By following these recommendations, we can all help to decrease water-resource pollution.

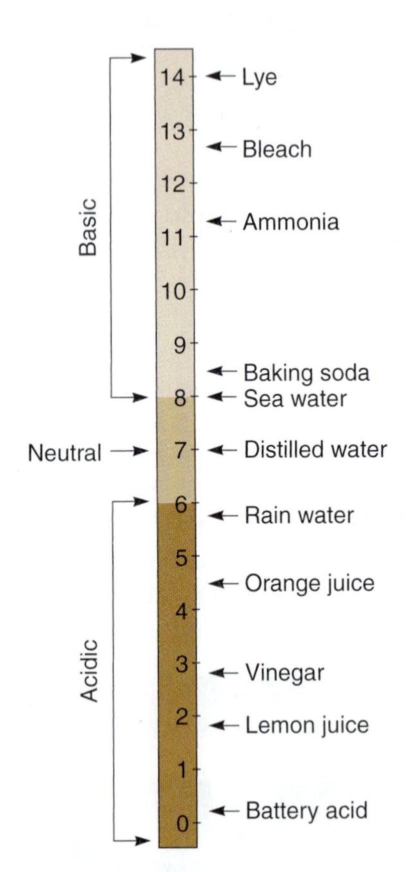

FIGURE 5–18 The approximate pH of some common substances. (Courtesy of Rick Parker.)

Water as a Living Environment

Water quality is important when water is used as a nutrient, but it is critical when it also becomes the living environment. In such an instance, water provides or supports all of the resources that sustain life. Its characteristics will determine whether or not an organism can survive.

High-quality water and plenty of it are the primary considerations for any animal that lives in a water environment. This is true for finfish, shellfish, and crustaceans as well as for birds and mammals. Water provides oxygen and food, serves as an excretory site, helps regulate body temperature, and may harbor disease-causing organisms. Understanding how water supports life is important if it is to be managed as an environment for living organisms.

Water pH

The **pH** of water is a measure of the number of hydrogen ions it contains. The pH scale spans a number range of 0 to 14, with the number 7 being neutral (Figure 5–18). The pH scale is logarithmic, so every one-unit change in pH represents a tenfold change in acidity. Test measurements above 7 are basic (or alkaline) and those below 7 are acidic. The farther a measurement is from 7, the more basic or acidic is the water. The **acid** and **alkaline** death points for fish are approximately pH 4 and below and ph 11 and above. Growth and reproduction can be affected between pH 4 and 6

Interest Profile

HOH—WATER!

Water—the stuff of life—covers 70 percent of the Earth's surface and ranges between 55 percent and 70 percent by weight of the bodies of plants and animals, depending on their stage of maturity. If all of the water were removed, a 200-lb. (91-kg) human would weigh only some 90 lb. (41 kg).

The water molecule is bipolar—that is, it has charged poles like a magnet—which gives it unique properties. Depending on the temperature, water exists in three forms. Between 32° and 212° F (0° and 100° C), or freezing and boiling, it is a liquid. It is a gas or vapor at temperatures above 212° F (100° C), and it is a solid (ice) at temperatures below 32° F (0° C).

Of all the naturally occurring substances, water has the highest *specific heat* (the amount of heat required to raise the temperature of a substance 1° C). This makes it a good coolant in biological systems and makes it resist rapid temperature changes.

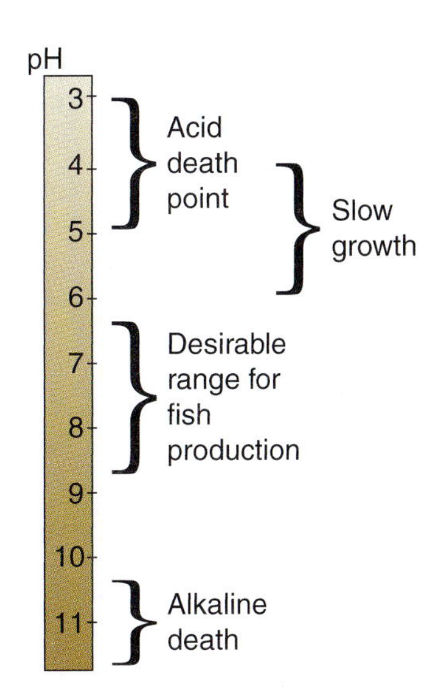

FIGURE 5–19 The effect of pH on fish survival. (Courtesy of Rick Parker.)

and pH 9 and 10 for some fishes (Figure 5–19). In addition, pH affects the toxicity of other substances such as ammonia and nitrite.

The pH of some ponds and shallow lakes may change during the course of a day and often may be between 9 and 10 for short periods in the afternoon. Fish can usually tolerate such rises that result when carbon dioxide, which is acidic, is used up by plants in photosynthesis. The most common pH problem for pond fish is when water is constantly acidic. The nature of the pond bottom and watershed soils is usually responsible. Water with a stable and low pH is only correctable by adding lime, which is basic, to the water.

Temperature

Water temperature helps to determine which species may or may not be present in the environment. Temperature affects feeding, reproduction, immunity, and the metabolism of aquatic animals. Drastic temperature changes can be fatal to aquatic animals. Not only do different species have different requirements, but also optimum temperatures can change or have a narrower range for each stage of life.

All species tolerate slow seasonal changes better than rapid changes. **Thermal stress** or shock can occur when temperatures change more than 2° to 3° F (1° to 2° C) in 24 hours. The **heat capacity** of water is exceptionally high, which makes it resistant to changes in temperature. This moderates the daily and seasonal climatic changes in temperature. Water temperature also is an important variable for many chemical tests and electronic measures for water quality, and sampling may require adjustments for temperature.

Temperature differences between the surface and bottom waters help produce vertical currents that move nutrients and oxygen throughout the water column. Temperature also influences the **solubility** of oxygen and the percentage of un-ionized ammonia in water.

Finally, water temperature affects many biological chemical processes. Spawning, for example, is triggered by temperature.

pH scale
aquatic animals, thermal stress
water, heat capacity

Salinity

Salinity is the measure of the total concentration of all dissolved ions in water. Sodium chloride (NaCl) is the principal ionic compound in seawater, but most lakes and inland ponds contain substantial concentrations of other ionic compounds (salts) such as compounds of sulfate and carbonate. Salinity also can be measured according to the density the salts produce in the water, the refraction they cause to light, or their electrical conductance. The result in all cases is reported in parts per thousand (ppt) salinity.

Standard seawater is 35 ppt or more. Well waters sometimes accumulate high amounts of dissolved ions because of the ionization of compounds of underground minerals or as a result of leaching from the high salt content of arid land surfaces. Dissolved ionic substances can be measured by electrical conductance. On laboratory reports, this may be shown as *specific conductivity*, which is reported as microohms per centimeter (cm) or EC $\times 10^6$. From such an electrical measure, tables can be used to derive tons per acre foot, parts per million, and grains per gallon, among other measures. Conductivities in natural surface water measure from 50 to 1,500 microohms per cm. Freshwater fish such as catfish, carp, tilapia, and trout tolerate some salinity and still grow and survive. Testing for water quality is an important part of ensuring that the fish are in the best environment possible.

INTERNET KEY WORDS

water, total dissolved solids
water, carbonates, bicarbonates

Total Dissolved Solids

Another measure that reports the presence of dissolved ionic constituents is total dissolved solids. This measurement is made by weighing the residue of an evaporated sample after passing it through filter paper. If the sample is not filtered, the reported value is *total solids* (TS) instead of *total dissolved solids* (TDS).

TDS sometimes serves as the only measure on which to make decisions regarding dissolved salts. As with salinity, 2,000 milligrams per liter (mg/l) (2 ppt) is known to adversely affect sensitive species or younger stages of some aquatic species. Catfish handle 6,000 to 11,000 mg/l salinity quite well, depending on acclimation. Besides seawater, water that flows through limestone and gypsum dissolves calcium, carbonate, and sulfate and results in high levels of TDS.

Major Dissolved Ionic Components

Helpful tests determine various ionic constituents of water. Calcium (Ca) and magnesium (Mg) compounds are preferred as major ionic components. The solubility of lime compounds is affected by the presence of sodium. Aquatic species are more affected by total ionic presence than individual ion concentrations.

Carbonates (CO_3^{2-}) and bicarbonates (HCO_3^-) are present in both surface and groundwater supplies at levels consistent with their solubilities. Measurements normally range below 300 mg/l and are considered harmless to fish life. Carbonate and bicarbonate act as pH **buffers**, helping water resist changes in pH.

Sulfate (SO_4^{2-}) and chloride (Cl^-) are expected to range higher than carbonate and bicarbonate in groundwater and surface waters where it comes into contact with appropriate rocks such as those composed of aluminum sulfate. Normal surface waters contain less than 50 mg/l of sulfate and chloride, but some well waters far from coastal areas reach 1,000 to 2,000 mg/l sulfate and several times that amount in chlorides.

Nitrate (NO_3^-) is generally nontoxic to fishes and can be expected to occur at less than 2 mg/l in natural surface water. Fish can tolerate several hundred mg/l. In some recycled waters or where feeding causes enrichment, nitrate could climb to several mg/l.

Phosphate (PO_4^{3-}), fluoride, and silicate are minor constituent anions. Like nitrate, phosphate is usually present in slight amounts (less than 0.1 mg/l) in natural surface and well water. Aside from promoting unwanted growth of algae in ponds, it is considered harmless.

Fluoride concentrations in surface water would be considered normal at less than 0.5 mg/l, high at 1 to 2 mg/l, and rare at more than 10 mg/l. Fish react differently to fluoride according to overall water conditions and species. In some cases, 3 mg/l causes major losses, but normal populations have been recorded in lakes where concentrations reach more than 13 mg/l.

Silica is rather nonreactive and harmless and is normally present in pond waters at less than 10 mg/l.

FIGURE 5–20 Obtaining a water sample that reflects the condition of all the water is difficult. (Courtesy of Rick Parker.)

water, sodium absorption ratio
water, total alkalinity, total hardness

Potassium (K^+) usually represents an extremely low percentage of surface water cations. Calcium (Ca^{2+}) and magnesium (Mg^{2+}) are typically greater and vary from site to site in proportion to sodium. Sodium (Na^+) is much more soluble than the other cations and can range into the thousands of mg/l; the others are limited to hundreds of mg/l. Water with up to 5 or less mg/l calcium is considered very low and 10 mg/l or less is still considered to be low in calcium. Problems to fish life from maximum cation amounts are typically associated with intolerance by fish to extremely high total dissolved solids (salts).

Some waters fed by wells have calcium and magnesium ions dissolved to perhaps twice that considered possible for natural surface water and yet will maintain living populations of fish and other aquatic animals. Excess sodium has a detrimental influence on attempts to adjust pH by liming and can cause it to have little positive effect. Often it is difficult to obtain water samples that reflect the conditions of various waters and conditions (Figure 5–20).

Sodium Absorption Ratio

A test for sodium absorption ratios is used to determine how much of an alkali hazard is posed by irrigation water. It compares concentrations of sodium with calcium and magnesium to assess the potential for sodium buildup in cropland.

Sodium Percentage

Another irrigation water test, sodium percentage or sodium hazard, compares the amount of sodium to all cations present. The effect of sodium on aquatic microplant production is not well understood. On crops, the sodium percentage should be less than 60.

Total Alkalinity and Total Hardness

Total alkalinity is a measure of the basic substances of water. Because in natural water these substances are usually carbonates and bicarbonates, the measurement is expressed as mg/l of equivalent calcium carbonate. In some cases, such as with groundwater and many western ponds, sodium carbonate is the predominant basic substance. These basic substances resist change in pH (buffering); where an abundance of calcium and magnesium bicarbonate is dissolved, the pH will stabilize between 8 and 9. If an abundance of sodium carbonate is present, the pH may exceed 9 or 10. Some laboratory forms report carbonate (CO_3^{2-}) and bicarbonate (HCO_3^-) in addition to total alkalinity. These are typically derived from alkalinity measurements by multiplication by standard conversion factors.

Total hardness is the measure of the total concentration of primarily calcium and magnesium expressed in milligrams per liter (mg/l) of equivalent calcium carbonate ($CaCO_3$). Calcium and magnesium are usually present in association with carbonate as calcium carbonate or magnesium carbonate. Total hardness relates to total alkalinity and indicates the water's potential for stabilizing pH. Waters can be high in alkalinity and low in hardness if sodium and potassium are dominant. Figure 5–21 lists **water hardness** classifications as ppm of $CaCO_3$.

Fish do best when hardness or alkalinity measures are between 20 and 300 mg/l. Levels below 20 mg/l can result in poor production. Cases in which total hardness is considerably below the measure for

Water Hardness	As $CaCO_3$ (ppm)
Soft	0 to 20
Moderately soft	21 to 60
Moderately hard	61 to 120
Hard	121 to 180
Very hard	>180

FIGURE 5–21 Water hardness levels. (Courtesy of Rick Parker.)

total alkalinity are also undesirable. Liming increases total hardness and total alkalinity.

Lime Requirement

The lime requirement test determines the amount of liming needed to neutralize acid because of cations that have been leached or drained from cropland. Sediment (mud) is removed and analyzed to determine the amount of ground limestone needed to bring the pH of soils to an acceptable level and hold it there for two to five years. For aquaculture, the test determines the amount of lime necessary to raise the pH of the bottom mud to 5.8 and raise the water hardness to acceptable levels. The lime requirement of the bottom mud must be satisfied before lasting effects can be expected in the water column. The lime requirement is reported in pounds per acre and may amount to one to several tons of lime per acre. Liming efforts may be futile if the presence of sodium is great or the muds are extremely acidic.

Suspended Solids and Turbidity

Suspended solids (unfiltered residue) should measure less than 2,000 mg/l in muddy pond waters (Figure 5–22). Many times this amount is needed to directly affect fingerlings and adult fishes. Muddiness can affect natural food production at suspended solid levels of 250 mg/l by shutting out sunlight. Such water is described as being **turbid**, and it interferes with reproduction of some fish at levels of less than 500 mg/l. Sometimes turbidity is used as a measure of suspended solids and is given in turbidity units. Turbidity units are derived by light transparency. Turbidity in fertile surface waters largely results from the presence of organic material, particularly algae. An expected measurement for water is 2 units, whereas algae-rich waters measure 200 units.

FIGURE 5–22 A special type of instrument called a Secchi disk can be used to determine turbidity.

Trace Elements

Trace elements include ionic constituents of water that dissolve to a small extent. Some elements are routinely included in reports of irrigation water and soil tests because of their effects on plants. Boron (B) is most common in this category. More generalized water reports include certain metals that could have toxic effects, and some laboratories offer metal analyses as special order items.

Aquatic organisms of all types are sensitive to metal poisoning when their concentrations reach a certain level in the water column. Exact levels of tolerable metals in solution that are considered safe for aquatic life are the subject of much discussion and disagreement. Certain fish groups tend to be more sensitive than others to particular metals. Copper, for example, is more toxic to rainbow trout than to channel catfish. The

INTERNET KEY WORDS

water, suspended solids, turbidity
water, trace elements, metals

toxicity of metals is influenced by the hardness of the water. A metal may poison fish in extremely soft water at a rate that would have to be increased tenfold to produce the same effect in hard water. Because the concentration of a metal required to produce toxicity may differ according to the overall water chemistry, clear statements on the toxicity of the various metals are difficult to find.

Trace elements normally present in unpolluted surface waters at concentrations of less than 1 mg/l include the following:

aluminum (Al)	lead (Pb)
arsenic (As)	manganese (Mn)
barium (Ba)	mercury (Hg)
beryllium (Be)	molybdenum (Mo)
cadmium (Cd)	nickel (Ni)
chromium (Cr)	selenium (Se)
cobalt (Co)	silver (Ag)
copper (Cu)	zinc (Zn)
iron (Fe)	

The toxicity of trace elements is affected by water hardness, dissolved organics, and suspended clays. Toxic action is also influenced by the form of the element—for example, as free ions or bound in organic compounds. Although trace elements can be toxic, some are essential to the health of aquatic life. Except for aluminum, arsenic, barium, and iron, these elements should be considered potentially harmful when present in concentrations higher than 0.1 mg/l.

Dissolved Gases

Dissolved gases determine the basic suitability of water for fish survival. These gases include oxygen (O_2), carbon dioxide (CO_2), nitrogen (N_2), ammonia (NH_4 and NH_3), hydrogen sulfide (H_2S), chlorine (Cl_2), and methane (CH_4). Dissolved gases are usually not found on water-analysis report forms because the way in which samples are collected and shipped can cause gas measurements to be inaccurate. Ammonia is the most stable of the group, and if a sample is processed within a day after collection, it should measure fairly accurately. Other measures are best taken at the water site using appropriate meters or chemical test procedures.

INTERNET KEY WORDS

water oxygen levels

Dissolved Oxygen

Aquatic life requires **dissolved oxygen** (DO). It varies greatly in natural surface water and is characteristically absent in groundwaters. Most aquatic animals need more than a 1 ppm concentration for survival. Most aquatic animals need 4 to 5 ppm to avoid stress. Concentrations considered typical for surface water are influenced by temperature but usually exceed 7 to 8 mg/l (ppm). In ponds, dissolved oxygen fluctuates greatly because of photosynthetic oxygen production by algae during the day and the continuous consumption of oxygen from respiration.

Source	Gain (ppm)	Loss (ppm)
Photosynthesis by phytoplankton	6–20	
Diffusion	1–5	
Plankton respiration		5–15
Fish respiration		2–6
Diffusion		1–5
Respiration by other organisms		1–3
Total	7–25	9–29

FIGURE 5–23 Dissolved oxygen ledger for ponds.

Dissolved oxygen typically reaches a maximum during the late afternoon and a minimum around sunrise. Cloudy weather, rain, plankton die-offs, and heavy stocking and feeding rates result in low levels of dissolved oxygen, which can stress or kill fish. As Figure 5–23 indicates, the gains and losses of the dissolved oxygen in ponds are nearly equal.

Oxygen is only slightly soluble in water. Water may be frequently supersaturated with oxygen in ponds with algae blooms. For example, at sea level at a temperature of 77° F (25° C), pure water contains approximately 8 ppm of oxygen when 100 percent saturated. During afternoon hours, however, levels of 10 to 14 ppm in ponds with healthy algae blooms are not uncommon (Figure 5–24). As water warms, is raised to higher altitudes, or becomes more saline, its oxygen-holding capacity declines. Water saturated with oxygen at 59° F (15° C) contains approximately 9.8 ppm, whereas water at 86° F (30° C) is saturated at approximately 7.5 ppm.

Dissolved oxygen is measured with oxygen meters or chemical test kits, which give results in mg/l. Guidelines for oxygen management usually report that oxygen levels should be higher than 4 mg/l (ppm) to avoid stressing warmwater fish. Most fish experience significant oxygen stress at levels of 2 mg/l, and levels of less than 1 mg/l (ppm) may

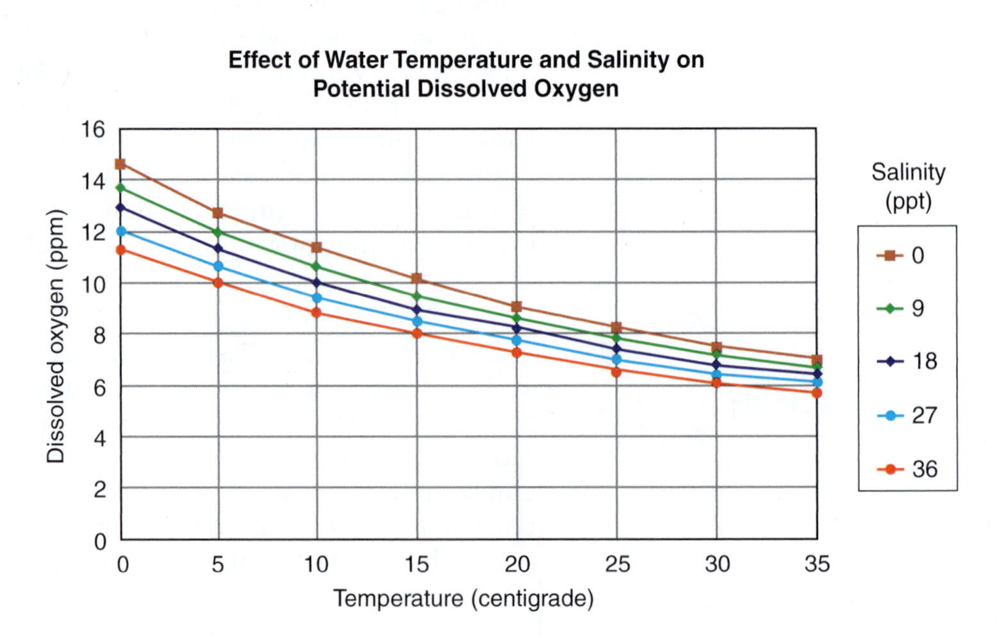

FIGURE 5–24 Increasing water's temperature or salinity reduces its oxygen-holding capacity. (Courtesy of Rick Parker.)

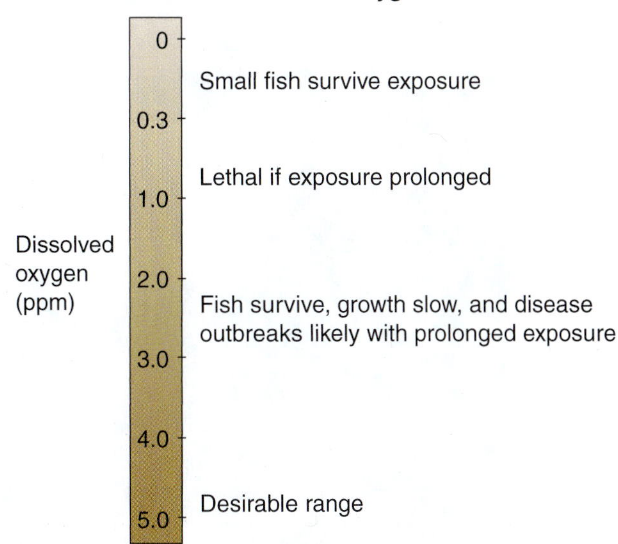

FIGURE 5–25 Fish need dissolved oxygen at 5 ppm. Lower concentrations stress fish and can be lethal. (Courtesy of Rick Parker.)

result in fish kills (Figure 5–25). Although these guidelines are accurate, fish actually respond to the percent **saturation** of oxygen rather than the oxygen content in water. A reading of 1 mg/l at 30° C (13.3 percent saturated) is a higher concentration than 1 mg/l at 15° C (10.2 percent saturated) and represents more available oxygen.

If dissolved oxygen in surface waters reaches low levels, fish will show deficiency signs including

- not eating and acting sluggish,
- gasping for air at the surface,
- grouping near flowing water,
- growing more slowly, and
- enduring outbreaks of disease and parasites.

Proper water management, particularly in the confinement systems typically seen with commercial aquaculture, prevents problems resulting from the depletion of dissolved oxygen. Management techniques include

- monitoring dissolved oxygen at critical times,
- avoiding overfeeding,
- keeping fish stocks at proper levels,
- avoiding overfertilization,
- controlling plant growth,
- using some form of **aeration** (see Figure 5–26), and
- keeping water circulating.

water, carbon dioxide levels
water, hydrogen sulfide pollution

Carbon Dioxide

Carbon dioxide (CO_2), a minor component of the atmosphere, is highly soluble in water. Most carbon dioxide in ponds and lakes occurs as a result of respiration. Levels usually fluctuate inversely to dissolved oxygen, being low during the day and increasing at night, or whenever

FIGURE 5–26 Pond aeration increases the dissolved oxygen. (Courtesy of Rick Parker.)

respiration occurs at a greater rate than photosynthesis (Figure 5–27). Carbon dioxide is present in surface water at less than 5 mg/l (5 ppm) concentrations but may exceed 60 mg/l (60 ppm) in many well waters and 10 mg/l where fish are maintained in large numbers.

Some aquatic animals, including fish, can endure stress and survive in CO_2 levels as high as 60 mg/l. Carbon dioxide interferes with the ability of the aquatic animal to extract oxygen from water, contributing to fish stress during periods of low oxygen. Aerating water to improve its oxygen content drives off excess carbon dioxide.

Adding quick lime ($Ca(OH)_2$) to water rapidly removes carbon dioxide without affecting oxygen content. This improves the ability of fish to use the available oxygen. Carbon dioxide acts as an acid in water, lowering pH as it increases in concentration. Carbonate buffers in water neutralize carbon dioxide and stabilize pH fluctuations within the range tolerated by fish. Waters low in alkalinity and hardness may experience extremes of pH because of their poor buffering against changes in carbon dioxide concentrations.

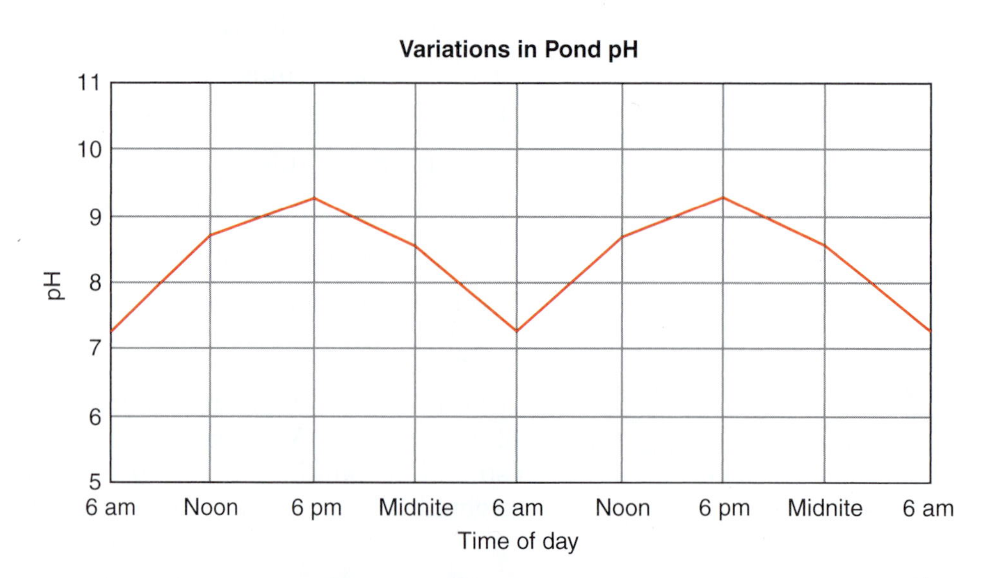

FIGURE 5–27 The CO_2 content of water affects its pH in a pond during the course of a day. Respiration raises the CO_2 and lowers the pH, whereas photosynthesis removes CO_2 and raises the pH. (Courtesy of Rick Parker.)

Hydrogen Sulfide

Hydrogen sulfide (H_2S) is toxic to fish and interferes with normal respiration. It is sometimes called rotten-egg gas because of its strong odor. It is present in some well waters but is so easily oxidizable that exposure to oxygen readily converts it to a harmless form. Vigorous aeration or splashing is usually sufficient to remove hydrogen sulfide from water. Its toxicity depends on temperature, pH, and dissolved oxygen. Any measurable amount that remains after providing reasonable aeration could be considered to be potentially harmful to fish life.

Hydrogen sulfide occurs in ponds as a result of the **anaerobic** decomposition of organic matter by bacteria in mud. In ponds, hydrogen sulfide can be released from anaerobic mud when the bottom is disturbed by certain fishnetting and harvesting activities. Liming ponds raises mud pH and reduces the potential for the formation of hydrogen sulfide. Toxicity is increased at high ambient temperatures and a pH less than 8 when the largest percentage of hydrogen sulfide is in the toxic un-ionized form. In aquaculture settings, potassium permanganate ($KMnO_4$) at concentrations of 2 to 6 mg/l (2 to 6 ppm) removes hydrogen sulfide from water and reverses the effects of its toxicity to fish.

Ammonia

Ammonia is present in slight amounts in some well and pond waters. As fish populations become concentrated in large numbers, ammonia can reach harmful levels. Any amount is considered undesirable, but stress and some death loss occurs at more than 2 mg/l (2 ppm). At more than 7 mg/l (7 ppm), fish loss increases sharply.

Ammonia is a waste product of protein metabolism in aquatic animals (Figure 5–28). In water, ammonia occurs either in the ionized (NH^{4+}) or

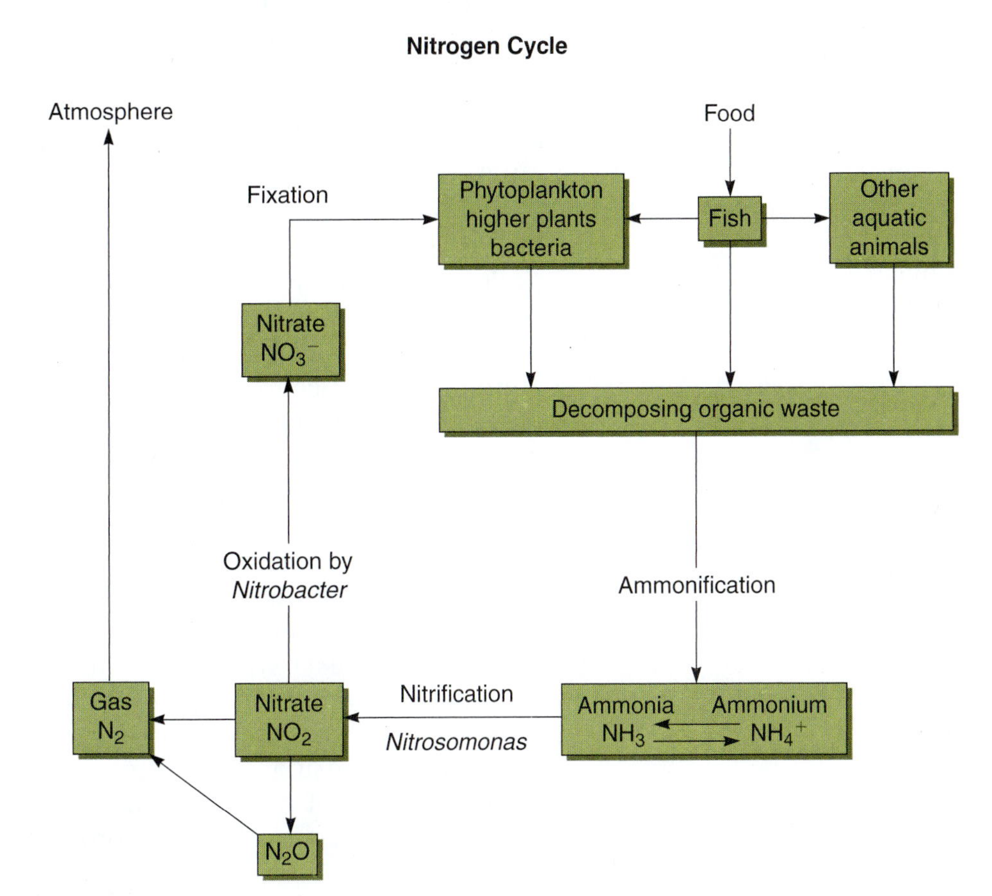

FIGURE 5–28 The nitrogen cycle. Foods and living tissue all contain nitrogen. Eventually the nitrogen is released and converted to several forms. (Courtesy of Rick Parker.)

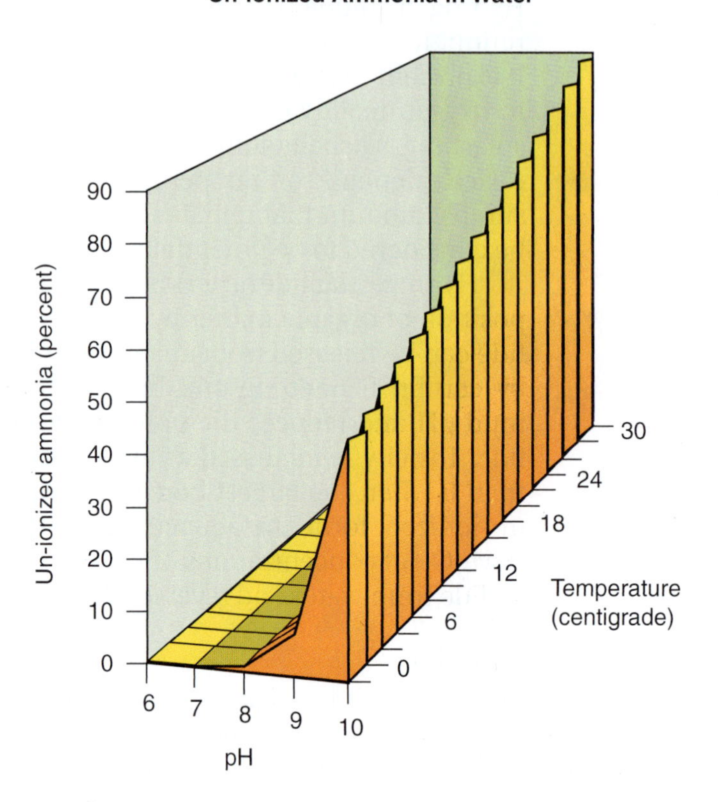

FIGURE 5–29 As temperature and pH increase, so does un-ionized ammonia. (Courtesy of Rick Parker.)

un-ionized (NH_3) form, depending on pH. Un-ionized ammonia is considerably more toxic to fish and occurs in greater proportion at high pH and warmer temperatures (Figure 5–29). For example, at 82.4° F (28° C) and pH 8, 6.55 percent of the total ammonia is present in un-ionized form. At pH 9, 41.23 percent of the ammonia is un-ionized. Un-ionized ammonia is stressful to warmwater fish at concentrations greater than 0.1 mg/l, and it becomes lethal at concentrations approaching 0.5 mg/l. Concentrations of 0.0125 ppm cause reduced growth and gill damage in trout.

ammonia pollution, fish
methemoglobinemia, fish

Test kits for determining ammonia in water measure total ammonia. To determine if a large percentage of the ammonia is in un-ionized form, pH is also measured. A pH above 8 in the presence of ammonia concentrations higher than 0.5 mg/l is cause for concern. Algae use ammonia as a nitrogen source for making proteins. Concentrations usually remain low in ponds with phytoplankton blooms. The greatest concentration of ammonia often occurs after plankton die-offs, at which time pH is low because of high levels of carbon dioxide, and the majority of ammonia is present in the relatively nontoxic ionized form.

Nitrite

Nitrite (NO_{2-}) is one of the basic products of organic matter decomposition. It acts as an intermediate stage in the conversion of ammonia to nitrate. These reactions occur in soils, mud, and water. Nitrite changes quickly to nitrate if oxygen is present. In ponds that are rich in nutrients, a temporary accumulation of this chemical in harmful amounts sometimes occurs. Nitrite measurements of more than 1 mg/l (1 ppm) should be suspect in causing fish deaths.

The binding of nitrite with the hemoglobin molecule gives blood a chocolate brown color. Fisheries professionals call this condition *brown blood disease*. In medical terms, it is known as *methemoglobinemia*. The toxicity of nitrite to fish is lessened by the presence of chlorides in water. The addition of salts, either calcium chloride or sodium chloride at rates of 20 mg/l for each 1 mg/l of nitrite, is a standard treatment for preventing nitrite poisoning in freshwater aquaculture ponds. Most warmwater fish can tolerate at least 0.4 mg/l of nitrite in freshwater without treatment if oxygen levels remain higher than 4 mg/l. Nitrite should be monitored frequently if a problem is suspected because its concentration may increase rapidly in pond water, especially during spring and fall, or when algae blooms suddenly die.

water, fish, chlorine, chloride
fish, gas bubble disease

Chlorine and Chloride

Chlorine is usually present at approximately 1 mg/l in municipal water supplies as a result of chlorination. Fish succumb quite easily at these levels. A common mistake is not recognizing the difference between chlorine and chloride. Chlorine (Cl_2), in gaseous form or as hypochlorites, is widely used to disinfect water supplies and to sterilize aquaculture equipment, tanks, and standing water in drained ponds. Tests to determine the chlorine or chloride content of water require completely different chemical procedures. The distinction between the two compounds and their significance in water is obvious but occasionally confused.

Chlorine acts as a powerful oxidizing agent, and it is toxic to fish at concentrations of less than 0.05 mg/l. Residual chlorine in municipal water supplies is normally between 0.5 and 2.0 mg/l. To be considered safe, water used for fish culture should not contain any residual chlorine. Chloride is a by-product of chlorine dissociation in water but is also widely associated with many other compounds that are highly water-soluble. Chloride occurs in a range of concentrations in water.

Fish are classified as freshwater or marine as determined by their physical adaptation to salinity. Chloride is important in regulating their physiological processes such as tolerance to salinity. Chloride is regarded as nontoxic within the tolerance range of each species.

Nitrogen and Methane

Nitrogen (N_2) and methane (CH_4) play critical roles only when present at abnormally high levels. For example, nitrogen may be driven to high levels in waters located below waterfalls or dam spillways where air mixes with the falling water. In some instances, the water dissolves nitrogen from the air, causing higher than normal concentrations of dissolved nitrogen in the water. Methane gas is a product that occurs in marshes and swamps as vegetative material decays. Sometimes it is called swamp gas. As it bubbles up to the surface, some of it becomes dissolved in the water. As total gas concentrations exceed 115 percent of normal, fish are affected by bubble formation in their blood, a condition called *gas bubble disease*.

Organic Material and Its Breakdown Products

In water, organic material consists of living organisms and various dissolved organic chemicals from their excretions. It also includes dead matter and associated decomposition products. Organic chemicals or tissues always include carbon and consist mostly of the other elements: oxygen, hydrogen, nitrogen, and sulfur. Organic material decomposes constantly into its ultimate breakdown products of carbon dioxide, sulfide, ammonia, nitrate, hydrogen ions, and water. Measuring organic matter can provide a general idea of the potential for oxygen deficiency in a body of water as organic matter decomposes.

Some organic materials such as pesticides will be toxic to fish at extremely low levels. Sophisticated analytical techniques such as chromatography are used to detect these specific organic constituents. Specific indicator tests include the following.

Total Organic Carbon Test

Total organic carbon is commonly reported on laboratory forms and given in mg/l. Natural surface water would be expected to contain 10 mg/l, and water that is exposed to high loads of organic material could build up to more than 30 mg/l. Water with decomposing plant life could also be expected to have high organic carbon. The measure of total organic carbon includes dissolved organic carbon and suspended material, which is commonly called *particulate organic carbon*.

Chemical Oxygen Demand

Chemical oxygen demand (COD) is a speedy and reliable estimate of organic load that is reported in mg/l. A normal measure would read less than 10. A measure of 60 would be considered rich.

Biochemical Oxygen Demand

Biochemical oxygen demand (BOD) is a standard test for organic material reported in mg/l per hour or per total test time. On a per-hour basis, 0.5 would be considered rich. BOD is usually measured over a five-day test period.

Chlorophyll

Measuring the photosynthetic pigment chlorophyll gives an estimation of plant life suspended in the water column. Unfertile ponds range as high as 20 micrograms per liter (mcg/l), and fertile ponds with rich phytoplankton blooms range from 20 to 150 mcg/l.

Oil and Grease

Most oils and grease float on the surface of a body of water. Whereas oily products are not natural releases from plants or other living agents of the pond, aquatic animals suffer when concentrations of oil and grease are above 0.1 mg/l . Sudden die-offs of fish can be expected when such concentrations exceed 100 mg/l. An oil slick is usually apparent in cases of pollution, but a sudden die-off of an algae population also will cause a slick from the oily materials produced by these plants.

pesticide, insecticide, fish

Pesticides

Pesticides are not a natural part of the environment. A knowledge of the historical application of pesticides in the watershed gives a clue to which chemicals to test for. Test laboratories offer scans of pesticides that are known to remain in the environment months and years after application. Newer pesticides disappear from detection soon after application.

Insecticides are more toxic to crustaceans than fish because of their closer kinship to insects. Pesticide groups include

- organochlorides such as dichlorodiphenyltrichloroethane (DDT) and toxaphene at 0.01 to 0.5 mg/l,

- organophosphates such as methyl parathion and malathion at 1 to 10 mg/l,
- carbamates such as carbaryl and methomyl at 0.1 to 10 mg/l,
- pyrethroids such as permethrin at 0.1 to 10 mg/l,
- fungicides such as benomyl and captan at 0.05 to 5 mg/l, and
- herbicides at 0.05 to 500 mg/l (aquatic and terrestrial toxicities vary greatly).

conserving water

The growth of the world's population has created many new demands on the supply of potable water. In many third world countries, the critical problem is developing safe water supplies for drinking and sanitation. In many developed nations, a water shortage means that water sources must be developed to serve new housing developments. In many cities, restrictions are placed on watering lawns and gardens during times of drought. Whatever the priority may be in your area, there is probably a water-conservation effort at some level of government.

In simple terms, there are only two ways to conserve water: Use it more efficiently or stop using it for certain less important purposes. There are many ways to use water more efficiently such as putting a brick in a toilet tank, setting sprinklers to irrigate lawns not sidewalks, and using aerators on sink faucets. Although these are simple remedies, they will conserve water in much larger amounts than we realize.

Water-Saving Tips

- Completely fill washing machines and dishwashers before each use.
- Reduce the water level in washing machines to match the size of the load.
- Rinse dishes in a sink instead of using running water.
- Compost most kitchen waste instead of running the garbage disposal.
- Make sure sprinklers water lawns instead of sidewalks and streets.
- Water yards during early morning hours.
- Wash produce in a sink or pan instead of using running water.
- Keep drinking water in the refrigerator instead of running water until it is cool.
- Throw tissues in the trash instead of flushing them in the toilet.
- Fix leaking faucets and toilets.
- Wash cars on the grass and use a hose nozzle that can be turned off during car washing or use a commercial car wash that recycles the water.
- Install low-volume toilets or put a brick or bottle filled with water in the toilet tank to reduce the amount of water per flush.
- Never thaw frozen food with running water.
- Use water-efficient equipment such as drip irrigation to water shrubs and trees.

- Teach family members to turn faucets all the way off.
- Install a low-flow showerhead.
- Soak pots and pans instead of running water while scraping them.
- Water lawns more deeply but less often.
- Adjust lawn mowers to a higher setting to reduce water loss by evaporation.
- Reduce the amount of grass in a yard by planting shrubs and ground cover plants.
- In arid regions, establish a Xeriscape landscape in a yard.
- Direct precipitation runoff to water trees and shrubs.
- Adjust the temperature of bathwater while the tub fills instead of running water down the drain until it is hot.
- Clean driveways with brooms instead of hoses.
- Use mulch around plants to reduce evaporation of moisture.
- Water plants with rinse water from produce.
- Check outdoor pipes, hoses, and faucets for leaks.
- Practice showering in less than five minutes.
- Regularly check pools and hot tubs for leaks.
- Do not let water run while lathering hands with soap.
- Water houseplants with leftover ice cubes.
- Weed gardens regularly to reduce water use by unwanted plants.
- Turn the water off while shaving, brushing teeth, or shampooing or using hair conditioner.
- Apply minimal amounts of fertilizer to reduce excess water demand.
- Make sure family members know where the water shut-off valve is located in case a pipe breaks or a water leak occurs.
- Water houseplants only when necessary. Never overwater them.
- Bathe pets outdoors where water is needed.
- Replace appliances with water-efficient models.
- Reuse towels to reduce the use of wash water.
- Bathe young children together.
- Insulate hot water pipes.
- Recycle water whenever possible.

Looking Back

Water is an essential nutrient for all plant and animal life. As a universal solvent, it dissolves and transports nutrients to living cells and carries away waste products. It stores heat and helps maintain a more constant temperature in the environment. It cleans the environment by diluting contaminants and flushing them away from vulnerable areas.

Two hydrogen atoms and one oxygen atom associate to form water, H_2O. In nature, water is not a pure substance. To some degree, it consists of many dissolved and suspended substances, including salts, organic matter, and dissolved gases. Marshlands contain plants that tolerate standing water and provide habitat for bacteria that are capable of breaking down materials that cause water pollution. Water quality can be improved by proper land management, careful water storage and handling, and appropriate use of water resources.

Much of the potable water supply is groundwater stored in huge natural aquifers that lay beneath much of the land surface. We must follow fundamental rules when we take water from aquifers. First, we must not pump groundwater at a rate that exceeds an aquifer's recharge rate. Second, we must not allow future priorities to divert recharge water to other uses. Third, we must not allow an aquifer to become polluted by sewer systems, chemicals, pesticides, or any other natural or human-produced contaminant.

Water quality is important when water is used as a nutrient, but it is critical when it also becomes the living environment. The characteristics of water resources determine whether or not an aquatic organism can survive. Technology makes it possible to test for many pollutants and to establish the degree of purity of water.

The demand for clean water is constantly increasing. There are only two ways to conserve it: Use it more efficiently or stop using it for low-priority purposes.

Self-Evaluation

Essay Questions

1. What is water, and why is it important to all living organisms?
2. Why is water considered to be an essential nutrient?
3. What are the most common substances that are found to occur naturally in water?
4. Describe the characteristics of water that make it suitable as a universal solvent.
5. Draw an illustration of a simple food chain that occurs in a water environment.
6. What are some of the relationships that are known to exist between land and water?
7. How is a wetland or marshland capable of removing pollutants from water?
8. In what ways are free water, gravitational water, and hygroscopic water different from each other?
9. What are some of the pollutants that are known to contaminate groundwater?
10. Give some examples of ways that water quality can be improved.
11. List four tests that are commonly used to evaluate the suitability of water as an environment for plants and animals.
12. Name three processes that are affected by water temperature.
13. Suggest six ways that we can conserve clean water in our homes and communities.

Multiple-Choice Questions

1. Most of the Earth's surface is covered with
   a. crops.
   b. farms.
   c. trees.
   d. water.

2. The bodies of plants, animals, and humans consist of approximately what percentage water?
   a. 10 percent
   b. 40 percent
   c. 70 percent
   d. 90 percent
3. Substances that dissolve readily in water to form simple ions are called
   a. salts.
   b. potable.
   c. organics.
   d. insoluble.
4. The universal solvent is
   a. gasoline.
   b. paint thinner.
   c. varsol.
   d. water.
5. The content of ocean water may be likened to
   a. fresh water.
   b. pure water.
   c. thin soup.
   d. varsol.
6. The process by which water changes from liquid form to a vapor or gas is
   a. precipitation.
   b. evaporation.
   c. distillation.
   d. condensation.
7. The function of a watershed is to
   a. protect potable water tanks from the weather.
   b. release a consistent flow of water throughout the year.
   c. shelter a water pump.
   d. protect water from contaminants.
8. Which of the following is not a function of a wetland?
   a. a safe dumpsite for leftover farm and garden chemicals
   b. natural water-treatment site for nitrates and phosphates
   c. migratory bird refuge
   d. sediment filtering from surface water
9. Groundwater that is unavailable for plant root absorption is called
   a. capillary water.
   b. free water.
   c. gravitational water.
   d. hygroscopic water.
10. Improved water quality can be achieved by
    a. appropriate use of water.
    b. careful water storage and handling.
    c. proper land management.
    d. all of the above.
11. Which of the following is not true about water?
    a. composed of two atoms of hydrogen and one of oxygen
    b. universal solvent
    c. freezes at 0° F
    d. covers 70 percent of the Earth's surface
12. A pH of 7 is
    a. basic.
    b. neutral.
    c. acidic.
    d. 7 ppm.
13. A measurement of the total concentration of ions in water is
    a. salinity.
    b. temperature.
    c. acidity.
    d. alkalinity.

Learning Activities

1. Collect a sample of drinking water from each of five sources as follows: a safe spring or well; bottled pure water; and faucets attached to (a) galvanized pipe, (b) plastic pipe, and (c) copper pipe. (Do not run off the water before obtaining the samples from the faucets.) Taste a small amount of each sample and describe the taste. Do the samples have different tastes? If so, what causes the differences?
2. Study the eating habits of one species of bird in your community. Describe the food chain that accounts for the survival of that species. What effect did the use of DDT as an insecticide have on that species of birds before it was banned from use?
3. Obtain a rain gauge and record the precipitation on a daily basis for several months. What variations did you observe from week to week? How do you explain the variations? What effects did these variations have on the community's agricultural activities?
4. Conduct an experiment to determine the effect of soil water on plant growth. Obtain three inexpensive pots of healthy flowers or other plants. Each pot must be the same size and type, have the same amount and type of soil, and contain the same size and number of plants. Use the following procedure:
   a. Mark the pots "1," "2," and "3."
   b. Plug the holes in the bottoms of pots 2 and 3 so water cannot drain from the pots. Leave pot 1 unplugged so it has good drainage.
   c. Add water to pot 1 as needed to keep the plant healthy for several weeks. Use it as a comparison specimen (called the "control").
   d. Add water slowly to pot 2 until water has filled the soil and is just level with the soil surface. All pores of the soil are now filled, and the soil is saturated. Weigh the pot and record the weight as "A."
   e. Remove the plug from the bottom of pot 2 and permit the water to drain out. After water has stopped flowing from the drain hole, immediately weigh the pot again. Record the weight as "B." The water that flowed from pot 2 when the plug was removed is the free or gravitational water. Weight of the gravitational water, "D," should be calculated and recorded using the formula $A - B = D$. Do not add any more water to pot 2 for two weeks.
   f. With the hole plugged in pot 3, add water slowly until the water is just level with the top of the soil. Do not remove the plug, and keep pot 3 filled with water to the saturation point for two weeks.
   g. Keep all three pots in a good growing environment for two weeks and record all observations.
   h. When the plant in pot 2 wilts badly because of lack of water, weigh the pot and record the weight as "C."

   Make the following calculations:

   Weight of the gravitational water, or (D) $= A - B$

   Weight of the capillary water, or (E) $= B - C$

   The weight of the hygroscopic water can only be determined by driving the remaining water from the soil by heating the soil in an oven and then weighing the completely dry soil.

TERMS TO KNOW

land
parent material
loess deposit
alluvial deposit
marine deposit
weathering
original tissue
humus
O horizon
mineral matter
A horizon
clay
silt
sand
tillable
topsoil
B horizon
subsoil
C horizon
bedrock
slope
texture
coarse-textured soil
medium-textured soil
fine-textured soil
drainage
flood hazard
erosion
topsoil thickness
subsoil thickness
anchorage
soil aeration
waterlogged soil
soil solution
cropland
shrink–swell potential
load-bearing capacity
decomposer
alluvial fan
erodability
saltation
suspension
surface creep
no-till
contour
cover crop

CHAPTER 6

Land and soil

Objectives

After completing this chapter, you should be able to

- explain the difference between land and soil
- explain the relationship between soil management and water quality
- describe the origin and composition of soils
- outline the processes by which soil is formed
- define the effects of weathering on parent materials
- explain the importance of organic matter in soils
- describe a mature soil profile
- explain how slope, soil texture, and drainage affect the productivity of soil
- identify the major uses that human society makes of soil and land resources
- list key factors that affect soil erosion by water and wind
- identify some proven measures used by farmers to control soil erosion by water and wind
- explain why soil erosion control should be important to every citizen

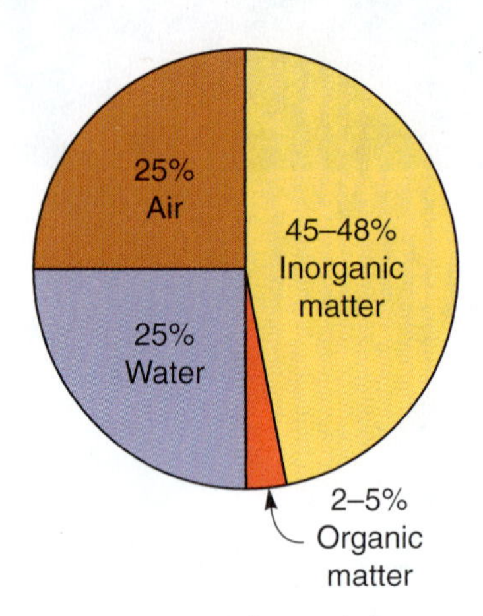

FIGURE 6–1 Physical components of a typical well-drained loamy soil. (Courtesy of Ed Plaster.)

Land and soil are among the world's most important natural resources because life as we know it cannot exist without them. Of equal importance with soil are sunlight and clean water. When soil, sunlight, and water are present, plant and animal life is possible. Without them, they do not exist. Energy from sunlight combines with nutrients and water obtained from the soil to form the simple sugars that nourish living organisms.

Land is the part of the Earth's surface that is not covered with water. It includes minerals deposited as rocks, gravel, and sand. It also includes organic materials. Soil is an important component of land. It is the surface layer of the Earth's crust consisting of minerals (inorganic) and organic material. Surface soil contains water and air (Figure 6–1), with soil particles ranging in size from quite coarse to extremely fine. Soil particles originate from a variety of materials to form soils of different types. It is from soil that nutrients are made available to all living organisms, including oxygen, water, carbon, and other nutrients. These materials are exchanged in the soil, which also provides the right temperature for the growth of organisms.

A close relationship exists between water and soil resources. Water has played an important role in the formation of soils, and the care and management of soils contributes to water quality. Conditions sometimes exist in nature that lead to excessive surface runoff. When this occurs, the presence of silt in surface water also causes water quality to decline. Where human activities are the cause of erosion or excessive overland flows of water, they must be corrected.

SOIL FORMATION

soil formation
parent material, soil formation

Soil is formed slowly. It results from natural forces acting on the mineral and rock portions of the Earth's surface. The process of soil formation has been going on ever since the Earth was first formed. Although the moon probably has its origins with a primitive earth, the powder on the surface of the moon is generally referred to as *lunar soil,* implying that "soil" formation may be a normal part of the development of any planetary body.

Productive soils develop on the Earth's surface as the atmosphere, sunlight, water, and living things meet and interact with the mineral world. If soil is suitable for plant growth to a depth of 36 inches or more, then the soil is regarded as "deep." Many soils of the Earth, however, are much shallower than this.

Plants attach themselves to the soil by their roots where they grow, manufacture food, and give off oxygen. Plants and animals of various sizes live on and in the soil, using carbon dioxide, oxygen, water, mineral matter, and products of decomposition.

Soils vary in temperature, organic matter, and the amount of air and water they contain. The kinds of soils formed at a specific site are determined by the forces of climate, living organisms, parent soil material, topography, and time (Figure 6–2).

Parent Material

In general, soil **parent materials** are those that lay under the soil and from which the soil was formed. There are five general categories of soil

Factors Affecting Soil Formation

Climate/location	Affects rate of weathering
Living organisms	Cause decay of organic material
Parent material	Influences fertility and texture
Topography	Affects distribution of soil particles and water
Temperature	Influences rate of weathering
Weathering	Causes soils to develop, mature, and age

FIGURE 6–2 Soil formation depends on many natural factors.

parent materials: (1) minerals and rocks, (2) glacial deposits, (3) loess deposits, (4) alluvial and marine deposits, and (5) organic deposits.

Minerals and Rocks

Minerals are solid, inorganic, chemically uniform substances that occur naturally in the Earth. Some common minerals for soil formation are feldspars, micas, silica, iron oxides, and calcium carbonates.

rocks, soil formation
glacial deposit, soil formation
loess deposit, soil formation

Whereas minerals are chemically uniform, rocks are not. Rocks, which are simply aggregates of minerals, are usually classified into three general groups: igneous, sedimentary, and metamorphic. *Igneous* rocks are formed by the cooling of molten materials pushed up to the Earth's surface. Lava and magma are two types of igneous rocks. *Sedimentary* rocks are formed by the solidification of sediment. We have all seen sandbars and mud flats along running streams. We have also seen rocks that look like layers of sand stuck together. Chances are that such rocks are sedimentary. They may have formed ages ago when sediment settled along a stream bed or on the ocean floor. Almost three-fourths of the Earth's surface is covered by sedimentary rocks. *Metamorphic* rocks are simply igneous or sedimentary rocks that have been reformed under great heat or immense pressure.

Glacial Deposits

During the four great ice ages, glaciers moved across vast areas of the northern hemisphere (Figure 6–3). As they moved, the glaciers scooped up massive amounts of surface material. They ground, pushed, piled, gouged, and eventually left behind great deposits of rocks, parent materials, and already formed soil materials. Much of the midwestern United States is covered by soils formed from glacial deposits.

Glacial period	Time
Wisconsin	10,000– 100,000 B.C.
Illinois	250,000– 350,000 B.C.
Kansas	600,000– 750,000 B.C.
Nebraska	900,000–1,000,000 B.C.

FIGURE 6–3 The four great ice ages. (Courtesy of Ed Plaster.)

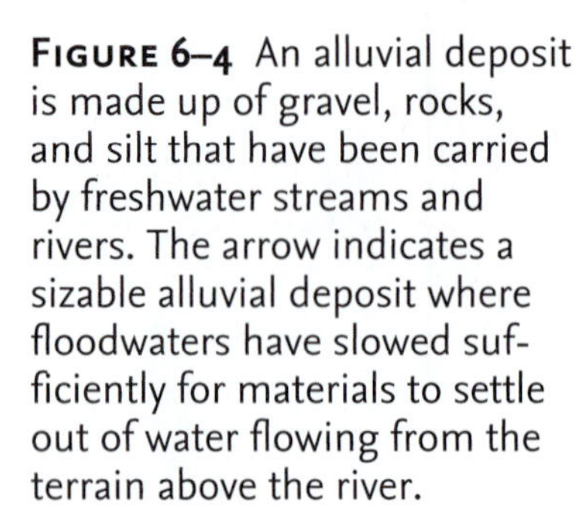

FIGURE 6–4 An alluvial deposit is made up of gravel, rocks, and silt that have been carried by freshwater streams and rivers. The arrow indicates a sizable alluvial deposit where floodwaters have slowed sufficiently for materials to settle out of water flowing from the terrain above the river.

Loess Deposits

Loess deposits are generally thought of as wind-blown silt. Much of the soil in the eastern Mississippi River valley are loess soils. The palouse soils of the Pacific Northwest are also loess soil deposits. Such soil deposits have the same general appearance as snow that has been blown into drifts by the wind.

INTERNET KEY WORDS

alluvial deposit, soil formation
organic deposit, soil formation

Alluvial and Marine Deposits

Both alluvial and marine deposits were created by waterborne sediments. **Alluvial deposits** were left by moving freshwater (Figure 6–4). **Marine deposits** were formed on ancient ocean floors. As water moves rapidly downhill, it picks up soil particles. As it reaches more level areas, it slows down and tends to "fan out." It is in the level fans that alluvial deposits are common. These alluvial fans are often in the form of floodplains or deltas.

Organic Deposits

In swampy and marshy areas, plant life may be extremely lush. As the plants die, the vegetation falls into the water where it decays slowly. Over the years, this partially decayed material begins to build up. It eventually gets thick enough to support plant life itself: It becomes a peat soil or a muck soil. Peat soils are made up of recognizable plant materials. Muck soils are more completely decayed so that plant parts are no longer recognizable. In the case of a lake that eventually fills with such materials, the peat or muck deposits may become quite thick, forming a soil in itself.

weathering

When minerals and rocks are exposed to the weather, they begin to break into smaller and smaller pieces. This is called **weathering**. The major weathering forces are temperature changes, water action, plant roots, ice expansion, and mechanical grinding.

Have you ever placed a cold glass into hot water? Did it crack? This may be an extreme example of heating and cooling, but similar actions

weathering, soil formation

take place in nature every day. The top of a rock can become hot in the sun while the bottom remains cold. Over time, such repeated heating and cooling causes rocks to crack into smaller and smaller pieces.

Some minerals are water-soluble, which means they dissolve when exposed to water. When a rock that contains some water-soluble mineral is thus exposed, parts of it dissolve. Caves are usually formed by this type of natural weathering action.

Have you ever seen a tree or shrub growing in the cracks of a large rock? Growing plant roots can exert great force. You have probably seen sidewalks or streets that have been cracked by tree roots. Once a crack forms in a large rock and soil starts to form in the crack, growing plants will soon follow, thus speeding up the weathering process.

Ice is another natural force that is important in weathering. As water changes into ice, it expands. One reason we put antifreeze in our cars is to prevent the water from freezing and cracking the engine. Bottles of liquids will sometimes shatter if they are left in the freezer until they are solid. By the same token, if a crack forms in a large rock it may sometimes fill with water. If it freezes there, the expanding ice can literally break the rock into pieces.

Rocks can be broken into smaller pieces by other forces, too. Sand, for example, may be blown against a large rock by high winds, causing both the sand particles and the rock to weather. As glaciers move rocks, the rocks grind against each other. As water moves soil particles and gravel, the pieces are ground together into smaller pieces.

The weathering process goes on continuously: day and night, year after year, century after century. Even as soil is being washed away from a field, the parent material beneath the soil is weathering.

The problem that faces modern land use is one of balance. If soil erosion takes place faster than soil formation, then the result can be a destroyed field. A badly managed topsoil can be destroyed in a few years, and it may take nature thousands of years to repair the damage through weathering and soil formation.

Soil Organic Matter

In most soils, the proportion of organic matter is relatively small (two percent to five percent). Its importance in the formation and productivity of the soil, however, is much higher than this small percentage would suggest. Soil organic matter consists of decaying plant and animal parts. As plants and animals die, their tissues are attacked by microorganisms such as fungi and bacteria, among others. The organic matter may be in two basic forms. **Original tissue** is that portion of the organic matter that remains recognizable. Twigs and leaves covering a forest floor are good examples. The other category is known as **humus**. Soil humus is organic matter that is decomposed to the point where it is unrecognizable (Figure 6–5). The brown color of some topsoil is a result of its humus content.

Figure 6–5 Humus consists of dried organic material.

A soil's organic matter serves many important functions:

- It stabilizes the soil structure by serving as a cementing agent.
- It returns plant nutrients to the soil, most notably phosphorous, sulfur, and nitrogen.

- It helps store soil moisture.
- It makes soil more tillable for farming.
- It provides food (energy) for soil microorganisms, which makes the soil capable of plant production.
- It makes the soil porous.
- It provides a storehouse for nutrients.
- It minimizes leaching.

characterizing soils

The Soil Profile

soil formation, humus
soil profile, A, O, B, C, horizon

Undisturbed soil shows four or more horizons in its profile. These are designated by the capital letters O, A, B, and C (Figure 6–6). The **O horizon** is on the surface and is composed of organic matter and a small amount of mineral matter. Organic matter originates from living sources such as plants, animals, insects, and microbes. **Mineral matter** is derived from nonliving sources such as rock materials.

The **A horizon** is located near the surface and is a combination of both mineral and organic matter. It contains desirable proportions of organic matter, fine mineral particles called **clay**, medium-sized mineral particles called **silt**, and larger mineral particles called **sand**. The appropriate proportion of these creates soil that is **tillable**, or workable with tools and equipment. With the presence of desirable plant nutrients, chemicals, and living organisms, the A horizon generally supports good plant growth. The A horizon is frequently called **topsoil**.

The **B horizon** is below the A horizon and is generally referred to as **subsoil**. The mineral content is similar to the A horizon, but the particle sizes and properties differ. Because organic matter comes from decayed plant and animal materials, the amount naturally decreases as distance from the surface increases.

Horizon	Name	Colors	Structure	Processes occurring
O	Organic	Black, dark brown	Loose, crumbly, well broken up	Decomposition
A	Topsoil	Dark brown to yellow	Generally loose, crumbly, well broken up	Zone of leaching
B	Subsoil	Brown, red, yellow, or gray	Generally larger chunks, may be dense or crumbly, can be cement-like	Zone of accumulation
C	Parent material (slightly weathered material)	Variable — depending on parent material	Loose to dense	Weathering, disintegration of parent material or rock

FIGURE 6–6 Characteristics of four soil horizons.

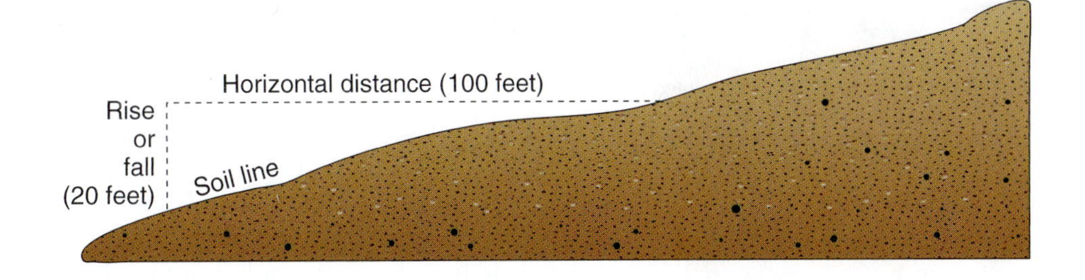

FIGURE 6–7 The slope of a soil surface can be determined by dividing the rise or fall by the horizontal distance. Here, slope = (20/100) × 100 = 20 percent. (Courtesy of Bill Camp.)

The **C horizon** is below the B horizon and is composed mostly of parent material. The C horizon is important because it stores and releases water to the upper layers of soil, but it does not contribute much to plant nutrition. It is likely to contain larger soil particles and may have substantial amounts of gravel and large rocks. The area below the C horizon is called **bedrock**.

The topsoil is the most productive part of the soil. It is here that most biological activity takes place. It is also here that most of the plant nutrients are available to plant roots. Thus, as you might expect, the A horizon is most valuable in the production of naturally occurring plants and cultivated crops.

Soil Physical Properties

soil texture, soil structure

The ability of a piece of land to produce agricultural crops is determined by six basic soil properties: (1) slope, (2) texture, (3) drainage, (4) flood hazard potential, (5) erosion, and (6) the thickness of both topsoil and subsoil.

Slope

Slope—the angle of the soil surface from horizontal—is the single most important factor in determining soil's productive potential. Slope is expressed as a percentage of rise or fall in a given horizontal distance (Figure 6–7).

Slope affects the productive potential of the soil in significant ways: It influences rainfall runoff rates, relates directly to the danger of soil erosion, and affects the use of farm machinery. Field size and shape may also determine the need for contour farming.

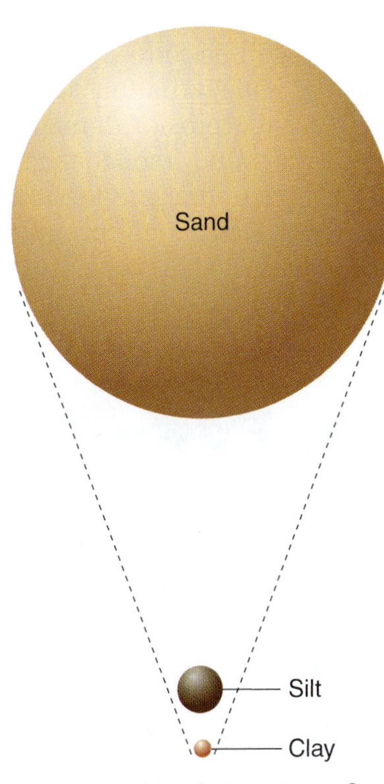

FIGURE 6–8 Relative sizes of sand, silt, and clay particles.

Texture

Texture refers to the proportions of sand, silt, and clay in the soil. Fine-textured soils are those with high proportions of clay. They tend to be sticky when wet and absorb surface moisture more slowly than other soils. Texture also refers to the size of soil particles (Figures 6–8 and 6–9).

The outstanding physical characteristics of the important textural grades are evident by the "feel" of the soil. **Coarse-textured soil** (sandy soil) is loose and single-grained with high proportions of sand. The individual grains can be readily seen or felt. Squeezed in the hand when dry, this soil will fall apart when pressure is released. Squeezed when moist, it will form a cast but crumble when touched. Sandy soils tend to drain more rapidly and, in extreme cases, may tend be more susceptible to drought conditions.

Name	Size, diameter in millimeters
Fine gravel	2–1
Coarse sand	1.00–0.50
Medium sand	0.50–0.25
Fine sand	0.25–0.10
Very fine sand	0.10–0.05
Silt	0.05–0.002
Clay	less than 0.002

Sizes of Soil Particles

FIGURE 6–9 Range of sizes of soil particles.

Medium-textured soil (loamy soil) has a relatively even mixture of sand, silt, and clay. The clay content, however, is less than 20 percent. (The characteristic properties of clay are more pronounced than those of sand.) A loam is mellow with a somewhat gritty feel, yet fairly smooth and highly plastic. Squeezed when moist, it will form a cast that can be handled quite freely without breaking.

Fine-textured soil (clay soil) usually forms hard lumps or clods when dry. It is usually extremely sticky when wet and quite plastic. When the moist soil is pinched between the thumb and fingers, it will form a long, flexible "ribbon." A clay soil leaves a "slick" surface on the thumb and fingers when rubbed together with a long stroke and firm pressure. The clay tends to hold the thumb and fingers together because of its stickiness.

Drainage

Soil **drainage** reflects the natural ability of the soil to allow water to flow through it. Well-drained soils allow excess water to move fairly quickly out of the plant-growing regions of the soil layers. Poorly drained soil holds excess water in their upper layers. Excessive wetness reduces the soil's productive potential by damaging plant root systems. Soils with poor drainage probably will have gray-colored or mottled subsoil or even topsoil.

Flood Hazard Likelihood

Flood hazard refers to the likelihood that a given area will receive flood damage. A field in a frequent flood plain does not have a good productive potential. Even though it may be capable of producing a good crop, such a field's long-range potential is lowered by the probability that the crop will be flooded in any given year.

Erosion

As a soil property, **erosion** refers to the degree the soil has already been damaged. Erosion may range from none to severe. A field that has been used for crop production, yet has suffered no or only slight erosion damage, can be used safely for further production. On the other hand, a soil that has suffered severe erosion should only be used for agricultural production with the utmost care. Its potential may be limited to permanent pasture or forest production. The appearance of large gullies or the loss of almost all of the original topsoil indicates severe erosion (Figure 6–10).

FIGURE 6–10 Erosion of topsoil is a serious environmental problem.

INTERNET KEY WORDS

thickness, topsoil, subsoil

Thickness

Topsoil thickness and **subsoil thickness** refer to the depths of those layers that are available for plant root production. A thin topsoil may limit plant growth and the potential for crop production. A combination of a thin topsoil and a thin subsoil will severely limit plant growth.

Medium for Plant Growth

INTERNET KEY WORDS

plant growth, nutrients, oxygen, water, anchorage

Soil has important functions in recycling resources needed for plant growth. In its most basic use, soil provides individual plants with four basic needs: a place to grow, water, oxygen, and nutrients.

Anchorage

In deep soil where roots grow freely, plants are firmly supported or anchored by their roots into the soil (Figure 6–11). This is called **anchorage**. Firmly anchored, they are able to grow as they reach for sunlight. When plants are grown in ways that deprive them of soil support, artificial support is often required. Landscapers may stake or "guy" a newly planted tree until the tree is firmly rooted, although staking weakens or even damages the trunk and is no longer recommended except in special cases. Poorly anchored trees can even cause serious safety or economic issues.

Water

The roots of a plant are its most effective structures for absorbing water because soil supplies nearly all the water a plant uses. For each pound of dry matter produced by growth, plants obtain between 200 and 1,000 pounds of water from the soil for photosynthesis, sap flow, and other uses. It is obvious that the water-holding capacity of a soil is important.

Oxygen

Except for some microscopic organisms, all living creatures, including plants, need oxygen. Plants release oxygen during photosynthesis

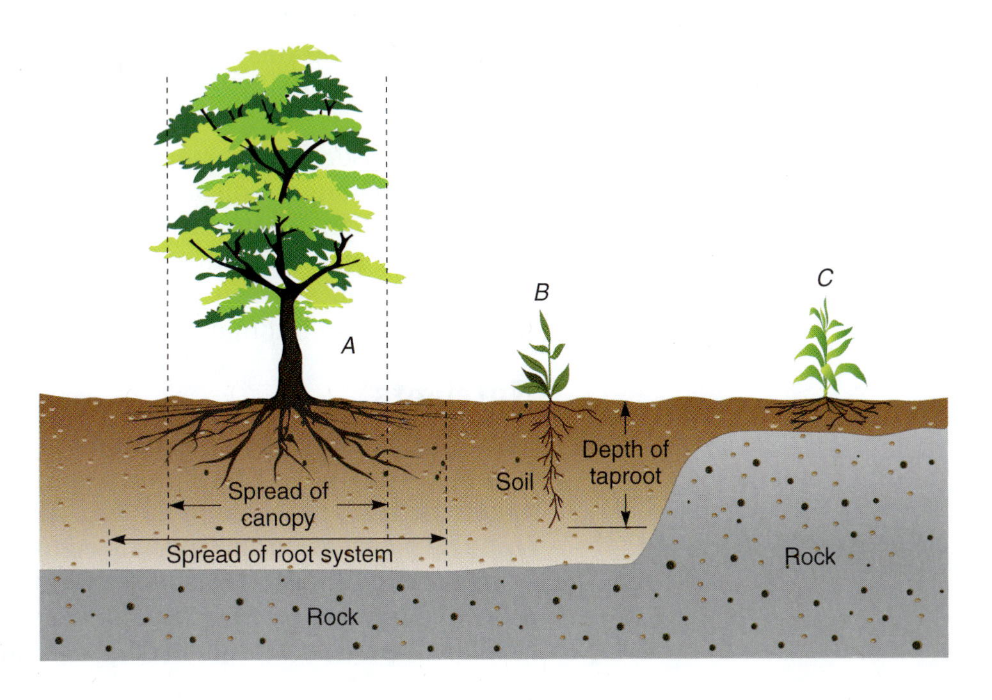

FIGURE 6–11 Roots that are able to grow freely in the soil provide anchorage to hold a plant upright. (Courtesy of Ed Plaster.)

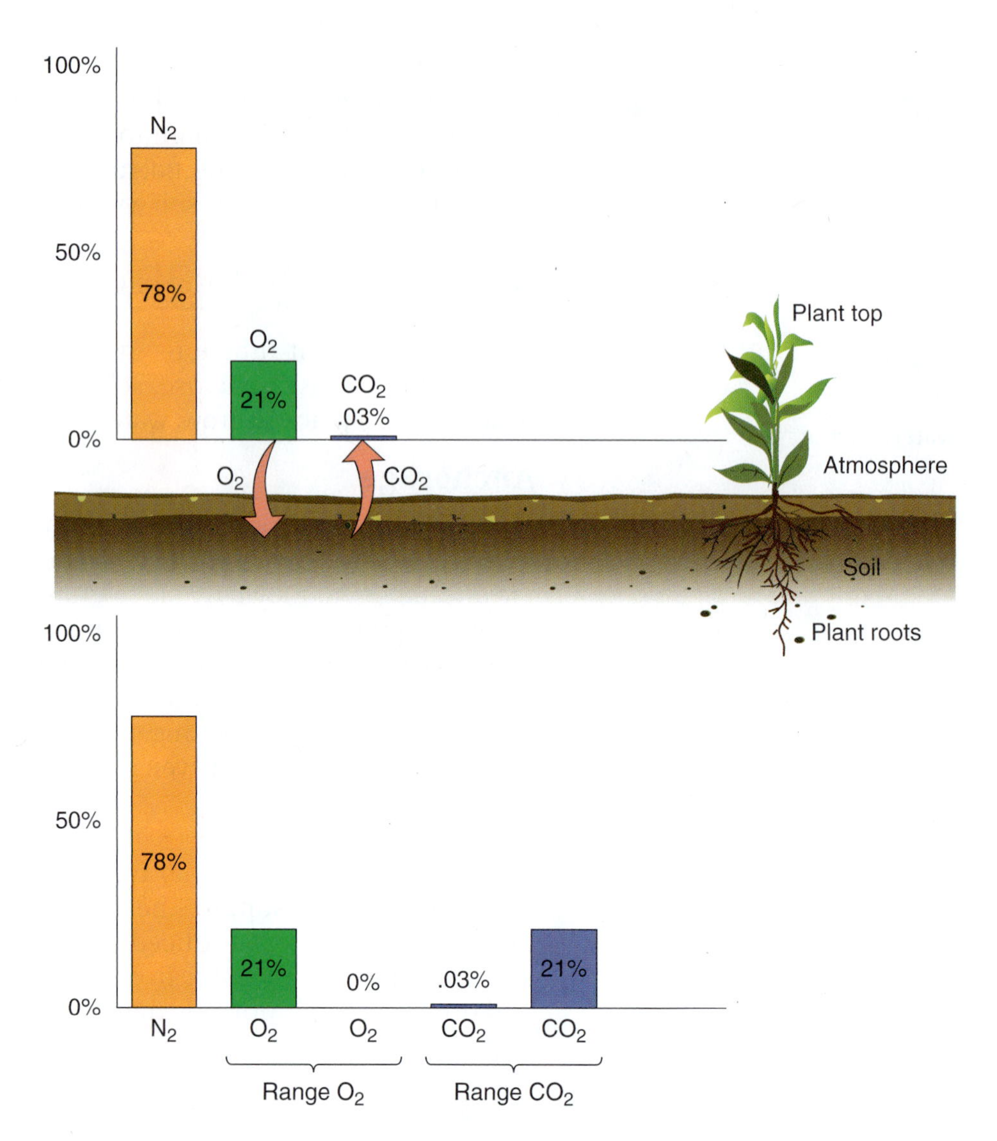

FIGURE 6–12 Soil air and aeration. Most of the gas in air and soil is nitrogen. Above the soil, air is approximately 21 percent oxygen. In the soil, the respiration of living things replaces oxygen with carbon dioxide. Aeration is thus the process by which carbon dioxide and oxygen are exchanged. (Courtesy of Ed Plaster.)

but consume it during respiration. The parts of a plant above ground, suspended in an atmosphere that is 21 percent oxygen, have all the oxygen they need (Figure 6–12). Plant roots and soil organisms located in the soil use up the oxygen and give off carbon dioxide. As a result, air in the soil contains less oxygen and more carbon dioxide than the atmosphere.

In the absence of factors that limit it, the process known as **soil aeration** exchanges soil and atmospheric air to maintain adequate oxygen for plant roots. Aeration varies according to soil condition. Saturated or **waterlogged soil** is an example of soil with poor aeration. The oxygen content near the surface of a well-aerated soil rarely drops below 20 percent but may approach zero in a saturated soil.

Nutrients

Of 17 nutrients usually considered to be needed by most plants, 14 are obtained from the soil. Carbon, oxygen, and hydrogen come from air and water; the rest are stored in the soil. Although leaves are able to absorb some nutrients, roots are specialized for this purpose. Root hairs absorb a **soil solution** that consists of plant nutrients dissolved in soil water by an active process that moves nutrients into root cells. The energy that drives this process is produced by respiration in the roots.

soiL uses

Human societies depend on soil to grow food, fiber, timber, and ornamental plants. Different agricultural uses require different soil management practices as described in the paragraphs that follow.

Cropland

cropland, soil
urbanization, land use

Cropland is land on which soil is worked and crops are planted, cared for, and harvested. Worldwide, the greatest acreage of cropland is devoted to *annual* crops, or those planted and harvested within one growing season. These include crops such as corn (Figure 6–13) and soybeans, fiber plants such as cotton, and horticultural crops like most vegetables. Annuals require yearly soil preparation. This activity gives growers a chance each year to control weeds and to work fertilizer and organic matter into the soil. Because the soil surface is bare much of the time, growers must be careful to keep soil from washing away.

Perennial forages, such as alfalfa, are in the ground for a few years. They may be harvested as hay to feed animals or be used for grazing. These crops cover the soil completely and so keep the soil from washing away. Because the soil is not worked each year, fertilization is different than for annual crops. Perennial crops also tend to build up and improve the soil.

Perennial horticultural crops include fruits, nuts, and nursery stock. Crops stay in the ground for three to as many as 20 years. Many crops are clean-cultivated to keep the ground bare and weed-free. Challenges to the grower of horticultural crops are controlling weeds, reducing erosion, preventing soil compaction, and keeping the level of organic matter stable.

Grazing Land

Much of the land in the United States is grazed by cattle and sheep. In the eastern half of the country, pasture is planted with perennial forage.

FIGURE 6–13 Agronomic crops occupy most of the world's cropland. (Courtesy of the U.S. Department of Agriculture's Natural Resources Conservation Service.)

FIGURE 6–14 Rangeland in Montana. (Courtesy of the U.S. Department of Agriculture's Natural Resources Conservation Service.)

In the western half of the country, which has a drier climate, most grazing is on rangeland (Figure 6–14). Range consists largely of native grasses and shrubs, with some non-native grasses planted through the existing vegetation. Partly because of the size of much rangeland, it is usually loosely managed.

Urbanization

One land use continues to grow at the expense of others: urbanization (Figure 6–15). This includes the building of cities, towns, factories, and roads. During the years 1982 to 1997, 25 million acres of rural land

FIGURE 6–15 Urbanization grows at the expense of other forms of land use. Here we see new subdivisions outside Las Vegas, Nevada. (Courtesy of the U.S. Department of Agriculture.)

were diverted to urban uses. During the same period, nonfederal rangeland declined by approximately 3 million acres, cropland decreased by 45 million acres, pastureland shrank by 18 million acres, forest lands increased slightly, and 34 million acres of crop and rangeland were enrolled in the Conservation Reserve Program (CRP). Some portion of CRP land will later be returned to its previous use.

Foundations

Before constructing a home, the builder tests the soil to a depth of several feet. People know that the structural soundness of a building depends not only on the skill of the builder but also on the soil under the house. Building foundations, for instance, will crack if the soil settles under the building. Even stricter requirements apply to soils for larger structures such as office buildings (Figure 6–16). In some communities, landscapers require an engineer's services in designing retaining walls to ensure they will hold firmly in the soil. Civil engineers also need firm soils that will settle little for the roadbeds of highways and foundations of bridges.

Examples of important soil properties for these engineering purposes include **shrink–swell potential** and **load-bearing capacity**. Many soils swell when wet and shrink as they dry. This causes walls to crack, foundations to be damaged, and pipes to break. Soils high in clay or organic matter have low load-bearing capacity. Foundations of buildings constructed on such soils may shift and crack. Roads and other structures built on such soils may also have structural problems. A major earthquake in 1989 brought down many buildings in San Francisco because most were located on loose "fill" soil that could not support structures when the ground began to shake.

waste disposal, landfills, soil logging, soil erosion

Waste Disposal

The treatment of human sanitary waste often relies on soil because it filters out some of the material, while microorganisms break down organic portions into less dangerous compounds. The common home

Figure 6–16 The foundation of a building under construction. The soil must have a good load-bearing capacity and low shrink–swell potential for the foundation to be sound.

septic system is an example. Sanitary or especially hazardous waste landfills require soils that will not allow hazardous materials to leach into the water table or run into neighboring streams or lakes. The search for landfill sites often arouses conflict in a community. Many people feel landfills cannot be entirely safe, and even those who agree landfills are necessary do not want them nearby.

Forest

Foresters probably disturb soil the least, but soil management is still a concern. When trees are harvested after many years of growth, logging equipment can tear up the vegetative cover and compact the soil. The result is increased soil erosion, and the soil surface may be damaged as a medium for the growth of newly planted seedlings. Other concerns of forestry include matching the tree variety to the soil type and ensuring good conditions for newly planted seeds or seedlings.

Recreation

Recreational uses of the soil surface are important. Visit an urban park and you probably will see children in the playground, softball teams on the field, and runners on jogging paths. Golf courses, parks, and campgrounds are examples of large areas used for recreation (Figure 6–17). The design of recreational facilities is a specialized skill that requires knowledge of soil properties.

Sports fields are probably the most demanding of all soil uses. To grow turf that withstands the punishment of football cleats or soccer shoes challenges even the best of managers. Soils in the best playing fields are highly engineered mixes of loam, specific sizes of sand, and other ingredients. They may even include a plastic mesh to hold the soil together. The fields generally have several soil layers, are carefully graded and drained, and are well maintained.

living organisms in soil

Those who manage playing fields must worry about sideways pressure, or shear, from shoes tearing the soil surface. Playing fields are designed to

FIGURE 6–17 Just as with farming, the design, building, and maintenance of golf courses and parks requires knowledge of soil.

FIGURE 6–18 Living organisms such as earthworms play important roles in breaking down such organic matter as leaves and grass. (Courtesy of PhotoDisc.)

have good shear resistance. Fields must be of a certain hardness to provide a proper playing surface and reduce injuries. They must dry quickly after a rain yet hold enough water to grow healthy turf grass. These and other considerations require knowledge of soil science.

SOIL AS A LIVING ENVIRONMENT

Soil is a living medium with a great variety of organisms living in it. Some of these organisms are large enough to see, but many are not. Among those that we are most familiar with are small mammals (such as mice, shrews, moles, and gophers), amphibians (such as frogs, toads, and salamanders), insects, snakes, earthworms, snails, slugs, millipedes, centipedes, spiders, sow bugs, and mites, among many others (Figure 6–18). In addition, many forms of microbes are too small to be seen.

Soils that become polluted are seldom capable of supporting the organisms that depend on the soil as a living environment. In some cases, bacteria are able to break down the pollutants and reclaim this important resource. In other instances, soil may become severely damaged to the point that no living thing can survive there.

Living organisms improve the soil in several ways. They excrete cellular or body wastes that become part of the organic content of soil. Microbes and the remains of larger plants and animals decompose or decay into soil-building materials and nutrients. The excretions and

SCIENCE CONNECTION PROFILE

LIVING SOIL

Soil is the most diverse ecosystem on Earth. The number of organisms per acre far outnumber that of any other place in the world. Most of us do not think about the abundant life under our feet as we cross a lawn. But just one square inch of this soil is teeming with busy creatures, most of which cannot be seen. Gardens and fields consist of fertile soil filled with living organisms. In one gram of this soil, there are:

- 3,000,000–500,000,000 bacteria (one-celled microscopic organisms)
- 1,000,000–20,000,000 actinomycetes (microscopic organisms that resemble both fungi and bacteria)
- 5,000–1,000,000 fungi (nonmicroscopic organisms such as mushrooms that get their food from dead material)
- 1,000–100,000 yeast (single-celled fungi, many of which are used in food production)
- 1,000–500,000 protozoa (small single-celled organisms such as the amoeba)
- 1,000–500,000 algae (one-celled organisms that contain chlorophyll)
- 1–500 nematodes (nonsegmented round worms)

Large numbers of slime molds, viruses, insects, and earthworms are also present. Some of these organisms are harmful to plants and animals, but most are **decomposers**, or organisms that break down and change once living material into the rich organic substances that add to the fertility of the soil.

The role these tiny creatures play in an ecosystem is irreplaceable. Imagine a world without decomposers. Leaves, dead animal carcass, and huge amounts of other dead organic matter would pile up. In a short period, material that is dead and has not decomposed would crowd out all life.

Soil is a unique substance. It provides living space for billions of organisms. It is the medium in which plants and other producers grow. It is also a place where once-living things are broken down and changed into the fertile organic material of the soil.

	Sheet and rill erosion 1982	Sheet and rill erosion 1997	Wind erosion (tons/acre/yr) 1982	Wind erosion (tons/acre/yr) 1997
Cultivated cropland	4.5	3.5	3.7	2.9
Pastureland	1.1	1.0	0.1	0.1
Rangeland	1.2	1.2	4.7	4.4

FIGURE 6–19 Estimated average annual loss of topsoil in tons per acre to erosion in nonfederal lands of the United States. (Source: U.S. Department of Agriculture, 1997 National Resource Inventory.)

decaying tissues of soil organisms also form materials that hold soil particles together, helping to maintain soil structure, a key element in preventing soil erosion.

soil conservation

INTERNET KEY WORDS

soil decomposers
soil conservation

Soil erosion is a major environmental issue. Each year, almost 2 billion tons of soil wash or blow away from U.S. farmlands, a quantity equivalent to losing the full plow layer from 2 million acres of land. Most of the loss—approximately 1 billion tons—results from water erosion. The remaining 0.8 billion tons are lost in wind erosion.

Soil scientists follow the rule of thumb that one acre of land can afford to lose no more than one to five tons of soil each year. The average soil loss to water erosion on cropland is thought to be 3.1 tons per acre per year, which is close to the limit. Added to this amount, however, is an average soil loss of 2.0 tons per acre per year to wind erosion (Figure 6–19). Approximately 21 percent of cultivated U.S. cropland suffers soil losses greater than the acceptable limit for water erosion (Figure 6–20).

http://www.nhq.nrcs.usda.gov/CCS/squirm/skworm.html

Soil erosion is not a new problem. It is a natural process, as evidenced by the river channels and canyons that have been cut through the surface of the land over tens of thousands of years. Erosion is also evident in the formation of **alluvial fans** where streams enter from a gorge into

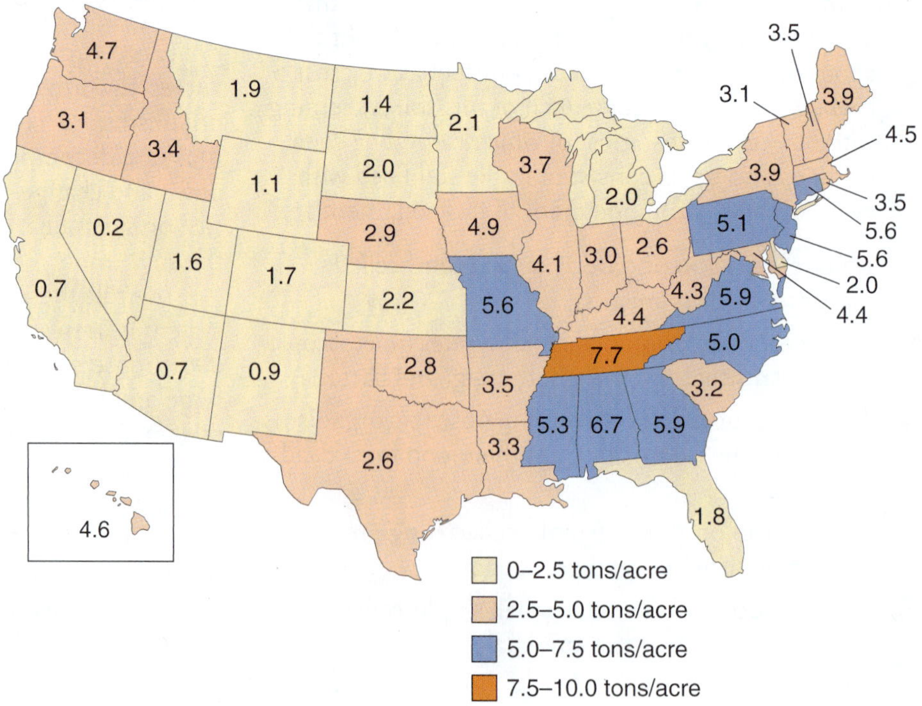

FIGURE 6–20 Average annual sheet and rill erosion in tons per acre by state on nonfederal land. (Source: U.S. Department of Agriculture, 1997 National Resource Inventory.)

INTERNET KEY WORDS

soil conservation

a plain or where a tributary stream joins with a main stream or river. Alluvial fans are composed of rocks, gravel, sand, and silt that have been carried to the area by streams of rapidly flowing water. They consist of huge soil deposits formed from materials that eroded from the land surface further upstream, were carried downstream, and were deposited when the water flow spread out and slowed down. Such events occur over long periods of time.

Erosion is a destructive force that has serious consequences for crop production, fish, and wildlife. Fish spawning grounds that are filled with silt may prevent developing fish eggs from getting the oxygen necessary to sustain life. Eggs become coated with silt particles, and the fish embryos soon die. Young fish also may be injured by water polluted with silt.

How Erosion Occurs

Water erosion follows three steps. First, the impact of raindrops shatters surface aggregates and loosens soil particles (Figure 6–21). Some of these particles float into spaces between soil particles, sealing the soil surface so water cannot readily infiltrate the soil. The scouring action of running water also detaches some soil particles. Second, detached soil grains move in flowing water and are carried down slopes. Third, the soil is deposited when the water slows down. These three steps are known as *detachment, transport,* and *deposition* (Figure 6–22).

Soil texture has two effects on soil erosion. First, it influences the infiltration rate of water. If rainwater infiltrates soil quickly, then less water runs off. With a lower volume of running water, less soil can be transported. Second, particles of different sizes vary in how easily they can be detached from the soil structure. Silt particles are most easily detached, so silty soils are vulnerable to water erosion.

INTERNET KEY WORDS

soil erodibility factors

Soil structure also influences infiltration: Like granules, good structural grades reduce runoff. The strength of soil aggregates is important,

FIGURE 6–21 Raindrop impact on soil. (Courtesy of the U.S. Department of Agriculture's Natural Resources Conservation Service.)

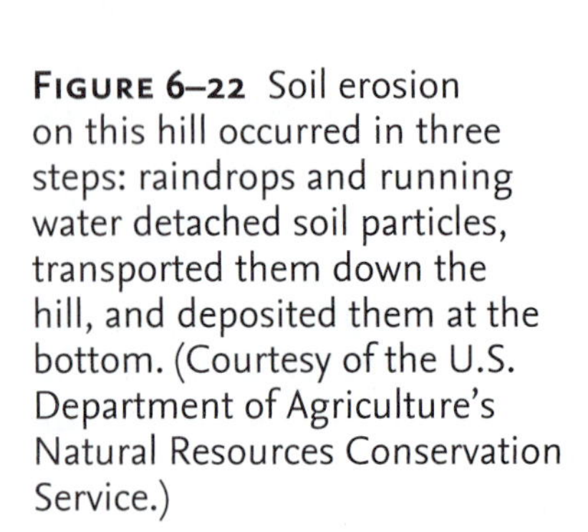

FIGURE 6–22 Soil erosion on this hill occurred in three steps: raindrops and running water detached soil particles, transported them down the hill, and deposited them at the bottom. (Courtesy of the U.S. Department of Agriculture's Natural Resources Conservation Service.)

too, because soil granules tend to resist the impact of raindrops. **Erodibility** is the tendency of soil aggregates to break apart, making the soil susceptible to erosion. Because of the importance of organic matter to soil structure, organic matter content has a strong bearing on erodibility. Compaction, loss of organic matter, and destruction of soil structure by tillage all reduce infiltration and increase the volume of water that is available to transport eroded soil. The combined effects of organic matter content, texture, and structure contribute to soil erodibility.

The length and grade of a slope also affect water erosion. On a steep slope, water achieves a high runoff velocity, increasing its erosive energy. On a long slope, a greater surface area is collecting water and thus increasing flow volume. On a long slope, running water can also pick up speed. For these reasons, long, gentle slopes can have the same erosive potential as short, steep slopes.

A rough soil surface impedes the downhill flow of water, slowing its velocity. When surface roughness takes the form of ridges across a slope and water pools behind the ridges, the volume of runoff water decreases. However, if enough water collects behind a ridge, it overflows the ridge and wears it away. Thus, roughness can fail to stop erosion during the heaviest rains, on long slopes, or when sealing of the soil surface stops infiltration. Surface roughness depends largely on tillage practices. The seedbed resulting from conventional tillage is smooth, whereas that from chisel plowing is rough. Tillage across slopes acts to impede downhill flow; tillage up and down the slope promotes downhill flow.

Bare soil is fully exposed to the erosive forces of raindrop impact and the scouring of running water. Soil cover reduces the energy available to cause erosion. A mulch or cover of crop residues absorbs the energy of a falling raindrop, lessens detachment of soil particles, and reduces sealing of the soil surface. Mulches also slow down runoff water. A complete crop cover such as turf or hay has the same effect; in addition, plant roots hold soil in place (Figure 6–23).

FIGURE 6–23 A complete vegetative cover almost eliminates soil loss. Here highly erodible land in Iowa is protected by a solid cover of native switchgrass. (Courtesy of the U.S. Department of Agriculture)

Figure 6–24 The lack of plant cover contributes to erosion on steep slopes. (Courtesy of Shutterstock.)

Crops that are grown with plants widely spaced, such as row crops or nursery stock, have a slightly different effect. As these crops close in between the rows, they form a canopy over the soil. This canopy intercepts rainfall and absorbs most of its impact. When water drips off plant leaves, the energy of the drops is less than free-falling raindrops, and erosion is reduced. It is important for erosion control that crops cover the soil surface as quickly and completely as possible. Unlike mulches, however, crop canopies have no effect on runoff speed or volume, so they have a less protective effect.

Erosion becomes more intense when the plant cover is removed on steep slopes (Figure 6–24). As the North American continent was colonized, most of the land considered suitable for tillage was developed for the production of crops. Some of this farmland is located on slopes that are highly vulnerable to erosion. Tillage practices that remove plant cover during the winter season leave the soil exposed to the forces of erosion from wind, heavy seasonal rains, and snowmelt.

Serious soil erosion often follows range, forest, or grassland fires (Figure 6–25). This results partly from the loss of the plant cover that protects the soil surface. Fire also breaks down the soil structure and causes it to be more easily damaged. Special conservation measures are necessary after fires that cover large areas. One important practice is to reseed the area with grasses and other cover plants to stabilize the soil as quickly as possible.

Wind erosion accounts for approximately 40 percent of the soil loss in the United States, mostly in the Great Plains states. Other areas with wind erosion problems include the muck and sandy soils of the Great Lakes states and Atlantic seaboards. Significant wind erosion occurs in some states, especially in the west (Figure 6–26). Dry areas with high winds are most likely to experience wind erosion. At greatest risk is soil that is kept bare because of clean-till summer fallow tillage practices.

FIGURE 6–25 Erosion in San Bernardino, California, as a result of firestorms. (Courtesy of the U.S. Department of Agriculture.)

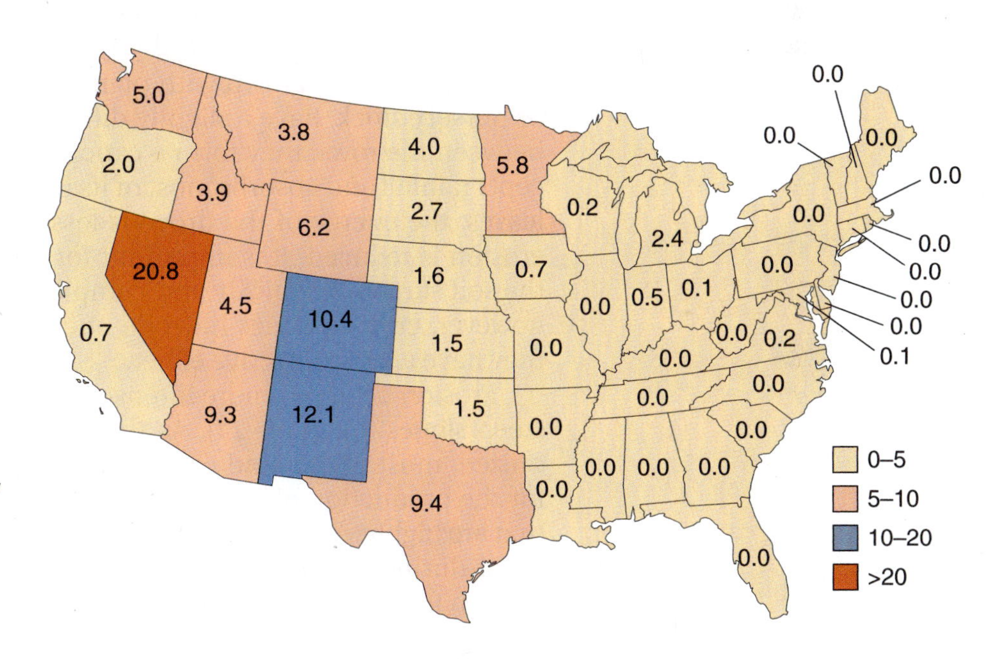

FIGURE 6–26 Average annual wind erosion in tons per acre by state on nonfederal cultivated land. (Source: U.S. Department of Agriculture, 1997 National Resource Inventory.)

Bare soil is the most vulnerable to wind erosion (Figure 6–27). An especially thin layer of still air covers the soil surface, but larger soil particles stick up above it. When wind reaches 10 to 13 miles per hour at a height of one foot above the surface, soil grains begin to move.

First, wind begins to roll soil grains in the size range of 0.004 to 0.02 inches (0.1 to 0.5 mm). These particles are fine to medium sands. Next, a sand grain is picked up in the air, rising as high as 12 inches. Wind then blows the sand grain several feet before it strikes the ground, where it may bounce up again or knock loose other particles. This process is called **saltation,** and it causes 50 percent to 75 percent of all wind erosion. In fact, more than 90 percent of all movement occurs within 1 foot of the soil surface.

INTERNET KEY WORDS

wind erosion, saltation

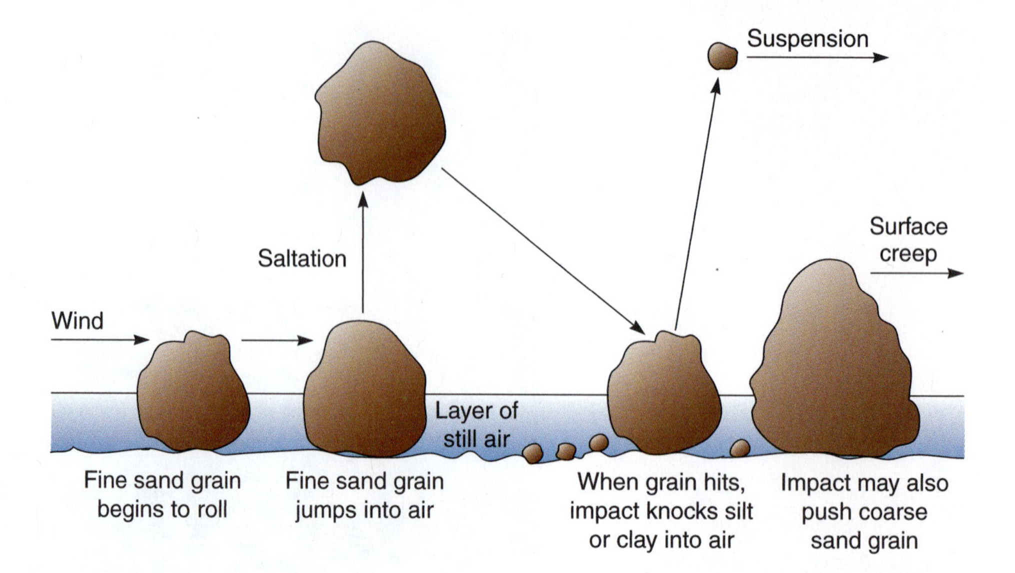

FIGURE 6–27 Saltation of fine sand triggers wind erosion. Fine sands are large enough to protrude above the layer of still air but small enough to be picked up by the wind. Surface creep and suspension both depend on the impact of fine sand in saltation. (Courtesy of Ed Plaster.)

Extremely fine silt and clay particles are too small to be picked up by the wind, but the impact of a sand grain moving by saltation may knock dust particles into the air. Once airborne, dust rises high into the air and is carried long distances. This process is called **suspension.** Silt particles move most easily, and the heavier sand grains tend to roll along the ground. This type of wind erosion is called **surface creep**.

Like water erosion, wind erosion removes the best soil first—the topsoil. It carries off fine soil particles, especially silt and organic matter. This shift toward a coarser soil texture reduces the soil's capacity to hold both nutrients and water. In addition, windblown soil particles "sandblast" young plants, tattering leaves and tearing away plant cells (Figure 6–28). Blowing soil can fill road or drainage ditches, affect the respiratory health of animals and people, and wear at paint and other surfaces.

FIGURE 6–28 Windblown soil "sandblasts" young crop plants. (Courtesy the U.S. Department of Agriculture's Natural Resources Conservation Service.)

FIGURE 6–29 Windbreaks are created to restrict water and wind erosion. (Courtesy of the U.S. Department of Agriculture.)

Preventing and Controlling Erosion

Soil conservation practices for farms include many practical ways of protecting soil surfaces and slowing the movement of water or wind across the soil. Examples of such practices are planting windbreaks, creating dikes along the contours of fields and hills, adapting no-tillage or minimum tillage farming practices, planting grass waterways, and planting high-risk fields to permanent cover crops (Figure 6–29).

Forest lands are protected from runoff following timber harvests by gouging holes in the forest floor to trap water so that it is held on or near the surface until it is absorbed into the ground. This practice effectively prevents excess water from flowing across the soil surface where it might cause erosion. Properly constructed logging roads prevent water from running down the road surfaces in large streams. The water is channeled off the road and into areas that have stable ground cover.

water quality
soil conservation, no-till,
cover crop

No-till means planting crops without plowing or disking the soil. This prevents damage to soil aggregates on the surface, thus reducing the risk of erosion. Another strategy is to plant alternating strips of close-growing crops (such as small grains or hay) alongside row crops (such as corn or soybeans). Farming on the **contour**—following the level of the land around a hill—is another effective strategy. Still another effective practice is to plant cover crops where regular crops do not protect the soil. A **cover crop** is a close-growing crop that is planted to temporarily protect the soil surface. It also is possible to reduce the volume of rainwater carried away by minimizing the amount of blacktop or concrete surface constructed. In low areas of fields, grass waterways can be used. Manure and other organic matter can be added to soil to increase water-holding capacity. Terraces on long or steep slopes also can be constructed (Figure 6–30). Finally, steep areas can be left covered with trees and sloping areas with close-growing crops.

Figure 6–30 Construction of terraces on sloping land (note the arrow) distributes water from precipitation along the terrace and helps to prevent soil erosion.

Best management practices for the control of water erosion include the following:

- Reduce raindrop impact to lessen particle detachment. This can be done by growing vigorous crops that fill in the canopy quickly. It is also wise to leave crop residues on the soil surface. Mulching and growing cover crops are also sound strategies.
- Reduce or slow water runoff. This lessens soil detachment by scouring and reduces the amount of soil that can be transported. Avoiding compaction, maintaining organic matter levels, and subsoiling help water infiltrate the soil. Contour practices and conservation tillage both help in reducing runoff.
- Carry excess water off the field safely by use of grass waterways or tile drain pipes.
- Till at right angles to the prevailing wind, leaving the soil surface rough and cloddy (Figure 6–31).
- Use conservation tillage or subsurface tillage equipment such as a rod weeder or a subsurface sweep to leave crop residues on the soil surface.
- Follow a crop rotation plan to increase organic matter, improve soil tilth, and improve the moisture-storing capacity of the soil.
- Keep the soil covered with vegetation as much as possible. Cover crops of winter grains work well in protecting soil over winter.
- Plant crops in strips (strip cropping) at right angles to the wind. Strips of soil in summer fallow may be partially protected by alternate strips of small grains or row crops.

FIGURE 6–31 Creating surface roughness and cross-wind ridges in Texas. This will help reduce wind erosion. (Courtesy of the U.S. Department of Agriculture's Agricultural Research Service.)

- Plant windbreaks of trees or large shrubs. Windbreaks shorten the field, reduce wind velocity, and capture blowing soil (Figure 6–32). Windbreaks should not be solid; they should block approximately 50 percent of the wind.
- Plant buffer strips as temporary windbreaks. For instance, tall wheatgrass barriers planted north–south and approximately 50 feet apart, reduce erosion by some 93 percent. The strips also capture blowing soil.
- Adopt contour farming methods on uneven ground.

FIGURE 6–32 Field windbreaks in North Dakota protect the soil against wind erosion. (Courtesy of the U.S. Department of Agriculture.)

- Keep off-road vehicles on established roads and trails.
- Establish permanent terraces on land that is prone to soil erosion.
- Plant the most critical areas with permanent grasslands or other vegetative cover.

Looking Back

Land and soil are among the world's most important natural resources because living things depend on soil for their existence. Living organisms need proper temperature, oxygen, water, carbon, and other nutrients. These factors are exchanged in the soil.

Soils are formed as parent materials break down to form soil particles. Weathering is a key force in soil formation. Soil forms in layers called horizons, with the A horizon containing topsoil, the layer in which plants grow and nutrients are exchanged. Soil provides anchorage, water, oxygen, and other nutrients to plants.

Soil provides sites for cropland, grazing, building sites, waste disposal, forest lands, and recreation. It also functions as a living environment for a large variety of organisms, both seen and unseen.

Soil is vulnerable to erosion, particularly from water and wind. Soil conservation must be practiced in order to preserve our soil resource. The use of best management practices must become a priority.

Self-Analysis

Essay Questions

1. Define the terms *soil* and *land* and explain how they are related.
2. How is water quality related to soil management?
3. Where do soils come from, and how are they formed?
4. What are some parent materials besides rocks and minerals?
5. Define weathering. What are the major weathering forces?
6. What are five reasons organic matter is an important component of soils?
7. Differentiate between original tissue and humus. Which gives topsoil its color?
8. What are the general soil horizons found in a soil profile? Describe each.
9. List six important physical properties of the soil. Why is each important?
10. How do slope, texture, drainage, and erosion of a particular soil affect its productivity?
11. What are some uses that human society makes of soil and land resources?
12. How does soil erosion occur? What roles do water and wind play?
13. List some ways that soil erosion can be prevented or controlled.

Multiple-Choice Questions

1. Which of the following soil parent materials is deposited by wind?
   a. loess deposits
   b. alluvial deposits
   c. minerals and rocks
   d. organic deposits

2. Which of the following is not a factor in soil formation?
   a. hydroponics
   b. gravity
   c. ice
   d. water
3. Which soil horizon is most important in supporting plant growth?
   a. O
   b. A
   c. B
   d. C
4. Which of the following soil particles is smallest?
   a. sand
   b. gravel
   c. clay
   d. silt
5. Which term is not considered to be a physical property of the soil?
   a. texture
   b. slope
   c. drainage
   d. soil solution
6. As a medium for plant growth, which of the following is not provided by the soil?
   a. energy
   b. nutrients
   c. anchorage
   d. water
7. Which of the following uses of soil and land continues to increase at the expense of all other uses?
   a. urbanization
   b. livestock grazing
   c. crop production
   d. forests
8. Of all the organisms living in the soil, which are most numerous?
   a. microorganisms
   b. mammals
   c. amphibians
   d. insects
9. Decay of organic matter is caused by
   a. large animals.
   b. microbes.
   c. rodents.
   d. water.
10. A destructive process that occurs in soils that are not protected against forces of flowing water or strong winds is
    a. denitrification.
    b. soil conservation.
    c. conservation of matter.
    d. erosion.
11. The tendency of soil particles to break apart as erosion occurs is called
    a. decomposition.
    b. illuviation.
    c. weathering.
    d. erodibility.

Learning Activities

1. Dig a pit in the soil or clear off a road bank so that you can identify the soil horizons.
2. Collect samples of soil with varying amounts of sand, silt, and clay. Moisten the samples and feel them. Can you tell the differences? Try to form ribbons of soil by pressing the samples between your thumb and index finger. Can you see the differences? An extremely coarse-textured soil will not even form a stable ball, but a finely textured soil will stick together firmly and allow you to form a fairly long ribbon with your fingers.

3. Research a soil conservation practice that you think is most useful in your area. Use the Internet or resources other than this textbook to prepare a report. Provide enough detail to fully describe how the practice conserves soil.
4. Prepare a demonstration of the effects of slope on erosion. Fill several trays with soil of the same type and texture. Raise one end of each tray to a different height to represent different slopes. Release a measured amount of water over the surface of each tray, making sure that the water is released at the same rate in each tray. This may be done by pouring the water through a large can with holes poked in the bottom. Collect the water that runs off the end of each tray using a plastic bag or other container. Filter the silt out of each water sample and weigh the dried filter to determine the amount of soil that was eroded from each tray. Create a graph that summarizes the amount of erosion that occurred at each slope.
5. Invite a soil technician or scientist to discuss soil conservation practices recommended for your region. A field trip might be arranged to view some of the practices at the sites where they have been implemented.

TERMS TO KNOW

- forest
- forestland
- timberland
- tree
- shrub
- forestry
- evergreen
- conifer
- softwood
- deciduous
- hardwood
- pulpwood
- heartwood
- cambium
- annual ring
- xylem layer
- sapwood
- phloem
- inner bark
- silviculture
- silvics
- stand
- natural regeneration
- artificial regeneration
- direct seeding
- germination
- seedling
- forester
- logging
- seed tree method
- shelterwood method
- coppice method
- environmental impact statement
- Environmental Protection Agency (EPA)
- biological value
- strata
- canopy
- understory
- shrub layer
- herb layer
- forest floor
- particulate matter
- biomass
- biomass power
- riparian zone
- silt load
- disease
- biotic disease
- abiotic disease
- growth impact
- terminal growth
- radial growth
- alfisol
- spodosol
- ultisol
- soil conservation

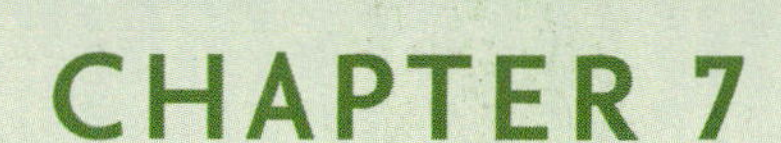

CHAPTER 7

Forests

Objectives

After completing this chapter, you should be able to

- explain the relationship between forestland and timberland
- describe the characteristics of trees known as evergreens, conifers, softwoods, and hardwoods
- name and describe the forest regions of North America
- identify important types and species of trees
- explain the significance of the cambium in a live tree
- define the terms *silvics* and *silviculture*
- describe a situation that might justify artificially renewing a forest rather than allowing natural regeneration to occur
- apply principles of good woodlot management
- list reasons why a harvest plan should be developed before any trees are cut
- discuss the different timber-harvesting methods and explain when each method is appropriate
- describe some of the roles that state and federal governments play in timber harvests
- identify ways in which forest resources are valuable
- explain the major functions and significance of watersheds
- identify some destructive elements from which forests sometimes require protection
- describe ways in which fire may be both beneficial and harmful to forests
- distinguish among the three soil orders that are of significance to North American forestry
- explain why a forest watershed is usually superior to a watershed without trees
- explain the relationship between soil erosion and surface water pollution

FIGURE 7–1 A forest contains trees, shrubs, and other plants, as well as animal life. (Photo courtesy of the U.S. Department of Agriculture.)

A **forest** is much more than a population of trees: It is a complex association of trees, shrubs, and plants that all contribute to the life of the community (Figure 7–1). Within its boundaries are found many other living organisms and natural resources.

The United States contains 731 million acres of forest: 483 million acres in timberland and 248 million acres in other forestland. This is approximately one-third of all land in the United States. The U.S. Department of Agriculture (USDA) defines **forestland** as land that is at least 10 percent stocked by forest trees of any size. **Timberland** is defined as forestland that is capable of producing in excess of 20 cubic feet per acre per year of industrial wood; the land also must not have been withdrawn from timber utilization by statute or administrative regulation.

A **tree** is a woody perennial plant with a single stem that develops many branches. Trees vary greatly in size, but they usually grow to 10 feet in height or more. A **shrub** is a woody plant that is smaller than a tree with a bushy growth pattern and multiple stems; it does not reach more than 15 feet in height. A productive forest is one that is growing trees for lumber or other wood products on a continuous basis.

Forestland may include parks, wilderness land, national monuments, game refuges, and other areas where the harvesting of trees is not permitted. When we consider that there are 860 species of trees in the United States, **forestry**—the management of forests—is obviously an important part of the economy. An unbelievable variety of products come from trees (see Figure 7–2).

Trees in U.S. forests are divided into two general classifications: evergreen and deciduous. **Evergreen** trees do not shed their leaves on a yearly basis. The most commercially important evergreens are mostly **conifers**; they produce seeds in cones, have needle-like leaves, and produce lumber called **softwood**. **Deciduous** trees shed their needles or leaves every year and produce lumber called **hardwood**.

trees, evergreen, deciduous, examples

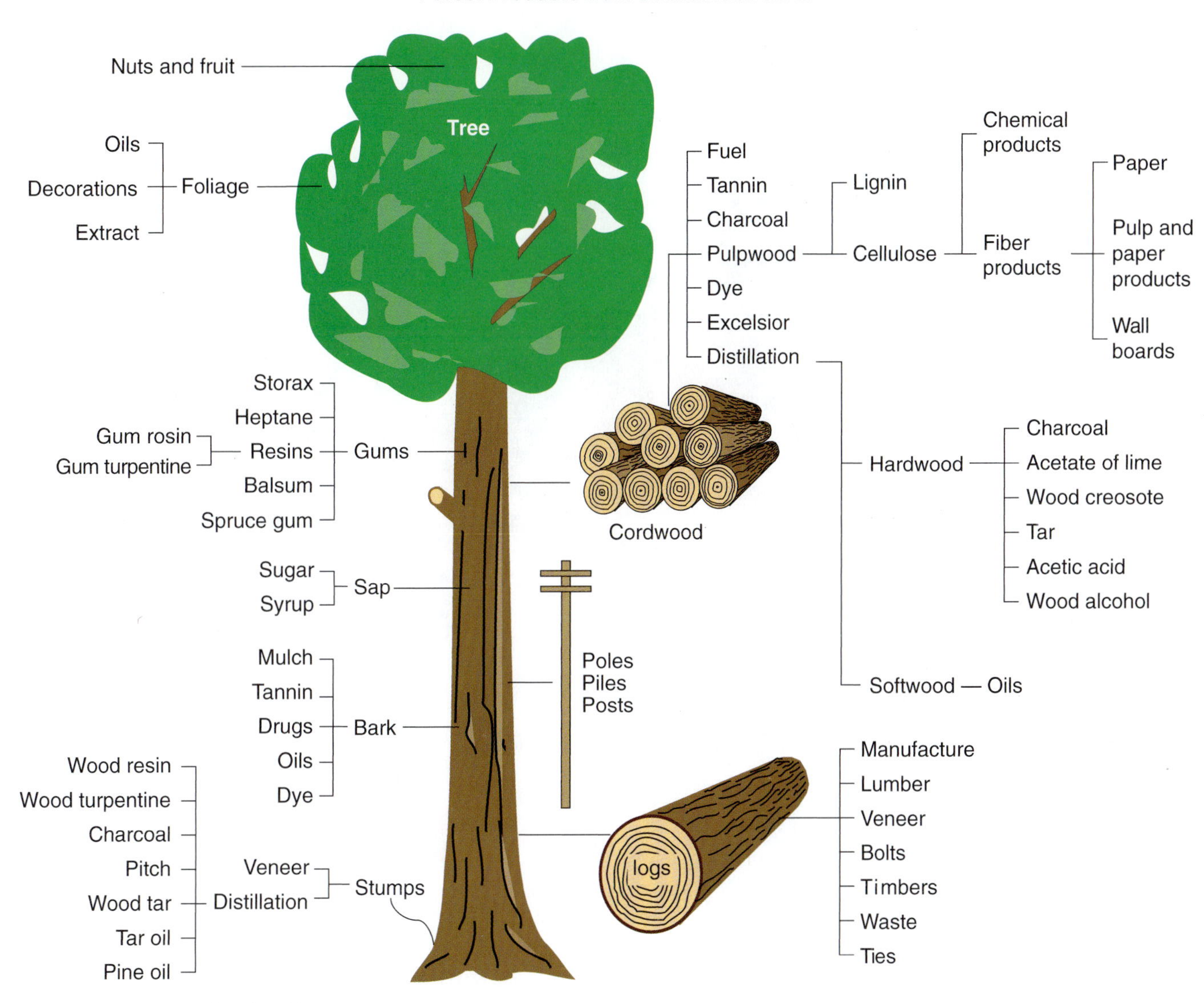

FIGURE 7–2 Many kinds of commercial products are obtained from trees and wood by-products.

forest regions map

Forest Regions of North America

Forestry educators and other forestry professionals generally recognize eight major forest regions (Figure 7–3): the northern coniferous forest, the northern hardwoods forest, the central broad-leaved forest, the southern forest, the bottomland hardwoods forest, the Pacific Coast forest, the Rocky Mountain forest, and the tropical forest. Some experts also include Hawaii's wet forest and dry forest (Figure 7–4).

Northern Coniferous Forest

The northern coniferous forest is the largest region in North America, extending across Canada and Alaska. It is characterized by swamps, marshes, rivers, and lakes and a cold climate. It contains vast regions of softwoods. Some areas along the U.S.–Canadian border contain mixtures of softwoods and hardwoods. The most dominant type of tree is the evergreen, and large amounts of **pulpwood** are harvested in this region. The most important species include the white spruce, Sitka spruce, black spruce, jack pine, black pine, tamarack, and western hemlock.

FIGURE 7–3 The forest regions of North America include those on the map and the bottomlands hardwood forest (which is mostly overlapped by the southern forest along the Atlantic and Gulf coasts).

northern coniferous forest
northern hardwood forest

Northern Hardwoods Forest

The northern hardwoods forest region reaches from southeastern Canada through New England to the northern Appalachian Mountains. It blends with the northern coniferous forest to the north and the central broad-leaved forest to the south. The region extends westward beyond the Great Lakes region and is populated by several important hardwood species, including beech, maple, hemlock, and birch trees.

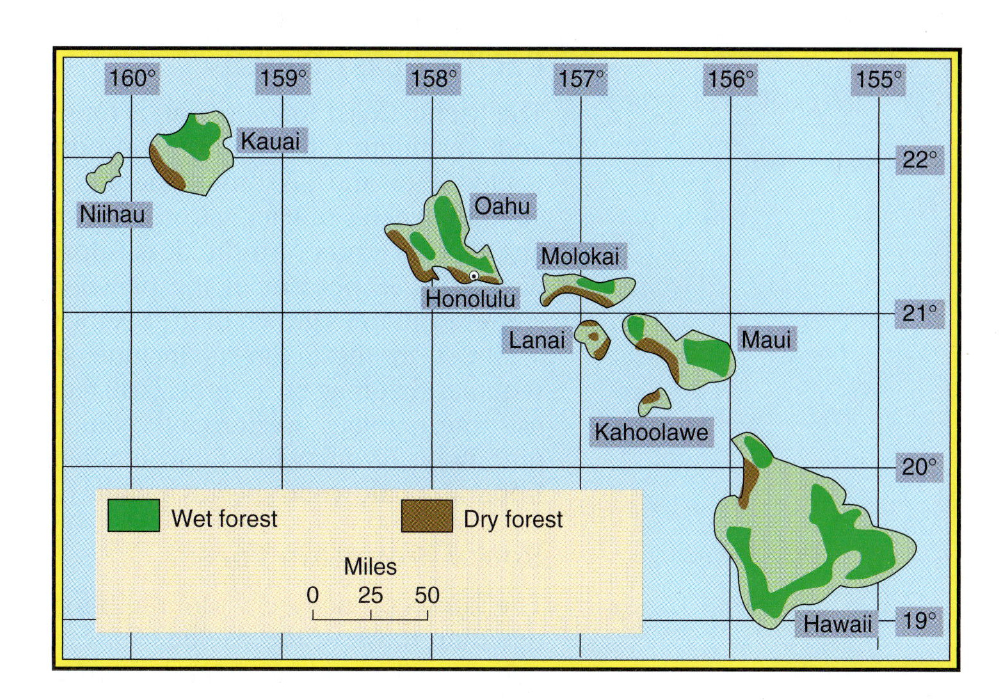

FIGURE 7–4 The forest regions of Hawaii. (Adapted from material provided by the U.S. Forest Service.)

Central Broad-Leaved Forest

The central broad-leaved forest region is located mostly east of the Mississippi River and south of the northern hardwoods forest. It consists of an arbitrary grouping of several distinctly different forest subgroups. It is a farming region in which most of the land has been cleared to produce cultivated crops. In contrast with some other regions, little of the forested area is owned by the federal government. High-quality wood is produced in this region, and much of it is used to construct high-quality furniture. Hardwoods of lesser quality are used for construction and to make industrial pallets.

central broad-leaved forest

This forest contains more varieties and species of trees than any other forest region. It is composed mostly of hardwood trees. Hardwoods of commercial importance in this forest region include oak, hickory, beech, maple, poplar, gum, black walnut, cherry, ash, cottonwood, and sycamore. The conifers that are of economic value in this region include Virginia pine, pitch pine, shortleaf pine, red cedar, and some hemlock.

Southern Forest

The southern forest region in the southeastern part of the United States extends south from Delaware to Florida and west to Texas and Oklahoma. It is the region with the most potential for meeting future U.S. lumber and pulpwood needs. The most important trees in the southern forest are conifers, including the Virginia, longleaf, loblolly, shortleaf, and slash pines. Oak, poplar, maple, and walnut are hardwood trees of economic importance.

Bottomland Hardwoods Forests

Bottomland hardwoods forests occur mostly along the Mississippi River. They contain mostly hardwood trees and are often among the most productive of the U.S. forests because of the high fertility of the area's soils. Oak, gum, tupelo, and cypress are the major hardwood species found here.

INTERNET KEY WORDS

Pacific Coast forest

Pacific Coast Forests

The Pacific Coast forest region is found in northern California, Oregon, and Washington. It is the most productive of the forest regions in the United States and has some of the largest trees in the world. Approximately 48 million acres of Pacific Coast forest provide more than 25 percent of annual U.S. lumber production. Approximately 19 percent of the pulpwood and 75 percent of the plywood produced in the United States comes from trees grown in the Pacific Coast forest region.

Trees in these forests include 300-foot-tall redwoods and giant sequoias that may be as large as 30 feet in diameter. Douglas fir, ponderosa pine, hemlock, western red cedar, Sitka spruce, sugar pine, lodgepole pine, noble fir, and white fir are conifers that are important in this region. Important hardwood species include oak, cottonwood, maple, and alder.

Rocky Mountain Forest

The forests of the Rocky Mountain forest region are much less productive than those of the Pacific Coast region. This region is divided into many small areas and extends from Canada to Mexico. Approximately 27 percent of the lumber produced in the United States comes from the region's 73 million acres.

INTERNET KEY WORDS

water cycle, forest relationships

Most of the trees of commercial value in the Rocky Mountain forests are the western pines: western white pine, ponderosa pine, and lodgepole pine. Spruce, fir, larch, western red cedar, and hemlock also grow there in small quantities. Aspen is the only hardwood of commercial importance in the region.

Tropical Forest

The tropical or subtropical forests of the continental United States are located in southern Florida and southeastern Texas. They compose the smallest forest region in the United States. The major trees—mahogany, mangrove, and bay—are unimportant commercially but critically important ecologically.

Hawaiian Forest

The wet forest region of Hawaii produces ohia, boa, tree fern, kukui, tropical ash, mamani, and eucalyptus. Most of these woods are used in the production of furniture and novelties. Hawaii's dry forest region produces koa, haole, algaroba, monkey pod, and wiliwili. None of these is of commercial value.

Important Types and Species of Trees in the United States

INTERNET KEY WORDS

tree species

Trees may be described in terms of the lumber they produce. The lumber's usefulness is evaluated by characteristics such as hardness, weight, tendency to shrink and warp, nail- and paint-holding capacity, decay resistance, strength, and surface qualities.

Softwoods

Commercially important U.S. softwoods, or needle-type evergreens, include Douglas fir, balsam fir, hemlock, white pine, cedar, southern pine, ponderosa pine, and Sitka spruce (Figure 7–5).

FIGURE 7–5 Some of the softwood tree species that are commercially important for wood products.

Douglas Fir

Douglas fir is probably the most important species of tree in the United States today. It typically grows to a height of more than 300 feet and a diameter of more than 10 feet. Approximately 20 percent of the annual U.S. timber harvest is Douglas fir. One-hundred-year-old stands of Douglas fir can produce 170,000 board feet of lumber per acre, which is five to six times the production of most other softwood species. Douglas fir is popular as construction lumber and for the manufacture of plywood, a construction material made of thin layers of wood that are glued together.

INTERNET KEY WORDS

Douglas fir
balsam fir
hemlock, eastern, western
cedar, eastern red, eastern white, western red
white pine, eastern, western
southern pine
ponderosa pine
Sitka spruce

Balsam Fir

Found in the forests of the Northeast, balsam fir trees have soft, dark-green needles and a classic triangular shape when grown at low densities. They are often used as Christmas trees. Balsam fir lumber is used mostly for framing buildings.

Hemlock: Eastern and Western

Eastern hemlock is strong and often used for building material, although it can sometimes be brittle and difficult to work. Eastern hemlock grows over most of the northern coniferous forest range.

Western hemlock grows in the Pacific Coast forest region, where yearly rainfall averages 70 inches. Western hemlock lumber is strong and one of the most important sources of construction-grade lumber. It is also important for pulpwood.

Cedar: Eastern Red, Eastern White, and Western Red

Because it is resistant to decay, eastern red cedar is used for fence posts. It is also used to line chests and closets because its odor repels many insects. White cedar is a swamp tree with decay-resistant wood that is often used for shingles and log homes. Western red cedar resembles redwood in appearance and is used where decay resistance, rather than strength, is important.

White Pine

White pine lumber is soft, light, and straight-grained. It has less strength than spruce or hemlock but is more popular as a wood for cabinetmaking. Eastern white pine grows from Maine to Georgia. Western white pine is found in the Rocky Mountain forest region.

Southern Pine

The southern pine category includes longleaf pine, shortleaf pine, loblolly pine, and slash pine. The southern pines grow in the southern and south Atlantic states. Lumber from southern pines is used for construction, pulpwood, and plywood.

Ponderosa Pine

A large tree, the ponderosa pine grows as high as 130 feet and reaches four feet in diameter. It is widely distributed in the western United States. The wood is heavy and can be brittle, but it is reasonably free of knots and other defects. Its most valuable use is for constructing wooden windowpanes and doors.

Sitka Spruce

Growing from California to Alaska, Sitka spruce trees attain a height of 300 feet and a diameter of 18 feet. Lumber from Sitka spruce is of especially high quality—strong, straight, and even-grained. Sitka spruce is also used in large quantities to make pulpwood.

Hardwoods

Hardwoods come from deciduous trees. The most commercially important species in the United States include birch, maple, poplar, sweetgum, oak, ash, beech, cherry, hickory, sycamore, black walnut, and willow (see Figure 7–6).

birch
maple
poplar
sweetgum
oak, white, red

Birch

Easily recognized by their white bark, birch trees grow in areas where summer temperatures seldom exceed 70° F. Birch lumber is dense and fine-textured and used for furniture, plywood, paneling, boxes, baskets, and veneer, as well as for many small novelty items. Veneer is a thin sheet of wood glued to a cheaper species of wood for use in paneling and furniture making.

Maple

Maple lumber is classified as both hard and soft. Hard maple lumber is heavy, strong, and dense and used for butcher blocks, workbench tops, flooring, veneer, and furniture. Soft maple is only 60 percent as strong as hard maple but used in the same applications. Some species produce sweet sap that is made into maple syrup.

Poplar

Poplar grows over most of the eastern United States. Although it is classified as a hardwood because of its deciduous structure, its lumber is soft, light, and usually knot-free. Poplar lumber may be white, yellow, green, or purplish and can be stained to resemble most of the fine hardwoods. Poplar is used for furniture, baskets, boxes, pallets, and building timbers.

Sweetgum

The sweetgum tree is easily recognized by its star-shaped leaves and distinctive ball-shaped fruit. Sweetgums grow to as high as 120 feet and to 3 to 5 feet in diameter. Its lumber has interlocking grain and is used for house trim, furniture, pallets, railroad ties, boxes, and crates. The gum that comes from wounds in the tree's bark can be used as natural chewing gum or as a flavoring or perfume.

Oak: White and Red

There are two general types of oak in the United States: white and red. White oak lumber is hard, heavy, and strong. Its pores are plugged with membranes that make it nearly waterproof. It is used for structural timbers, flooring, furniture, fencing, pallets, and other uses where wood strength is required.

Red oak is similar to white oak, except that it is highly porous and not resistant to decay; for outside use, it must be treated with wood preservatives. Its chief uses include furniture, veneer, and flooring.

aspen
ash
beech
cherry tree lumber, uses

Aspen

Aspen trees grow in the Northeast, Great Lake states, and the Rocky Mountains. Aspen grows rapidly, and its lumber tends to be weaker than most construction-grade timber. It is also used for pulpwood.

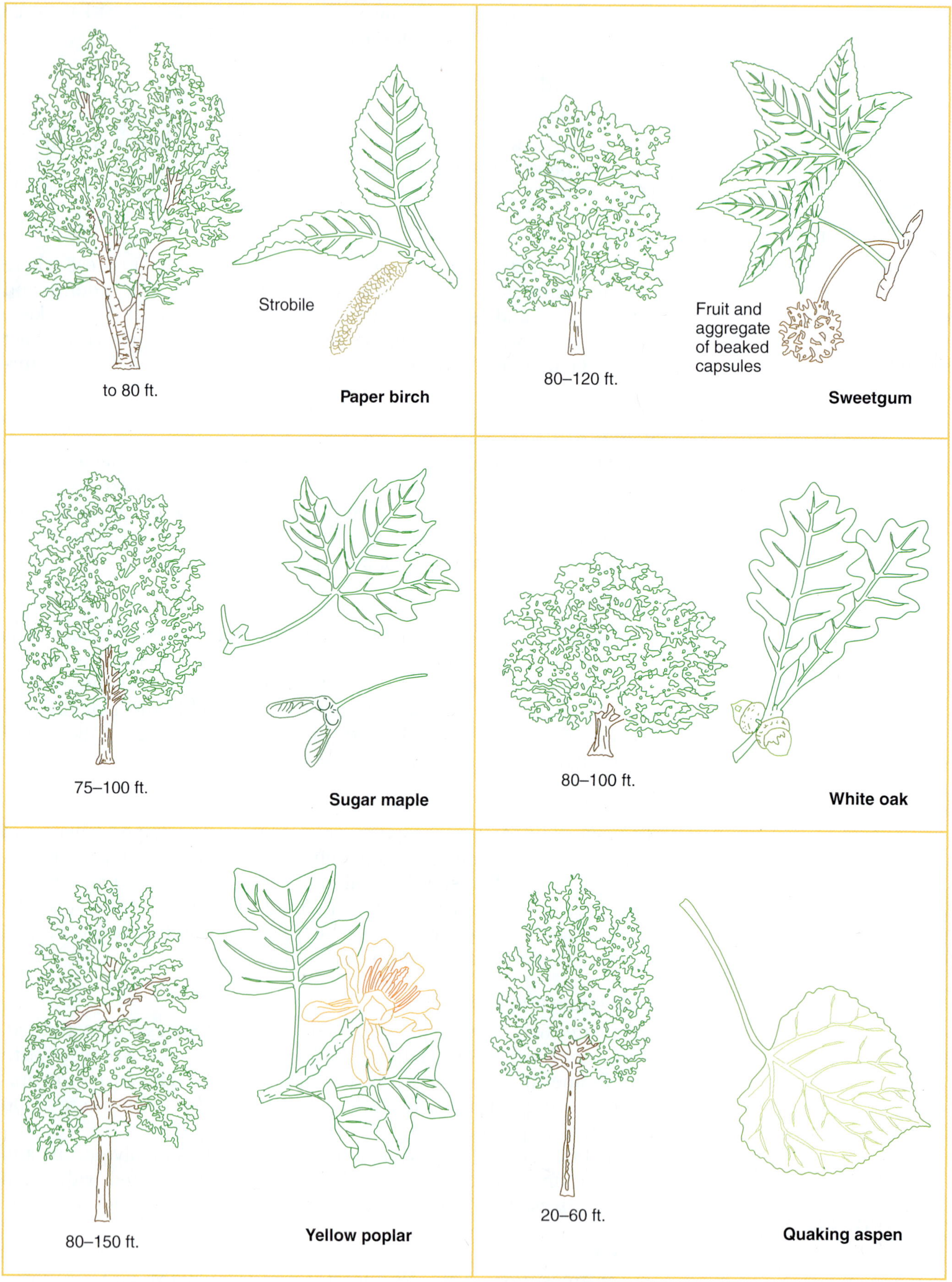

Strobile
to 80 ft.
Paper birch

Fruit and aggregate of beaked capsules
80–120 ft.
Sweetgum

75–100 ft.
Sugar maple

80–100 ft.
White oak

80–150 ft.
Yellow poplar

20–60 ft.
Quaking aspen

FIGURE 7–6 Species of hardwood trees that are commercially important for wood products.

FIGURE 7–6 *(continued)*

Ash

Ash lumber is heavy, hard, stiff, and highly resistant to shock. It also has excellent bending qualities and is popular for use in handles, baseball bats, boat oars, and furniture. It resembles oak in appearance.

Beech

Grown in the eastern United States, beech is heavy and hard and noted for its shock resistance. It is difficult to work with, however, and is prone to decay. Beech is used in veneer for plywood, and for flooring, handles, and containers.

Cherry

Cherry can be found from southern Canada through the eastern United States. Cherry wood is dense and stable after drying. It is desirable and popular in the production of fine-quality furniture. Because it is expensive and in limited supply, it is used mostly for veneer and paneling, although it can be used for other woodworking purposes.

Hickory

Hickory grows best in the eastern United States. Its lumber is hard, heavy, tough, and strong: In fact, it is somewhat stronger than Douglas fir when used as construction lumber. Other uses for hickory include handles, dowel rods, and poles. It is also popular as firewood and for smoking meat.

hickory
sycamore
black walnut
black willow

Sycamore

Sycamores grow from Maine to Florida and west to Texas and Nebraska. The wood is used for flooring in barns, trucks, and wagons because of its strength and shock resistance. Other uses include in boxes, pallets, baskets, and paneling.

Black Walnut

A premier wood for the manufacture of fine furniture, black walnut grows from Vermont to Texas. The wood has straight grain and is easily machined with woodworking tools. Because walnut is slow growing and desirable, it is often made into veneer to get more use from its chocolate brown **heartwood**, the inactive core that gives a tree its strength and rigidity. It is also the source of black walnut nut meats.

Black Willow

Most of the black willow of commercial value is grown in the Mississippi River valley. It is soft and light and has a uniform texture. Black willow is used mostly in construction for subflooring, sheathing, and studs, but some is also used for pallets and interior components in furniture. It is sometimes a low-cost substitute for black walnut because it has a similar brown appearance when finished.

Many other domestic softwoods and hardwoods grow in the United States, and many are important in local areas. The types discussed here are but a sampling, rather than a definitive list, of the most commercially important trees.

Tree Growth and Physiology

Trees use carbon dioxide (CO_2, a carbon and oxygen molecule) from the air and water (H_2O, a hydrogen and oxygen molecule) from the soil to manufacture simple sugars in their leaves. The leaves then use

age, annual rings

additional carbon, hydrogen, and oxygen atoms to convert simple sugars into complex sugars and starches. Nitrogen and minerals from the soil are then used to manufacture proteins, the building blocks for growth and reproduction.

A tree typically starts from a seed. For instance, oak trees grow from seeds called *acorns*, pine trees start from seeds in pinecones, and peach trees grow from peach seeds. Trees may also sprout and grow from stumps or other tree parts. When a seed germinates, a shoot grows upward to form the top growth, and roots grow downward and outward to form the root system. Both roots and shoots extend themselves by growth at the tips through cell division and elongation. At the same time, tree roots, stems, and trunks grow in diameter by adding cell layers near their outer surfaces (see Figure 7–7). This growth layer in a tree root, trunk, or limb is called the **cambium**. The outward growth of the cambium in one year creates an **annual ring** that can be seen in the cross section of a root, trunk, or limb.

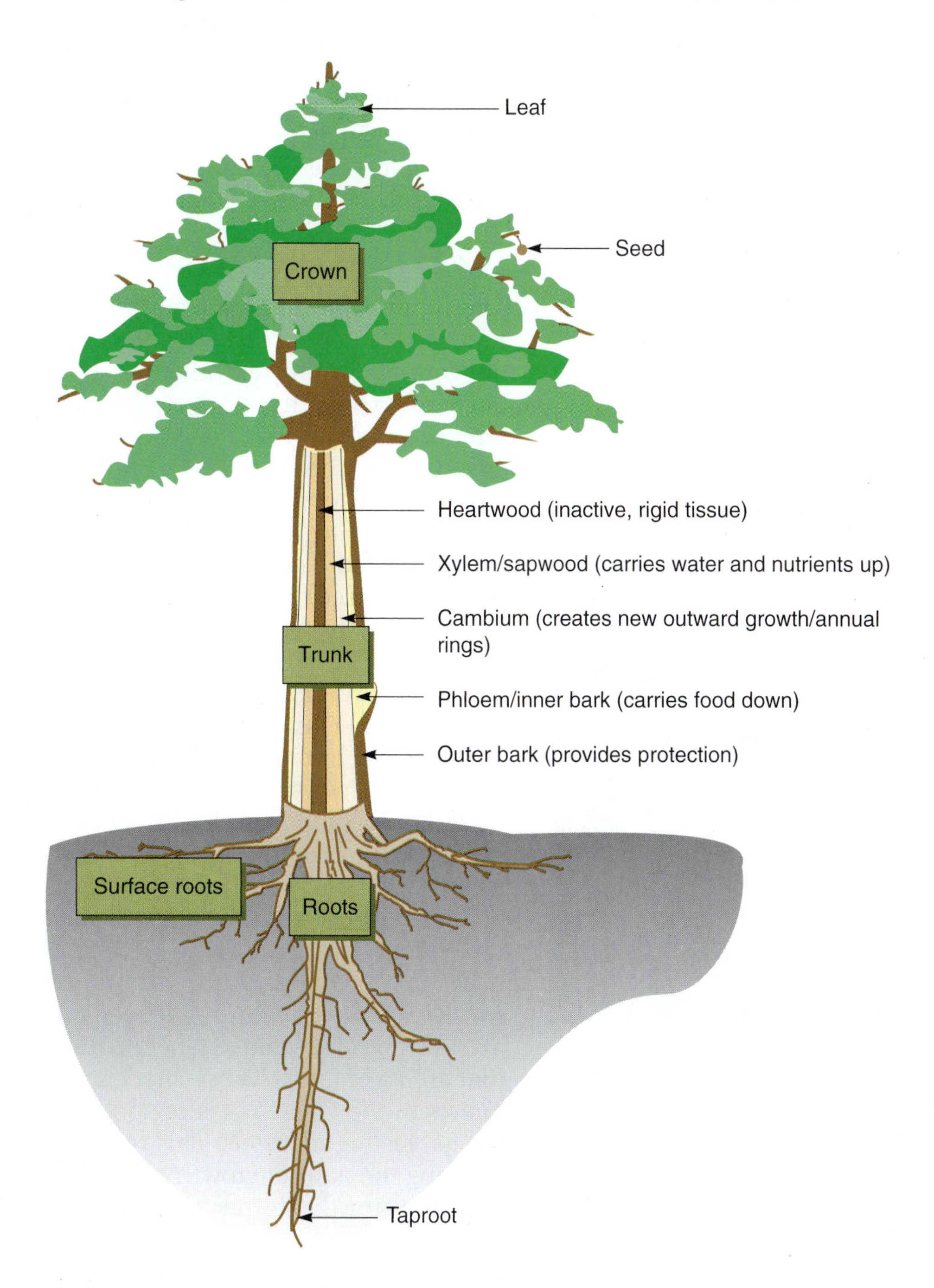

FIGURE 7–7 The major parts of a tree.

Water and minerals are taken in by the roots and transported up to the leaves through a layer of cells called the **xylem layer** or **sapwood**.

The xylem is located just inside of the cambium layer. Just outside the cambium is another layer of cells called the **phloem**, or **inner bark**, which carries food manufactured in the leaves to the stems, trunk, and roots. Each year, the tree grows new cambium, xylem, and phloem tissues, and the older sapwood becomes heartwood.

Forest Management

INTERNET KEY WORDS

woodlot management
silviculture, silvics
natural, artificial, regeneration, forests

The art and science of tree production is called **silviculture**. It is a specialized area of study within the larger field of forestry. It is based on an understanding of **silvics**, which is the study of forests and forest relationships and includes plant, soil, and animal interactions with trees. Silviculture is practiced on the assumption that a forest environment can be managed so that it is favorable to tree production. Of course, there are many degrees of intensity with which forests are managed. Many of our national forests are managed with much less intensity than privately owned commercial forests. Any cultural practice that manipulates the forest environment to achieve specific goals in the forest is a form of silviculture.

Forest Reproduction and Regeneration

The forest reproduction processes known as *sexual reproduction* and *vegetative reproduction* are both important methods for regenerating forests. Sexual reproduction produces seeds that are capable of becoming young trees. Vegetative reproduction, also known as *asexual reproduction*, occurs when young trees are produced from leaf, root, or stem tissues (see Figure 7–8). Both kinds of reproduction are important in forest regeneration.

A population of trees that has been established in a forest environment is called a **stand**. When a tree population is few in number and widely scattered, it is considered a poor or weak stand, whereas a population of healthy trees that are properly spaced in the forest is considered a strong or vigorous stand. These terms will be used throughout this text to describe the characteristics of specific populations of forest trees.

FIGURE 7–8 Some tree species such as poplars are regenerated vegetatively by planting "suckers" or young stems and branches in moist soil where they generate roots.

Forests are usually considered to be naturally renewable without any need for human intervention. Although this is generally true, forests do not always produce the kind of trees that are wanted or needed. Some kinds compete well with other forest plants, and some do not. For example, the eastern white pine forests tended to be replaced by oak forests after they were harvested. The stage in the biological succession on these sites favored the climax species of trees such as oak.

When a particular kind of tree is desired in an area, it may be necessary to change the environment to favor its growth. For this reason, silviculture is widely practiced on private forestlands. Some species of trees reproduce and regenerate naturally when conditions are created that are favorable to those species. For example, regeneration of aspen forests in the Great Lakes region can often be accomplished by clear-cut harvesting methods. This removes the shade from the soil surface, allowing the natural growth of young aspens. Conversely, shaded areas favor the growth of other kinds of forest plants and trees.

interest profile

SEED DISTRIBUTION BY SQUIRRELS AND BIRDS

Squirrels and birds play important roles in distributing the seeds of trees and other plants to new locations (Figure 7–9). Squirrels actively harvest seeds by gathering acorns, pinecones, and other kinds of nuts to be eaten during the winter season. Some of the seeds that are hidden in the debris of the forest floor are forgotten and may eventually germinate and grow. Most of these seeds are not distributed far beyond the trees on which they grow, but squirrels play a role in planting them beneath the vegetative cover on the forest floor.

Birds distribute seeds over wide areas. Some seeds are carried in flight and dropped in distant locations. Others may be overlooked in the shell or husk materials that surround most seeds from trees. Some seeds have hard seed coats that can survive a bird's digestive process. These seeds are distributed in the feces of birds.

FIGURE 7–9 Squirrels and birds play important roles in distributing tree seeds to suitable new locations. (Courtesy of the U.S. Fish and Wildlife Service. Photo by Jon Nickles.)

Two types of forest regeneration occur following timber harvests: natural and artificial. **Natural regeneration** occurs on a forest site when young trees begin to grow without having to be planted. Sometimes, seeds have been dispersed in the area by the wind or by wild animals. Some hardwoods grow from the roots or stumps of the harvested trees. In some instances, advance regeneration occurs because of existing seedlings and saplings that survived the harvest. Natural regeneration of forest trees depends on several important growth conditions such as the availability and dispersion of fertile seed in the area, the availability of soil moisture, warm temperatures, the condition of the soil, favorable weather, favorable light intensity, and freedom from diseases and harmful insects.

Artificial regeneration occurs when seeds or seedlings are planted at the harvest site (Figure 7–10). This method often results in a uniform stand of trees that are evenly dispersed throughout the area. The forest manager also has control over the species of trees that make up the new forest planting. Seedlings can be selected from superior parent stock that has the potential to increase yields. Because artificially regenerated trees are all the same age and at the same growth stage, they are easier to manage. Tree plantations are usually planted in rows. This makes it possible to use mechanical equipment within the plantation for weed control, thinning, and pruning, among other purposes.

Direct Seeding

Planting tree seeds to generate new forest growth is called **direct seeding**. Seeds can be either planted directly into the soil surface using mechanical equipment or dispersed in the area by aircraft. Aerial seeding is a good method to use following fires. The soil surface in a burned area is

FIGURE 7–10 Forest renewal is often accomplished by planting seeds or seedlings soon after a mature forest has been harvested.

usually free of debris, and large areas can easily be planted using this method. When direct seeding methods are used, it is important to take advantage of good moisture conditions to ensure a good stand of trees. Drought conditions will prevent seed germination.

The use of high-quality seed is important in forest regeneration. Tree seeds should be collected only from superior trees (Figure 7–11). Seeds must be collected at the right stage of maturity and carefully cleaned to eliminate damaged or shrunken seeds. They should be stored in a cool, dry, and dark location to avoid untimely germination.

Seeds are sometimes damaged by extreme temperatures, drought conditions, and other environmental factors, which can reduce their fertility. Seeds should always be tested before use to make sure they will sprout and grow, a process called **germination**. Seed laboratories are available in

FIGURE 7–11 Seed tree farms are often maintained to ensure a supply of high-quality seed for forest regeneration.

Figure 7–12 Seeds should be tested for germination before planting to ensure a good stand.

INTERNET KEY WORDS

seed testing

most states to provide seed testing services (Figure 7–12). These laboratories issue certification tags that are attached to each container of seed that was part of the tested seed lot. Certified seeds reduce the risk of establishing an inadequate population of young trees in the plantation, one of the greatest problems associated with the direct seeding of trees.

Adequate amounts of seed must be planted to ensure that the established seedling population is properly spaced in rows. Enough young trees must survive and require later thinning to remove weak, damaged, and deformed trees from the stand. Allowances must be made in the seeding rate for seed that is eaten or destroyed by birds, squirrels and other rodents, and insects. Some seeds are killed by molds and fungi, which are nearly always present in forest environments. Seeds can be protected from seed-eating insects, birds, and animals by coating them with protective fungicides, insecticides, and repellents.

Because not all of the planted seeds will grow, more seeds than are expected to grow must be planted. The seeding rate is different for each kind of tree because the seeds of different species are of different sizes. The volume or weight of the seed that is needed to plant each acre is much less with small seeds than for large seeds because there are more seeds per pound for small seeds. The seeding rate is also affected by the way the trees will be used at harvest. Trees that are used for pulpwood can be spaced more closely between and within the rows than would be desirable for timber production.

Woodlot Management

The proper management of a wooded area or woodlot involves more than just harvesting trees and removing unwanted species. A woodlot is a small, privately owned forest, and the management of the many woodlots affects the health of their regional environments. The production of trees for harvest is a long-term investment, and mistakes in management take a long time to correct.

woodlot management

Factors that must be considered in woodlot management include soil, water, light, type of trees, condition of trees, markets available, methods of harvesting, and replanting. A woodlot must be protected from fire, pests, and domestic animals if it is to maximize its yield. Woodlot grazing by cattle and sheep usually destroys all small seedlings in the forest and also eliminates most of the woodlot floor coverage. Livestock may also strip the bark from trees, causing them to die. Even though woodlots do not provide much food for grazing animals, it is still wise to exclude livestock from them.

Restocking a Woodlot

Natural seeding is the least expensive method of replacing trees harvested from a woodlot. Sources of seed for the desired species must be available in the forest, and conditions must be right for seed germination to take place. If seed from natural sources is not available, then seeds from other sources may be planted on the forest site.

A surer method of restocking a woodlot is to plant trees of the desired species rather than rely on seeds to do the reforestation. In most cases, **seedlings** (young trees started from seeds) are planted during late winter and early spring, before the new season's growth begins (Figure 7–13). Woodlots can be planted with one species or a mixture of several compatible tree species.

Timber Management

Management of a forest is much more involved than just sitting back and watching it grow. Proper care is important if the forestry enterprise is to be successful. Trees that are of no commercial value should be removed as soon as possible to eliminate competition for light, moisture, and nutrients. Because these "weed" trees are removed when they are small, there is seldom a market for them, and they are left on the woodlot floor to decay. If weed trees are of sufficient size, however, they may be used for firewood. When all the trees of a forest are nearly the same age (typically15–30 years old), they often need to be thinned. Trees should be thinned any time their crowns or branches occupy less than one-third of the overall height of the woodlot. Usually, this amounts to approximately one-fourth to one-third of the trees in a woodlot.

Trees that are being grown for lumber are often pruned of side branches to produce a better-quality log that is free of knots. Only rapidly growing trees, however, should be pruned. Branches should be pruned flush with the trunk of the tree, usually during the fall and winter when the trees are dormant.

Seedling

Unrestricted, conical root shape

FIGURE 7–13 To maximize the seedling's survival chances, a conical root shape must be maintained as the young tree is transplanted into the soil.

Planning a Harvest

A forest should never be harvested without a good harvesting plan to maximize yield over many years. It is usually wise to use the services of a **forester**—a person who studies and manages forests—when developing such a plan. A timber harvest, also known as **logging**, is a natural

FIGURE 7–14 Once a tree has matured, its health usually declines until it dies. It must be harvested in a timely manner before it dies and begins to decay. Failure to harvest in a timely manner results in dead or decaying trees that have no commercial value.

outgrowth of planning. Once a long-range forest-management plan has been implemented, proven harvesting practices will lead to quality timber on a timetable that maintains sustainable yields.

Most forest managers can predict, with reasonable accuracy, how long it will take for an even-aged timber stand to mature. They also know that once the trees have matured, a harvest must be planned and initiated to avoid timber losses (Figure 7–14). Heart rot and decay, for example, become serious problems in overage forests, and the increase in timber volume from growth can be more than offset by losses from rotting.

Planning for a timber harvest begins long before any trees are cut. One of the first decisions that must be made is the type of harvest that will be conducted. This chapter will now detail different timber-harvesting methods.

clear cutting, benefits, problems

Clear-Cutting

One of the most important considerations in planning a timber harvest is to determine what method will be used to regenerate the forest. If an even-aged stand is desired, it can be achieved by planning how the current stand of trees will be harvested. For example, an even-aged stand of eastern white pine can best be established by clear-cutting followed by site preparation and planting either seeds or seedlings to regenerate the desired stand.

Clear-cutting is an unpopular harvesting method among some political action groups because of lost scenic value and dangers to the environment, both real and perceived. If clear-cutting is the method of choice, then harvests in adjacent areas must be timed to ensure that forests of different ages are established (Figure 7–15). In the political world of the twenty-first century, it will be important to restrict the size of clear-cut harvests to reduce the potential for erosion. Harvest planning should also include selection-cutting methods near roads and

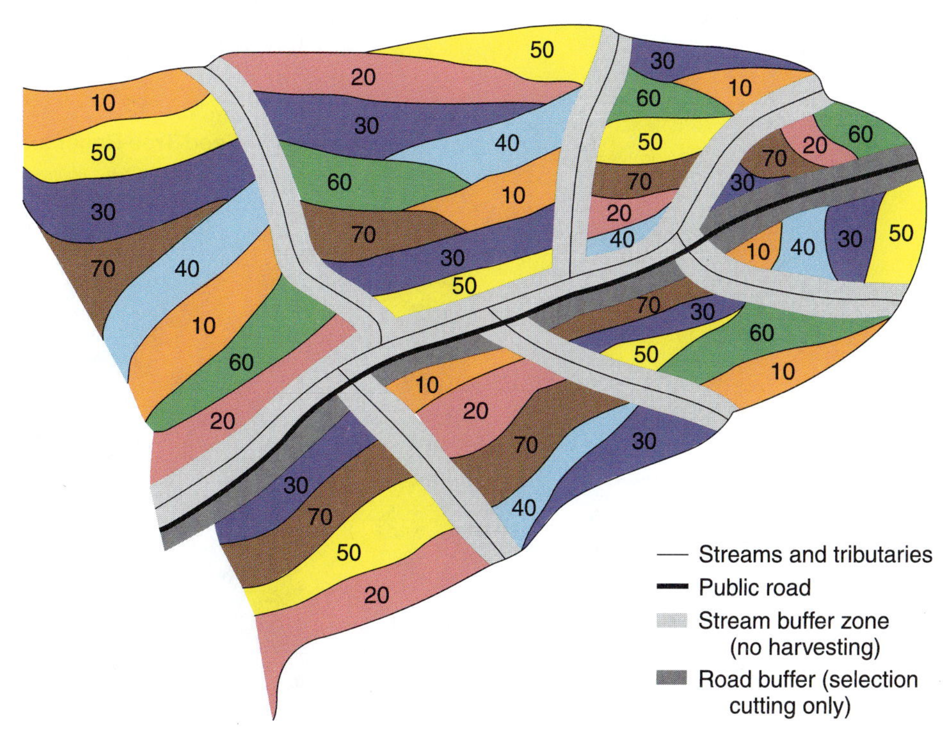

FIGURE 7–15 Clear-cuts should be set up parallel to contour lines of the terrain and isolated in relatively small pockets. This harvest plan allows trees to be harvested for sustained yields every 10 years on a 70-year rotation.

FIGURE 7–16 Unharvested buffer zones near roads and streams provide wildlife habitat and preserve the beauty of the forest environment.

unharvested buffer zones of trees near streams both to protect water quality and provide wildlife habitat (Figure 7–16). These kinds of compromises will be necessary if clear-cutting harvest methods are to be used in the future.

Seed Tree Method

The **seed tree method** of timber harvesting is part of a management plan in which mature trees of the desired species are protected from cutting in scattered locations throughout the forest (Figure 7–17). The purpose of such protection is to provide seed for forest regeneration. In most instances, site preparation is necessary to reduce competition from shrubs and established seedlings of less desired species. This method is effective in establishing even-aged timber stands with southern pine and western larch. The method, however, has limitations in situations where site preparation is not practiced.

FIGURE 7–17 The seed tree method protects mature trees scattered throughout the harvest area and provides seed for the next crop. The seed trees may be harvested once the young trees have become established.

FIGURE 7–18 The shelterwood method leaves enough mature trees in the harvested area to provide seeds along with shade and protection for the young seedlings.

seed tree harvest method
shelterwood harvest method
coppice harvest method

Shelterwood Method

The **shelterwood method** is a modified seed tree harvest method in which mature trees are left in the harvested area in sufficient numbers to provide shade and protection for seedlings (Figure 7–18). Once seedlings have become established, some of the mature trees are harvested, leaving an overstory that provides partial shade on the forest floor. The number of mature trees that are harvested in the first cutting depends on the seedlings' shelter needs. These needs vary from one species to another, and the species that eventually become dominant in the stand may be strongly influenced by the availability of shelter during critical periods of development.

When the new stand has become well established, the remaining mature trees are harvested. This method has proven to be effective where harsh environmental conditions make it difficult for young trees to survive. This method for establishing even-aged timber stands also tends to be viewed more favorably than clear-cutting because the landscape is never completely stripped of trees.

Coppice Method

The **coppice method** of forest regeneration is a silviculture system in which trees are clear-cut and the forest is regenerated from stump sprouts (Figure 7–19). It is included here because of its connection with clear-cutting as a harvest method. As with any clear-cut harvest, care must be taken to protect against soil erosion by planning the sequence of clear-cuts. Most of the trees that are generated by the coppice method are fast-growing trees such as oaks and aspens that are managed on short rotation periods between harvests. In most cases, these trees are used for fuel or pulpwood. This method also has great promise in the production of biomass.

Selection Cutting

As described in this textbook, selection cutting is not the same as the selective cutting that occurred as native forests were harvested. In those instances, only the high-value trees were harvested. Eventually, the only

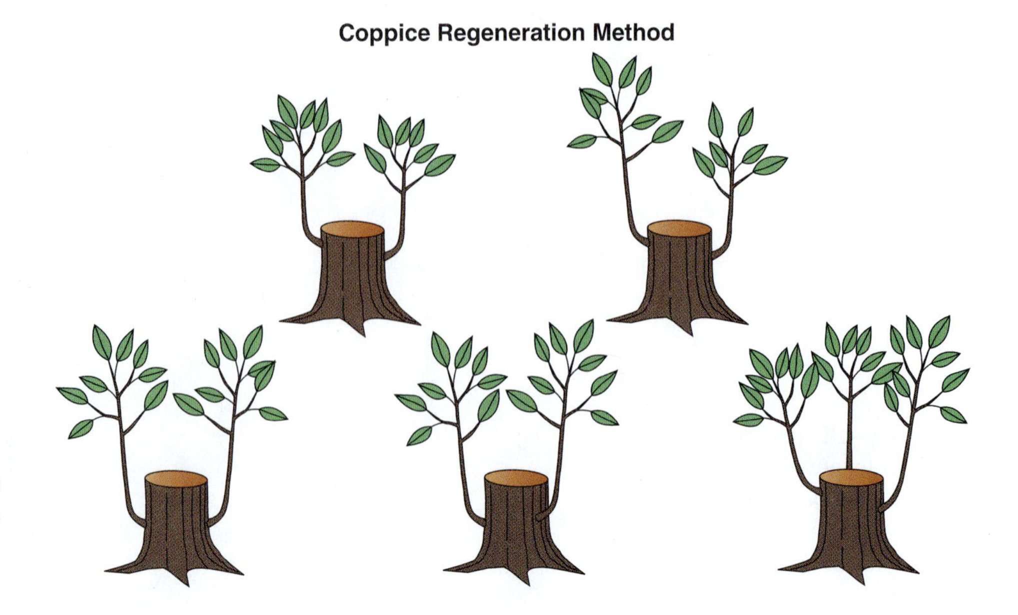

FIGURE 7–19 With the coppice harvest method, a new generation of trees sprouts from the stumps of harvested trees.

trees that remained were the least desirable varieties. Modern selection cutting identifies and harvests trees near the end of their productive lives (Figure 7–20). It is important to harvest these trees while they are still vigorous, and before they become victims of decay or disease.

process, selection cutting
salvage harvesting, forest
NEPA

Salvage Harvesting

Natural disasters occur regularly in forests, and salvage harvesting often follows. Trees that are dying from insect damage and diseases can still be used for lumber products if harvested before they are completely dead. Physical damage also sometimes occurs, including charred trunks from forest fires and broken trunks and limbs caused by high winds. Once damage has occurred, these trees need to be harvested as soon as possible. Waiting longer than two or three years to harvest damaged trees often results in poor-quality lumber.

FIGURE 7–20 During selection cutting, only mature trees that are near the end of their productive lives are marked for harvesting.

Sustained Yield

When selection-cutting practices are followed, harvest planning is a continual process because trees are removed from the forest at regular intervals as they mature. Regular inspections and forest inventories are required with this kind of harvest method to ensure that the correct amount of timber is harvested. When too much timber is harvested, the ability of the forest to produce at a sustained level gradually decreases. Eventually, the size of the trees diminishes, and larger numbers of trees are required to maintain the same harvest yields.

Sustained yields can be established and maintained by determining the forest growth and harvesting only as much timber as the forest can replace. Selective cutting requires a plan with a regular harvest rotation or a minimal tree size. Before each harvest, individual trees are evaluated, and a harvest decision is made based on the vigor and the potential future production of each tree. Growth rate is a good indicator of vigor, and individual trees may be retained in the forest instead of being harvested because they are still healthy and their potential for growth is high. Individual trees to be harvested are marked.

Environmental Impact Statements

The passage of the National Environmental Policy Act (NEPA) in 1969 required the filing of an **environmental impact statement**, a science-based study of the harvest area that specifically details the expected effects of human activity on the environment and wildlife in the area. Such studies must be completed before timber harvests may be conducted on federal lands. This piece of legislation has had major impacts on forest-management practices in the western area of the United States where large tracts of federal forestlands are concentrated.

An environmental study on the impact of logging includes the effects of road construction in the area, which can cause stream pollution with runoff water laden with silt; soil types and the expected effects of logging activity on soil stability; and the effects of logging on native plants and wild birds and animals. The presence of an endangered or threatened species in the area may require major modifications in a timber harvest plan or even stop all logging. The environmental impact statement also addresses the ways in which environmental damage and pollution problems will be prevented (Figure 7–21).

EPA, Environmental Protection Agency

FIGURE 7–21 Senior-level forest managers spend a great deal of their time researching and writing forest plans that limit negative environmental impacts on forest resources.

science connection profile

THE ENVIRONMENTAL PROTECTION AGENCY

The **Environmental Protection Agency (EPA)** was established in 1970 to protect and maintain the environment for future generations. The EPA is responsible for the enforcement of environmental laws designed to reduce air and water pollution from noise, radiation, pesticides, and harmful chemicals and materials. This agency has established water-quality standards and monitors the disposal of toxic waste as well as chemical residues in humans, wildlife, and food.

science connection profile

CONSERVING OUR BIODIVERSITY

Our biodiversity is at risk! Not far from the shores of Florida and Texas, the tropical rain forests of Central and South America are disappearing, just as they are in parts of Africa and Southeast Asia. Commonly called a *jungle*, a tropical rain forest is a hot, wet, green place characterized by an enormous diversity of life and a huge mass of living matter. Biologists estimate that more than half of the world's plant and animal species are found in these forests; unfortunately, many are not found anywhere else. The rain forests occupy only seven percent of the land area of the Earth, and their total area is diminishing at an alarming rate!

Abundant moisture and warm temperatures encourage tremendous plant growth year-round. Plants and moisture provide an excellent environment for animals and microorganisms to grow and develop. Rain forests are covered by a dense canopy formed by the crowns of trees as high as 150 feet. Under the shade of this canopy are smaller trees and shrubs that create a second layer called the *understory*. The two layers are typically woven together by strong woody vines, which may exceed 700 feet in length.

The third layer is the forest floor, which is dotted by palms, herbs, and ferns. With its shrubs and trees, the floor is home to a myriad of insects and animals. Annual rainfall may reach 400 inches, and the daily rate exceeds the rate of evaporation. Life is plentiful.

The soil beneath the forest, however, is poor, shallow, and erodible. As human populations expand into the rain forests, the foliage is cut and burned to make way for logging, grazing, and cropping. Because of its poor quality, the land often will support grazing or crop production for only a few years before it is worn out, depleted of fertility and soluble minerals, and eroded. Once the forest is destroyed, it may be lost forever. The human occupants then typically move on to virgin rain forest to slash and burn new tracts and repeat the cycle. It is estimated that one-half of the world's original tropical rain forests are gone. At the current rate of destruction, only one-quarter will remain in the near future.

Rain forests are our richest source of plant and animal species and biodiversity.

Because the rain forests are the storehouse for more than half of the genetic diversity in the entire world, species are being lost at an alarming rate. Many probably become extinct before they are discovered and recorded. In fact, scientists estimate that they have discovered and named only one-sixth of all the plant and animal species in the world's rain forests. Earth's biodiversity is indeed threatened by human hands.

FIGURE 7–22 Forest management has become much more difficult in recent years because of political and environmental conflicts over the appropriate uses of forest resources.

INTERNET KEY WORDS

rain forest, biodiversity
rain forest, destruction

The Forest as a Resource

Management of the forest ecosystem is a complicated and controversial profession in our politically charged world (Figure 7–22). The social and political sciences have become as important in forest management as the biological sciences that are the basis for modern forest management practices. Forest management now must address several different values: economic, biological, habitat, climate moderation, energy, livestock, recreation, and watershed.

Economic Value

Thirty percent of the land area in the world is forestland, and forest products are critically important to the economies of developed countries. The importance of North American and world forests goes far beyond the production of wood products such as paper, cardboard, lumber, plywood, and structural beams. Forests also provide solvents, medicines, fuels, and many other high-value products that are important for our health and comfort (Figure 7–23).

FIGURE 7–23 Forests are important sources of some of our most useful products, many of which are made of or extracted from wood.

Forest products are important natural resources to the U.S. economy. In contrast to a nonrenewable resource such as coal that is permanently gone once consumed, a forest is a renewable resource, meaning that it can be regrown following harvest (Figure 7–24). Many North American forests have been harvested at least twice since European settlers arrived. Good management practices should extend the production of our forests well into the future.

Biological Value

All of the living organisms that are found in forest environments contribute to the value of a forest. In fact, the **biological value** of a forest takes into account the worth of all of the life forms found there and includes the value of important natural functions of forests such as effects on climate, watersheds, water temperature, soil erosion, and

FIGURE 7–24 A forest is a renewable resource that replaces itself when conditions are favorable to new growth.

INTERNET KEY WORDS

forest, economic value
forest biological value

wildlife. Forests and other forms of plant life restore oxygen in our atmosphere through photosynthesis. They also provide the plant materials from which many of the medicines in use today have been derived, and many more plants with medicinal value and potential for new medical cures may yet be identified. We are likely to find many new plant materials in our forests that are valuable to society. Forests also function as huge biological filter systems that clean the environment by removing impurities from air and water. They also function in the elemental cycles (such as the carbon and nitrogen cycles) and in the water cycle.

Habitat Value

A forest consists of an area in which trees are the dominant living organisms. Sometimes a forest consists of a single species, but many forests are made up of more than one kind of tree (Figures 7–25 and 7–26) as well as many plants other than trees. Several layers of vegetation called **strata** are found in a forest (Figure 7–27). Taller trees form a ceiling or **canopy** at the highest levels and receive the most sunlight.

FIGURE 7–25 Some forests consist almost entirely of a single species of trees.

FIGURE 7–26 The majority of forests are made up of different kinds of trees.

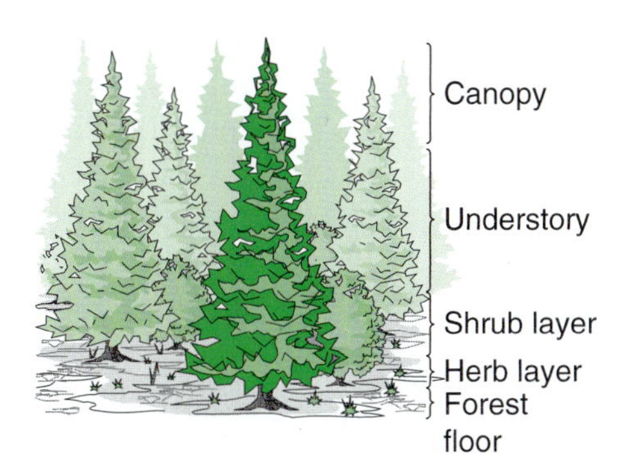

FIGURE 7–27 Different species tend to become dominant in different strata or vegetative forest layers.

The area beneath the canopy is filled with smaller trees that make up the forest's **understory**.

The lower strata of a forest is often inhabited by small woody plants and thus is called the **shrub layer**. The shortest plants such as ferns, grasses, and flowering plants are collectively called the **herb layer**. The **forest floor** is composed of a layer of decaying plant materials that acts as a mulch to preserve soil moisture. Each stratum is inhabited by living organisms that are adapted to live in that particular environment.

For many people, the value of a forest lies in the wildlife that can be found there. Forests provide many of the essential elements for the survival of a wide variety of creatures. Many of them—including insects, mammals, birds, amphibians, reptiles, and fish (Figure 7–28)—obtain

FIGURE 7–28 A forest provides food sources such as forage, seeds, fruits, and tender twigs for the wild animals that live there. (Photo courtesy of Clare Harkins.)

FIGURE 7–29 Forests provide ideal nesting materials for birds and small mammals.

food directly from plants or from other animals found in the forest. Whether the food source is meat, berries, nuts, seeds, shrubs, or forage plants, the amount available is a limiting factor in the size of each animal population living in the environment.

The other basic need of an animal is a habitat that provides shelter from weather and climate and protection from its natural enemies (Figure 7–29). A great variety of forest animals and birds find this among trees, shrubs, and other plants. Birds and animals also use the materials found in forests to provide dens and nesting places (Figure 7–30). Forests also tend to isolate animals and birds from humans and human civilization, allowing them to avoid disturbances during critical periods in their lives such as birds' nesting and fledgling phases when eggs and young birds are vulnerable to predators. Young mammals are also vulnerable to predators, partly because of their natural curiosity. Some also lack the mobility and caution that develop later in life (Figure 7–31).

FIGURE 7–30 Trees are sources of both food and shelter for birds such as woodpeckers and flickers.

FIGURE 7–31 A newborn elk calf is most vulnerable to predators in the first few hours after it is born. Its safety depends on forest cover and camouflage coloring. (Photo courtesy of Robert Pratt.)

FIGURE 7–32 Trees are important to fish habitat in streams. They provide places for fish to hide and help lower water temperatures by shading the water from direct sunlight.

Climate Moderation Value

Forests have some influence on the climate in a local area. During summer months, they provide shade and cooler temperatures (Figure 7–32). In winter, they are comparatively warmer than surrounding areas. Air movement is restricted by dense forest vegetation, reducing chill factors and restricting moisture evaporation from the forest floor. Large amounts of water vapor, however, are lost from plant leaves through transpiration, resulting in increased air humidity.

Flooding occurs whenever melted snow or runoff rainwater exceeds the rate at which water can be absorbed into the soil. Frozen soil prevents water from being absorbed down through the soil profile. By breaking up hard soil layers, deep-rooted plants allow more water to infiltrate the soil. Other smaller and more fibrous roots tend to hold soil particles in place when water moves across the soil surface. In these ways, healthy forest watersheds help protect against erosion and lessen the flooding of lowland areas.

Energy Value

Wood is the most important source of heat in many countries. It is used for cooking and for heating houses and other buildings that require supplemental heat. Energy is released from wood in the form of heat when combustion or burning occurs. Most of the wood that is used to heat homes is harvested and dried before it is used. Wood as fuel has the advantage of being a renewable resource. It tends to be low in sulfur compared to coal, so it produces little pollution such as acid precipitation in rain or snow. It does, however, produce pollution from **particulate matter**, which consists of tiny particles in smoke. This problem is greatest when damp wood is burned because combustion tends to be incomplete.

Some wood fuel is obtained as a by-product of wood manufacturing in the form of wood pellets. This product consists of waste lumber that has been ground into small chips and extruded through a pellet mill for use in wood-burning stoves and furnaces. This process efficiently uses wood products that otherwise would be wasted.

Some commercial products also use wood as a heat source. The most common of these are industries that convert wood to coke. Coke is a wood product that is obtained by heating wood to temperatures above its combustion temperature using large ovens from which oxygen is excluded. The product that is obtained from this process is capable of burning at extremely high temperatures. The charcoal briquettes that are used in home barbecue grills are one type of coke. This fuel is used in processes that require high temperatures and clean-burning fuels.

Biomass includes vegetation and waste materials that contain significant amounts of vegetable matter. Forests comprise the most important source: 42 percent of the Earth's total available biomass. Other sources include agricultural crops, crop residues such as straw and fodder, crop-processing wastes, animal wastes, and solid wastes from cities and towns. Significant amounts of energy can be obtained from these renewable biomass sources.

Trees and shrubs that are grown as energy crops can be cultivated in dense plantings with rows of plants that are much more narrowly spaced than in forest plantings. Fast-growing varieties of trees and herbaceous plants are produced using intensive management practices such as fertilization, weed control, and more frequent harvesting. Such crops are well adapted to land that is not suitable for agricultural crops because of poor soils or steep slopes.

The wood from harvested biomass plants may be chipped, dried, and burned to produce steam to generate electricity (Figure 7–33). Electricity from this source is called **biomass power**. Other uses for biomass materials include the manufacturing of paper products, construction materials, and ethanol fuels. The production of biomass as an energy crop has been practiced in woodlots for many years, but modern biomass production is much more intensive. These crops are usually harvested every three to seven years.

Livestock Value

Livestock grazing is a practice that allows for harvesting forages into forest environments. Vast forest areas provide habitat for forage plants, and domestic livestock species such as cattle and sheep are able to

FIGURE 7–33 One form of biomass consists of wood chips that are dried and burned as an energy source to generate electricity.

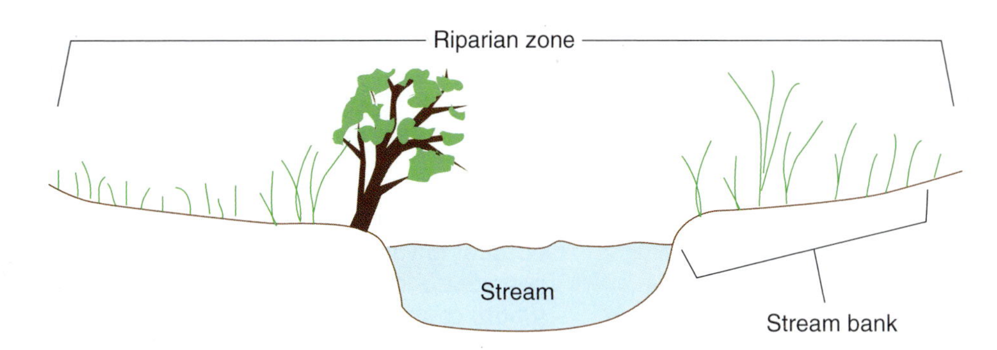

FIGURE 7–34 The riparian zone includes the area adjacent to the bank of a river or stream.

convert these plant materials to meat. A well-managed system of livestock grazing removes vegetation before it becomes old and unpalatable and allows for new plant growth. The regrowth of succulent forage on forest rangelands also contributes to improved nutrition for large game animals such as deer, elk, and moose that use these areas for winter range.

The untimely use of a resource often becomes abuse of the resource. Just as riding a motorcycle across dirt trails or hillsides when they are wet leads to soil erosion and fishing in prime areas during the spawning season may significantly reduce wild fish populations, grazing livestock in sensitive areas contributes to damaged rangelands. The **riparian zone** is one of these sensitive areas. It consists of the land adjacent to the bank of a stream, river, or other waterway (Figure 7–34).

INTERNET KEY WORDS

riparian zone

Any time the riparian zone is overgrazed, it is likely to contribute to severe damage to the natural plant cover in the area. This often leads to erosion of soils from the banks of streams and lakes, causing surface water contamination. The amount of soil that is suspended in flowing water is the **silt load**. Silt fills in lakes and reservoirs, destroys fish spawning areas, and kills young fish when water is muddy for extended periods of time (Figure 7–35).

Ranchers and herders who hold permits to graze domestic livestock on forestlands must do so in responsible ways. Only those who wisely use our natural resources should enjoy the privilege of continued use. Those who abuse them should no longer be allowed the privilege. A single

FIGURE 7–35 Silt consists of small soil particles that become suspended in the water of flowing streams. It is the greatest single source of water pollution.

FIGURE 7–36 Forests continue to hold appeal for people who enjoy outdoor environments, and many people use forests for recreation. (Photo courtesy of Joni Conlon.)

abuser of a natural resource is likely to damage the trust relationship that the multiple-use management system needs to work effectively.

Recreation Value

Many resources are available in our forests that have recreational value for people. Lakes and streams provide opportunities for fishing, boating, and other water sports. Many people enjoy hiking, mountain biking, horseback riding, photography, picnicking, harvesting mushrooms and wild fruits, camping, fishing, and big game hunting (Figure 7–36). All of these recreational activities draw people into our forests in large numbers. Outdoor recreation has become a huge industry, and the citizens of many communities located in or near forestlands derive much of their income from the sales of food, services, or supplies related to recreation. Entire industries are devoted to the production of recreational products such as tents, camp stoves, boats, and fishing supplies.

Many people also enjoy forest environments as places where wild animals can be observed in their natural settings. The thousands of visitors who take vacations to our national parks and monuments each year are evidence of the value that citizens place on forests as wildlife habitat. It is a difficult to measure intrinsic value, but this is the most valuable of all the forest uses to people who place great value on wild places and natural environments.

Watershed Value

An area where precipitation is absorbed into the soil to form groundwater is called a *watershed* (Figure 7–37). It is an area in which water

FIGURE 7–37 A watershed may be an area in which rainwater and melting snow are absorbed to emerge as springs of water or artesian wells at lower elevations.

FIGURE 7–38 A healthy watershed is important in stabilizing stream flows.

from rain and melting snow is absorbed to emerge as springs of water or artesian wells at lower elevations. Each watershed is separated from other watersheds by natural divisions or geological formations, and each is drained to a particular stream or body of water. Watersheds are valuable because they act like huge sponges, soaking up water from precipitation and melting snow and releasing it slowly. A watershed that is forested is superior to other watersheds because trees tend to slow the melting of snow, allowing it to be absorbed into the soil instead of running off the soil surface.

Vast expanses of forests and other vegetation are needed to regulate a uniform flow of water in rivers and streams. Precipitation on land surfaces can cause severe soil erosion unless plants are available to slow the flow of water over the land. Slow-flowing water infiltrates into the soil and comes out of the earth purified of most contaminants. Trees aid in water absorption by reducing the depth to which soils become frozen. This allows water to seep into the soil sooner than it does in unprotected areas. The snow usually lasts several weeks longer in forested areas than it does in open areas. Extending the length of the snowmelt reduces spring runoff and helps to maintain constant stream flows (Figure 7–38).

Protecting Forests

forests, insects, disease
forest, pest management
biotic, abiotic, diseases

Insects, pests, and diseases have always been part of the forest environment. Insects and diseases cause more damage to existing forests than fires (Figure 7–39). Pests cause trees to be weak and deformed by consuming leaves, damaging bark, and retarding tree growth. They kill billions of trees each year. However, controlling diseases and insects in forestlands is difficult and expensive, but removing dead, damaged, and weak trees helps reduce disease and insect problems. Not only are weak trees attacked, but healthy ones are also sometimes the favorite targets of pests. Prompt action in dealing with outbreaks of insects, pests, and diseases will do much to ensure a healthy forest.

Pests

There was no real effort to manage disruptive pests in the forests until long after the European settlers arrived in North America. Until the last century or so, the forests of this continent seemed too vast to raise

FIGURE 7–39 When insects and diseases become epidemic in a forest, they can kill many trees during periods of drought or stress.

concerns over potential shortages, and even the damage to the forests that was caused by insects, pests, and diseases seemed insignificant. Today we see these forests as finite resources that may be depleted if careful oversight and management practices are not used to protect and maintain them.

Tree Disease Agents
Biotic diseases
Bacteria
Fungi
Viruses
Microplasmas
Parasites
Nematodes
Abiotic diseases
Drought conditions
Temperature extremes
Poisonous effects
Nutrient deficiencies
Acid precipitation

FIGURE 7–40 Tree diseases are caused by both living (biotic) and nonliving (abiotic) factors.

Diseases

All living organisms are affected by destructive disorders that interfere with their health or cause illness. When a particular disorder occurs that can be traced to a specific cause with consistent symptoms, it is called a **disease** (see Figure 7–40). A disease has predictable symptoms, and the causes of the disease are the same in each occurrence. Those that are caused by living agents of infection such as bacteria, fungi, viruses, micoplasmas, parasites, or nematodes are described as **biotic diseases**. A disease that is caused by a nonliving factor or condition is an **abiotic disease**.

Insects

Insect interactions with trees include both good and bad effects. Some forest insects are pollinators and help ensure that fertile seed will be

Interest Profile

INTEGRATED PEST MANAGEMENT

An insect-management plan that uses a variety of control methods is the most acceptable form of pest control because it does minimal damage to nontarget insect species. Integrated pest management is the concept behind controlling harmful insects or other pests while providing some protection for useful organisms. It involves the use of some chemical pesticides in emergency situations, but it also relies on natural enemies and other biological strategies to control harmful pests. Integrated pest management does not attempt to kill all of the harmful insects because such control also kills natural enemies of the pest species. Natural insect enemies also must have a small population of the harmful pest on which they can depend for food.

FIGURE 7–41 Some insects cause terminal damage by eating the meristem tissue. Other insects cause terminal damage by destroying the phloem and xylem of the immature branch tissue. (Photo courtesy of the Boise National Forest.)

forest fire, benefits, problems

FIGURE 7–42 Fire can cause major damage to timber resources when it gets out of control. It is important to plan for fire protection in forests that are highly vulnerable. (Photo courtesy of Boise National Forest.)

produced by trees and other plants. Some insects are predators, eating other insects and, in some cases, playing important roles in controlling harmful insects. Other insects are parasites that infect and control populations of harmful insects. A delicate balance thus sometimes exists between predatory and parasitic insects and the insects that damage the forest. A similar balance exists between insect populations and the diverse tree species that make up a forest. A forest composed of a mix of different kinds of trees is less vulnerable to insects than a forest composed of a single variety of tree.

Insect damage to forest resources is indisputable, and it causes greater loss of timber than all forest fires combined (Figure 7–41). Insect damage is measured in terms of **growth impact**, which calculates timber losses caused by reduced growth rates and tree deaths. Forestry professionals have estimated as much as 123 million cubic meters of timber have been lost in the past 50 years because of insect damage.

Insect damage to every forest ecosystem is an ongoing destructive force that continues every day. Damage occurs to terminal buds, causing trees to lose **terminal growth** (vertical growth). It also occurs beneath the bark and inside the woody tissues of trees, interfering with **radial growth**, or growth that increases the diameter of a tree. Massive insect infestations sometimes occur in forests, but the greatest insect losses result from day-to-day reductions in tree growth.

Fire

Fire is both beneficial and harmful to forests. Controlled ground fires are used as management tools to burn the debris that accumulates on the forest floor. When done regularly, such burning reduces competition between young trees and shrubs and keeps accumulations of combustible materials at minimal levels. This reduces the number and intensity of wildfires that feed off large amounts of debris.

Uncontrolled wildfires cause great damage to forest resources. Millions of dollars worth of timber are destroyed by fire each year (Figure 7–42). In addition to actually killing some trees, fire slows the growth of others and damages some so severely that insects and diseases are able to destroy them. An extremely hot fire also burns the organic matter on the forest floor, which takes nutrients away from the trees and exposes the soil to erosion.

To help prevent fires, debris should be removed from around trees. Weeds, brush, and other trash around the edges of a woodlot should be removed. Prohibiting human use of an area for activities such as hunting and fishing during dry periods may also reduce the potential for fire. The construction of permanent firebreaks is another useful tool for fire control.

A plan for dealing with fire is important in minimizing damage should a fire occur. State and county foresters can assist in developing fire-prevention and fire-control plans. These strategies include planning for water storage and setting procedures for notifying proper authorities and obtaining appropriate equipment and help.

In some instances, fires are allowed to burn. For example, nature is usually allowed to take its course in the national parks and monuments because these areas have been preserved as natural environments. Timber harvests are not allowed in these areas, and fire serves

Interest Profile

TO FIGHT OR NOT TO FIGHT

Before humans began managing our forests, this was nature's job. Fire was a natural occurrence. When it erupted, it would burn underbrush, ground litter, and trees. Often, these fires would not kill the larger mature trees but instead kill the smaller trees that were in competition with these older trees for nutrients. The giant sequoia or ancient redwoods that stand today in some parts of our country would not have grown as tall or as thick as they have if fire had not thinned the population of competing trees. Some types of trees actually need fire to reproduce. The seeds of the lodgepole pine cannot germinate without first going through the intense heat of fire.

In the early 1900s, people in the United States began fighting forest fires as routine practice. For almost 100 years, few forest fires were allowed to burn. The result now is thicker, more densely growing forests than were present a century ago. There are more trees, and they are much closer together in today's forests. Now when fires start, the damage is much more severe. Fires burn hotter, faster, and with more intensity than in the past. Now they demolish forests, kill wildlife, destroy ecosystems, burn homes, and threaten human life. Although fire was once nature's way of cleaning and thinning forests, it now can wipe out all life in its path.

The job of keeping our forest healthy turns to the forest industry. Some types of logging practices will thin trees and lower the amount of combustible material available for fire. The wood is then sold for construction, papermaking, and many other uses.

Still, some people strongly oppose both logging and fighting fires. They would prefer that whole forests burn and the lives of forest creatures be lost. However, as Dr. Patrick Moore, cofounder and former president of Greenpeace, noted in the following statement to the *Wall Street Journal*, "The root of the problem is that when we protect our forests from wildfires, over time they become susceptible to disease and to catastrophic wildfires as fuel loads build up. The only way to prevent this is to actively remove dead trees and to thin the forest. The active management of these forests is necessary to protect human life and property, along with air, water and wildlife." Individuals who are truly concerned with the health of our forests understand that thinning our forests by logging and fighting dangerous forest fires is the best way to preserve this natural resource for generations to come.

to cleanse the environment from time to time. There are also occasions when fires are allowed to burn on public lands. In some instances, this because of a lack of resources to fight the fire. In other instances, the forest may be benefited by fire. A forest that is badly damaged by insects or diseases may have little useful timber remaining, and fire may serve a useful purpose.

Forest Soils

Soil is the surface layer of the Earth's crust that consists of particles of different sizes and origins from a variety of materials. *Soil texture* is a measurement of the proportion of mineral particles of different sizes that are found in a sample of soil. The smallest of the mineral particles is clay (less than 0.002 millimeter in diameter). Silt includes mineral particles between 0.002 and 0.05 millimeter in diameter. The largest soil particles are sand grains ranging between 0.05 and 2 millimeters in diameter.

Organic matter or humus is important in soils because it provides food to soil organisms, releases nutrients to plant roots, increases the water-holding capacity, and increases the movement of air through

Soil Orders		
Alfisols	**Spodosols**	**Ultisols**
Oak	Spruce	Loblolly pine
Hickory	Fir	Shortleaf pine
Northern hardwoods (New York)	Eastern white pine	Oak
	Northern hardwoods	Hickory
Aspen	Aspen	
Birch	Birch	
Ponderosa pine	Western hemlock	
Lodgepole pine	Sitka spruce	
	Longleaf pine	
	Slash pine	

FIGURE 7–43 Of the ten soil orders, only the alfisols, spodosols, and ultisols are of real significance to forestry.

the soil. It also tends to insulate the soil, causing temperatures to be more constant. Forest soils tend to build up deposits of litter on the forest floor. As the litter decomposes, it creates humus, which is also added through the decomposition of small roots.

INTERNET KEY WORDS

soil orders
alfisols
spodosols
ultisols

Soil Orders

The U.S. Department of Agriculture has identified 12 soil orders and the types of forest cover that are usually associated with them. This soil classification method is based somewhat on color, texture, and the mineral content of a particular soil. Of the 12 soil orders, three are of real significance to forestry in North America: alfisols, spodosols, and ultisols (Figure 7–43). The remaining soil orders are restricted mostly to agricultural uses.

The **alfisols** are high in calcium, magnesium, sodium, and potassium. The B horizon (see Chapter 6) tends to be high in clay content. Alfisol soils are slightly acidic. They are best suited to such trees as the oaks, hickories, aspens, birches, and ponderosa and lodgepole pines.

The order of soils known as **spodosols** are products of cold, damp climates and coarse silica parent material. The subsoil is illuvial, meaning that it is composed of materials (humus, aluminum, and iron) that have leached into the B horizon from the A and E horizons. These are acid soils that tend to be light in color. They support the growth of such forest species as spruce, fir, white pine, northern hardwoods, aspens, birches, western hemlock, Sitka spruce, and longleaf and slash pines.

The third soil order that is of importance to the forest industry is the **ultisols**. These soils are found in warm, humid climates, and they generally show evidence of heavy weather action as evidenced by illuvial deposits in the subsoil. The B horizon is composed of clay and reddish-colored iron deposits and silicate-based clays, and the E horizon is yellow in appearance. These soils support mostly loblolly and shortleaf pines along with oaks and hickories.

Erosion

Many forest soils have lost large amounts of topsoil as a result of soil erosion, the number one source of water pollution in North America. It damages wildlife populations by polluting water supplies, killing young fish and aquatic animals, and filling reservoirs and lakes with deposits of soil.

Forestlands are most susceptible to soil erosion following forest fires that have destroyed the plant cover. Fires also kill the roots of many plants on the forest floor, and they damage the topsoil by breaking down the structure of the soil granules. Soil particles that have endured extreme heat no longer adhere to other soil particles, and the pores in the soil tend to become clogged with small soil particles, reducing the soil's permeability and water-absorption capacity. These damaged soils become a sort of powder that is easily washed away by heavy rainfalls. Heavy flows of water from rapidly melting snow also contribute to severe erosion.

Harvest practices that disturb the soil surface over large tracts of land can also contribute to soil erosion. This is especially true when all of the trees are harvested from a large area through clear-cutting or when soil surfaces have been disturbed to construct loading sites or roads. Special practices must be used in damaged areas to allow precipitation to be absorbed by the soil and to prevent it from flowing overland across soil surfaces. One frequently used method to prevent erosion is to create an uneven surface that will trap and hold excess water until it is absorbed by the soil.

Soil erosion is also prevented in disturbed areas by using straw bales, plastic sheeting, and other types of silt traps to slow the flow of water, allowing heavy silt particles to settle to the bottom of the flow. These structures are also used to disperse the flow of water over a larger area, improving the prospects that the water will be absorbed into the soil instead of flowing across the soil surface. **Soil conservation** is the practice of protecting soil from erosion caused by strong winds or flowing water. Any action or method that contributes to the protection of soil surfaces is considered to be a soil conservation practice.

Looking Back

A forest is a complex association of trees, shrubs, and plants that all contribute to the life of the community.

The United States has a total of 731 million acres of forestland. Trees in U.S. forests are divided into two general classifications: evergreen and deciduous.

Conifers are evergreen trees that produce seeds in cones, have needle-like leaves, and produce lumber called *softwood.* Deciduous trees shed their needles or leaves every year and produce lumber called *hardwood.* Forestry professionals generally recognize eight major forest regions in the United States. The art and science of tree production is called *silviculture.* Forests are considered to be naturally renewable, but plantings allow for control over tree varieties and improve equipment access. An *environmental impact statement* is a science-based study of the harvest area that details the expected effects of human activity on the environment and wildlife in the impact area. Sustained yields can be established and maintained by determining a forest's growth and harvesting only as much timber as the forest can replace. The social and political sciences have become as important in forest management as the biological sciences, which are the basis for modern management practices. Forests contribute value to society in many ways. The biological value of a forest takes into account the worth of all of the life forms that are found there. Prompt action in dealing with outbreaks of insects, pests, and diseases will do much to ensure a healthy forest. Fire is both beneficial and harmful to forests. Of the 10 soil orders, three are of real significance to forestry in North America: the alfisols, spodosols, and ultisols. Soil conservation is any action or method that contributes to the protection of soil surfaces.

Self-Evaluation

Essay Questions

1. In what ways does soil erosion affect surface water pollution?
2. How is a tree different from a shrub?
3. What are the eight forest regions of North America?
4. Which of the forest regions ranks first in timber production?
5. Which tree species are the most important softwood trees in the United States? Which are the most important hardwood trees?
6. How does a tree transport water and nutrients to the leaves, and how are the food materials transported from the leaves to the roots, trunks, and limbs?
7. What is the difference between natural and artificial regeneration of forests?
8. Why is it important to develop a harvest plan before cutting down any trees?
9. What are some reasons why the clear-cutting harvest method has become controversial?
10. Explain the principle of sustained yield as it relates to timber harvests.
11. What role does the Environmental Protection Agency (EPA) perform in forest management?
12. Explain how integrated pest management works in managing harmful forest insects.
13. Distinguish between biotic and abiotic diseases.
14. What are the three most important classes of forest soils, and what characteristics separate them into distinct classes?

Multiple Choice

1. A woody perennial plant with a single stem and many branches is a
   a. bush.
   b. scrub.
   c. shrub.
   d. tree.
2. A tree that does not shed its leaves on a yearly basis is
   a. evergreen.
   b. coniferous.
   c. deciduous.
   d. hardwood.
3. Approximately 75 percent of the wood for plywood is harvested from the __________ forest region.
   a. northern coniferous
   b. central broad-leaved
   c. Hawaiian
   d. Pacific Coast
4. The forest region that contains more varieties and species of trees than any other is
   a. central broad-leaved.
   b. northern coniferous.
   c. Pacific Coast.
   d. southern.

5. The most important commercial species of trees in the United States is
   a. oak.
   b. Douglas fir.
   c. redwood.
   d. black walnut.
6. Water and minerals are taken in by the roots and transported up to the leaves of a tree through a layer of cells called the
   a. phloem.
   b. xylem.
   c. heartwood.
   d. cambium.
7. The scientific management of forests is
   a. silviculture.
   b. arboriculture.
   c. pomology.
   d. olericulture.
8. Forest regeneration that occurs when seeds or seedlings are planted at a harvest site is called
   a. artificial regeneration.
   b. germination.
   c. natural regeneration.
   d. vegetative reproduction.
9. A timber harvest method in which all of the trees over a minimum size are removed is the
   a. shelterwood method.
   b. coppice method.
   c. clear-cut method.
   d. seed tree method.
10. The ____________ value of a forest takes into account the worth of all of the life forms that are found there.
   a. biological
   b. political
   c. numerical
   d. social
11. A form of electrical power that is obtained from trees and foliage that have been chipped, dried, and burned to supply steam is called
   a. hydropower.
   b. solar power.
   c. nuclear power.
   d. biomass power.
12. Insect damage to forests is measured in terms of
   a. growth impact.
   b. radial growth.
   c. terminal growth.
   d. terminal counts.
13. A ____________ soil order is high in concentrations of calcium, magnesium, sodium, and potassium.
   a. alfisol
   b. ultisol
   c. spodosol
   d. humus
14. A destructive process that sometimes occurs in soils that are not protected against forces of flowing water or strong winds is
   a. denitrification.
   b. conservation of matter.
   c. erosion.
   d. soil conservation.

Learning Activities

1. Working with your teacher, invite a person who has expertise in insect identification to visit the class. Extension educators, foresters, and urban foresters in many areas are competent to lead discussions about the ways in which insects affect trees and forests. Ask the visitor to use photos, slides, or actual insect collections to raise the interest levels of the students. Assign pairs of students to bring a picture or illustration of a particular insect to class. Make a collage on the classroom bulletin board with the pictures and illustrations contributed by the students.

2. Look up "rain forests" on the Internet or using the school library. Find one organism that is used to benefit humans (besides building products). Write a one-page report describing the location where the organism is found, its importance to humans, and its survival expectations if forests are destroyed.
3. Obtain a coring tool from your teacher and find a large tree. Core the tree by following your teacher's instructions, then count the rings in the core to determine the tree's age.
4. Visit a local forest and identify the types of trees growing there.
5. Working with your teacher, invite a professional forester to visit the class to speak about managing the forest in ways that protect the environment.

TERMS TO KNOW

rangeland
range
grassland
forb
biome
carrying capacity
overgrazing
Taylor Grazing Act of 1934
decreaser
increaser
invader
undergrazed
grazing capacity
animal equivalent unit (AEU)
hydrology
hydrophyte
hydric soil
jurisdictional wetland
ecological wetland
marsh
floodplain
swamp
bog
prairie pothole
vernal pool
natural wetland

CHAPTER 8

Grasslands and wetlands

Objectives

After completing this chapter, you should be able to

- explain the importance of rangeland and its careful management
- discuss the history of the rangelands in United States
- define and discuss the three major types of grasslands in the continental United States
- describe the types of vegetation prevalent in the grasslands of the continental United States
- explain the relationship between grazing and grassland management
- explain what a wetland is
- describe the difference between on-site and off-site wetland identification procedures
- explain the three common types of wetlands
- list and explain the major causes of wetland areas
- explain the major wetland-management practices
- list the major governmental programs and agencies that regulate wetlands

rangeland

For the vast majority of Americans who live in the eastern or midwestern states and in large cities, the word **rangeland** is almost meaningless. The vastness of the original rangelands in this country and the significance of their contribution to the development of our nation may well be surprising, but the size and importance of the world's rangelands on a worldwide basis today can hardly be overstated.

In its broad sense, **range** refers to the land area that provides food, particularly in the form of forage and browsing for animals. Using that definition, almost all land area must be considered a range for something. As we use it here, the term is more narrowly defined. For our purposes, rangeland—also known as *range*, **grassland**, and *prairie*—refers to the land areas of the world that tend to be naturally covered by grasses, grass-like plants, **forbs**, and shrubs as the primary vegetation instead of trees. Naturally, that excludes the world's true deserts and forested areas.

Many parts of the world are considered rangeland. Natural rangelands occur in the form of tundra, meadows in openings within forests, marshes, wetlands, vegetated areas above the tree lines on mountains, shrub lands, and grasslands. Of these, by far the most important to human development have been the grasslands.

Human-generated rangelands consist of pasture land and forage cropland produced on areas that were originally covered by forests or natural grasslands. Of the total land area on Earth (approximately 32.2 trillion acres), some 26 percent (approximately 8.3 trillion acres) are managed as permanent pasture. On the North American continent's total land area of some 5.3 trillion acres, 17 percent, or approximately 0.9 trillion acres, are managed as permanent pasture.

Before the advent of agricultural settlement across the continental United States, the tree lines of eastern forests reached about as far westward as a line drawn roughly between Dallas, Texas, and Chicago, Illinois. West of that lay open grassland stretching as far as 1,500 miles to the Rocky Mountains. In your history books, the general area is frequently referred to as the Great Plains, a term that describes a geographical and physical region rather than a **biome**, a group of ecosystems within a region that have similar types of vegetation and climatic conditions.

Today, rangeland makes up approximately one-third of all the land in the United States, some 770 million acres. Of that, approximately 200 million acres are in Alaska. Little of Alaska's rangeland is used for pasture for livestock, however, so almost all of it is left for wildlife to graze. In contrast, Canada has only 70 million acres or so of rangeland suitable for livestock grazing.

In the southwestern United States, as much as 85 percent of the land area is considered rangeland, and as much as 90 percent of that rangeland is managed as pasture for grazing livestock. Of the rangeland in the continental United States, approximately two-thirds is privately owned or owned by local or state governments. Of this amount, the most productive and accessible rangelands are generally privately owned. Land that is less productive or that is less accessible has tended to be retained by the various levels of government. The remaining one-third of the total rangeland is owned by the federal government and managed by various federal agencies. The Bureau of Land Management (BLM) manages most of that federal land, leasing much of it to ranchers for grazing livestock. Other rangeland is managed by the U.S. Forest Service and the U.S. National Park Service.

Grasslands

In general, the grasslands in what is now the continental United States consisted of three broad types: tallgrass prairies, transition prairies, and shortgrass prairies. More than 25 percent of the land area was covered by rangeland when the first European settlers landed on the continent's eastern coast (Figure 8–1).

tallgrass prairie, shortgrass prairie

The most striking of the grasslands was the tallgrass prairie, which covered much of the area between the current states of Indiana to Nebraska and southward to Texas. The land that produced the tallgrass prairies was extremely rich. The topsoil was deep and abundant in plant nutrients. Historically, the climate produced a fairly regular and adequate supply of rainfall, with winters that varied from cold to extremely harsh.

The result of such a combination of adequate rainfall and rich soil produces a lush growth of prairie grasses that might reach 6 feet in height at the peak of a good growing season. It also makes for highly productive agricultural land, some of the richest farmland in the world. When the settlers came, they recognized the richness of the land and cleared it for farm production to feed a growing and hungry nation. Because of this, the tallgrass prairies are almost gone today (Figure 8–2).

The shortgrass prairies stretched from central Canada through parts of the Dakotas to central Texas. In general, the soils of that region were also rich in plant nutrients with deep topsoils, as were the tallgrass prairies farther east. However, this area has a much lower average annual rainfall and it is much less dependable than that of the tallgrass prairie region. At the peak of a good growing season, these grasses might reach a height of as much as 3 feet (Figure 8–3).

Between the shortgrass and tallgrass prairies were the transition prairies. This region experiences more average rainfall than the shortgrass prairie, but precipitation tends to be less dependable than that of the tallgrass areas. The maximum height of the grasses falls between 3 and 6 feet in good growing years, but good years are fewer and farther between.

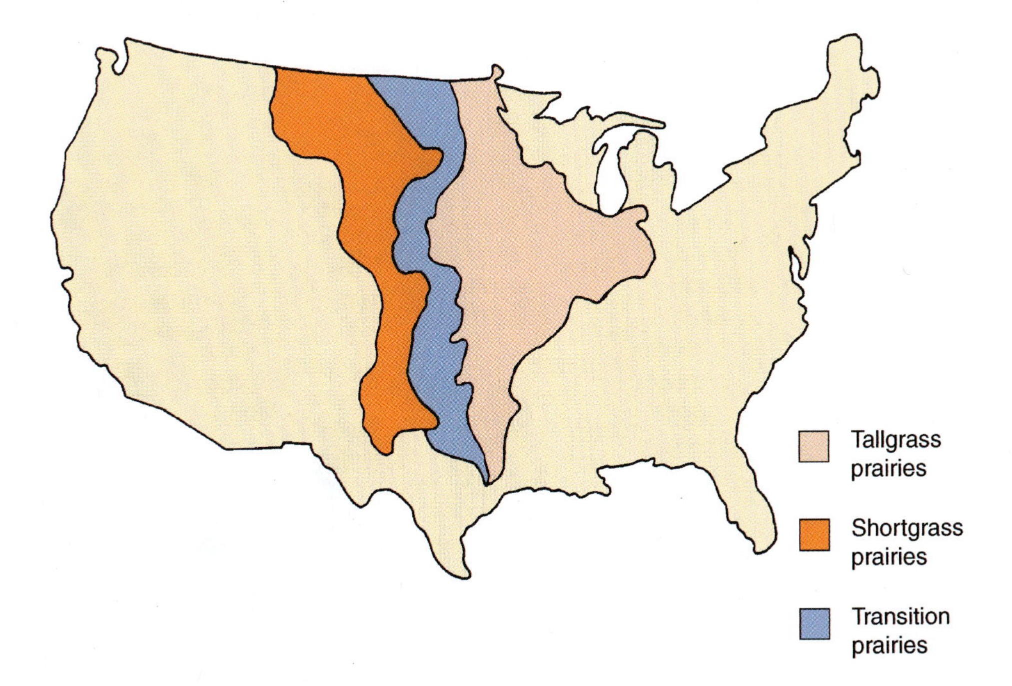

Figure 8–1 Major grasslands were dominant in what is now the continental United States before European settlement.

FIGURE 8–2 This tallgrass prairie is located in the Flint Hills area near Riley, Kansas. (Courtesy of Shutterstock.)

At least parts of all three original grassland types still exist, although the vast majority of the land in all three types has been plowed under at one time or another. Substantial areas of shortgrass prairie remain, primarily because it has less agricultural productive potential. Virtually all of the shortgrass region requires irrigation to make cultivated farming profitable. Almost all of the original tallgrass prairie land has been converted to farm production or other human developments, primarily because of its high productive potential.

In other parts of the world, the great grassland regions are known by different names. On the Asian continent and in Eastern Europe, the

FIGURE 8–3 This example of shortgrass prairie is in Custer State Park, South Dakota. (Courtesy of Shutterstock.)

great *steppes* stretch for thousands of miles. The African *veld* is vast and rich grassland. We have all seen television documentaries about the huge herds of herbivores in Africa. What these documentaries usually do not point out is the absolute reliance of those wild animals on the grasses of the veld. Parts of South America hold great regions of *pampas*. In Brazil, the grasslands are called *campo*. In other parts of the Americas, *savannah* is the term used to describe some grasslands. In Spain, it is the *protero;* in the Philippines, the *kogonales*.

History of the North American Grassland Ranges

Early History

Great Plains history

Before European explorers and settlers arrived, the center of this continent was a land of vast, uninterrupted grasslands (refer to Figure 8–1). Ecologists have long disagreed as to whether grasses are the true climax vegetation of the region. Many believe that much of the region could easily support forests; in fact, the rainfall is quite adequate for that in much of the grasslands. It may be that lightning-caused fires brought about the destruction of the forests that could have flourished there in prehistoric times. Later, the combination of natural fires and fires set by the Native Americans as a hunting tool caused the eventual elimination of the trees that would have been needed to provide seeds for reforestation (Figure 8–4).

Although the middle of the continent was covered with grasslands, spots occurred all the way from the Pacific Ocean to the Atlantic Ocean. Wherever fire destroyed forest, rapidly growing grasses would quickly replace the trees until new trees returned to dominate the smaller plants. Often the fires were caused by lightning or other natural causes, but just as often they were set by Native Americans for hunting or to establish grasslands to attract bison and other grazing animals. In these cases, the grasslands were temporary and did not truly belong to the great grassland biome.

These vast areas were virtually void of trees when the first European settlers arrived. At that time the only checks on the population of

FIGURE 8–4 This range is in the Flint Hills area near Riley, Kansas. (Courtesy of Shutterstock.)

herbivores were climate, disease, and predators, including the large cats, wild dogs, wolves, and Native Americans. The last group hunted just enough animals to meet its needs for meat for food, hides for clothing and shelter, and bones for tools. Their meager needs would hardly have affected the population levels of wildlife on the prairie. Still, their hunting methods could be quite destructive. To harvest enough wild game—much of it larger and faster than the hunters themselves—they often resorted to the use of range fires. A "killing party" would be set up at a location selected to intercept the fleeing animals. A second party would set a fire on the opposite side of the herd of bison or other herbivores. The frightened animals would then flee into the ambush. Another technique involved forcing animals to run over a cliff or into a deep ravine. Falling either to their deaths or at least to severe injuries and easier killing, the animals thus provided the needed food and hides.

Native American hunting, fire grassland animal life

When these range fires did not spread far, the technique was not too wasteful. When there was nothing to stop them, however, such fires sometimes burned out of control until they reached a river or other natural barrier wide enough to extinguish them or rain put them out.

That method may seem a remarkably destructive way to gather food, but there were several reasons it was done. The primary reason is that the hunters were on foot (horses did not appear on the continent until the Spanish introduced them in the sixteenth century). A group of hunters on foot might not be especially successful in harvesting deer, antelope, or bison in quantities large enough to provide a reliable food source for their people (Figure 8–5).

Remember also that the grasses were not actually eliminated by the fire. Immediately after a range fire, the group might be bare; but most of the nutrients in the plant matter remained after the fire and were returned to the soil. More important, many of the seeds and roots remained alive despite the heat from the fire. The new crop of grass would begin to grow with the next rain, and the range soon would become green and rich again. As a result, trees were virtually eliminated from most of the grasslands.

FIGURE 8–5 The bison thrived in the grassland regions of North America. (Courtesy of the U.S. Fish and Wildlife Service.)

Unlike grasses, trees take many years to grow large enough to produce seeds. Range fires killed trees that would otherwise have become dominant in the most productive grasslands. Where range fires were frequent and regular, the shrubs and forbs were also eliminated for the same reason. A rangeland without shrubs and trees to compete with grasses for sunlight, nutrients, and water, produces lush grazing and has a much higher **carrying capacity** for herbivores. Carrying capacity is defined as the population, number, or weight of a species that a given environment can support for a given time.

As noted, beyond the effect of human-caused and natural fires on the carrying capacity of the grasslands, disease, other predators, and climatic fluctuations took weak animals and became important in naturally "managing" animal populations when they exceeded carrying capacity. Whenever a population of grazers became large, predators multiplied more rapidly, which meant more predation and thus downward pressure on the grazers' population.

Climate—in the form of periodic droughts coupled with years of unusually heavy rainfall—also reduced grazing during some years and made for more abundant grazing during other years. Abundant years encouraged increased breeding, and sparse years caused high death rates among animal populations. The result, although seemingly cruel, was an effectively controlled population.

Even with the effects of predation, natural disasters, Native Americans, and disease, the grazing animals of the plains grew to millions of bison, antelope, deer, and other wild animals. The grasslands had a massive carrying capacity for herbivores. Early Western pioneers, for example, spoke of herds of "buffalo" that stretched from horizon to horizon. It was only with the coming of agriculturally based civilization that the plains came to be stretched beyond their ecological limit.

As the first Spanish explorers marched northward into this region in the sixteenth century, they found seemingly endless expanses of rich grazing land populated by millions of animals (Figure 8–6). These explorers brought donkeys as pack animals, cattle and other grazing

FIGURE 8–6 Pronghorns are well adapted to this shortgrass prairie. (Courtesy of PhotoDisc.)

grassland overgrazing, cause, effect

animals for food, and horses to ride. Some of their animals escaped or were released into the wild where they found plentiful and nutritious grazing with few natural enemies. They reproduced in large numbers, and the result was **overgrazing** by cattle and horses in many limited areas. In addition, it was only after the Spanish introduced horses to North America that Native Americans acquired them to ride and to use as draft animals, making hunting fires less necessary.

The earliest Westerners came to the region as exploiters, not as settlers. They were looking for gold and silver and to harvest the fur pelts of beaver and other wild animals for the use of wealthy Europeans. These earliest pioneers put little real pressure on the ecosystem of the plains.

After the Settlers Came

Later waves of English, French, Spanish, and American explorers moved westward to make a living from those same great grasslands and herds of herbivores. Large numbers of trappers and market hunters profoundly affected game populations, wasting many animals and causing herds to decline. They took some game animals for food, but wildlife was mainly exploited for furs, hides, and feathers.

Soon, however, other settlers moved westward from the American colonies along the east coast and northward from Mexico. At that point in time, agriculture became necessary for human survival in settlements. Most of the new settlers came with the specific intent of earning a living from farming.

The grasslands offered one particularly important source of potential wealth for enterprising newcomers. Tall, abundant grass on the open ranges offered free grazing for potentially millions of cattle. How else could one more surely make a profit than by putting cattle into areas where they could eat, drink, and multiply freely at no cost to their owners? The cattle could then be slaughtered for meat and hides, with only a minimum of financial investment by cattle ranchers. The biggest problem ranchers faced was getting their cattle to eastern and midwestern markets. Railroads offered a solution in the mid-1800s. As a result, many great fortunes were made by ranchers who were strong enough and ruthless enough to survive in the open-range cattle industry.

Much of American folklore comes from the time between about 1850 and 1890. Particularly during the two decades after the American Civil War, the open-range cattle industry became a critically important source of wealth in the United States. By this time, much of the tallgrass regions had been cleared for cultivation. The somewhat less productive, but still rich, lands of the shortgrass prairies remained mostly in their original condition.

This was also a time, coincidentally, when nature played a cruel trick on the people in these areas. For more than 20 years, there had been a climatic cycle in which unusually dependable rainfall and mild winters prevailed. Cattle ranchers pushed their herds to sizes that overtaxed the carrying capacity of the shortgrass prairies. Great ranches sprang up, and the cattle barons grew rich and powerful. Farmers who wanted to clear the land for plowing were unwelcome, and cattle ranchers kept them away, often by force. Fencing material was largely unavailable in the region, and in any case would not have been allowed by the cattle ranchers. After all, the grasslands were a vast region that belonged to no one in particular. The grass was a gift of nature to anyone who was able to take advantage of it. This was a huge common area, and the cattle industry grew up explicitly to exploit the wealth of the commons.

By the mid-1880s, however, the effects of overgrazing were beginning to be obvious everywhere. The abuse of the grasslands occurred for several reasons. First, the cattle ranchers were not ecologists. They had little understanding of the effects of their activities on the ecosystem. Beyond that, the motive of cattle ranchers was profit, not conservation. Thus, cattle populations were pushed to and even slightly beyond the ecological limits of the ranges, with the cattle ranchers hoping each year for good weather to make the grasses grow rapidly. The land began to dry out more than usual because of the loss of soil cover. The organic matter in the topsoil also became depleted, further decreasing the soil's water-holding capacity.

Calamity struck when the vast herds of cattle were beginning to be less healthy because of poorer grazing. The winter of 1885–86 was particularly cold and harsh. Millions of cattle died or became ill as a result of the unusual cold combined with the lack of adequate grazing. The following summer the grazing was still inadequate because of the effects of overgrazing in previous decades. The winter of 1886–87 followed with even colder and harsher conditions.

The open-range cattle industry was shaken to its foundations by this triple catastrophe. True, the ranchers had made such an event almost inevitable because of their management practices. At the same time, one cannot blame any individual rancher for the disaster. The calamity occurred because the prairie was a rich common area with value to many people. It belonged to no one but could be used by anyone who voiced a claim for it and was willing to stand behind that claim. Finally, it was an area in which no one was responsible for making and enforcing overall management decisions.

Barbed wire was invented in the 1870s but was violently resisted by cattle ranchers. To them, they were the ones who had taken the land from the Native Americans and who had labored to develop huge herds of cattle. They had "tamed" the land. Settlers who wanted to clear the land for farming were forced to build fences to keep cattle out of their fields against the wishes of the cattle ranchers. The cattle ranchers saw this development as an invasion of "their" land—which was not legally their land at all.

Resistance to the settlement and fencing of the open ranges soon became so violent that laws against cutting fences became extremely harsh. Not until the disasters of 1885–87 was the fate of the open ranges determined. After that, fences came to be largely accepted as necessary to the future of the cattle industry. Cattle ranchers came to understand that range restoration was not possible without fencing to control grazing. Without range restoration, the cattle industry was unlikely to improve and become profitable again.

Still, overgrazing continued even after the range was settled and fenced. This was true for several reasons, not the least of which were a general lack of understanding of the long-term effect of overstocking and farm and ranch families' need for income for short-term survival. In addition, much of the rangeland remained the property of the federal and state governments. That land remained essentially a common area under no one's direct management. Thus, the public land was still left available to ranchers to freely graze their cattle.

Taylor Grazing Act of 1934

The Taylor Grazing Act of 1934

The **Taylor Grazing Act of 1934** was designed to assert federal control to prevent continued overgrazing in the arid grasslands. When it was being considered by Congress, it was strongly opposed by much of the western livestock industry on economic grounds and by political conservatives in the name of states' rights.

Responsibility for the administration of the Taylor Grazing Act was finally passed to the federal Bureau of Land Management in 1946. Under that agency, the concept of the grazing district has become an effective tool for managing government rangelands. The BLM now controls grazing on more than 1.75 million acres of public lands.

Only ranchers who own land adjacent to grazing districts are eligible to lease land from BLM-managed ranges. Those who hold leases to BLM grazing rights can graze their livestock at low cost on public domain grasslands. As many as 9 million head of sheep and cattle are allowed on public grazing lands each year. Ranchers must abide, however, by strict management policies set by BLM officials based on the recommendations of government range conservationists. At the same time, the reality is that BLM decisions can often be swayed by political considerations under the pressure that can be applied by ranchers who hold these substantial rights.

INTERNET KEY WORDS

grass, decreasers, increasers, intruders

Types of Grassland Vegetation

Four major types of vegetation predominate in the rangelands: grasses, grasslike plants with fibrous root systems, forbs, and shrubs (many of which have taproot systems—that is, deep center roots). Each type of vegetation must compete for sunlight, moisture, and nutrients in the arid plains region. Grasses are particularly valuable from two standpoints. First, the complex root systems are highly adept at holding the soil in place and thereby preventing erosion while also preserving soil moisture. Second, grasses provide a higher-quality source of nutrition for grazing animals than do most other plants that can grow freely on rangelands (Figure 8–7).

FIGURE 8–7 This rangeland area near San Diego, California, has been invaded by chaparral and other tap-rooted shrubbery. The brushy conditions developed because of extreme undergrazing. (Courtesy of the U.S. Department of Agriculture's Natural Resources Conservation Service.)

The proportions of grasses vary substantially from year to year and even during a single growing season. Grazing rates particularly affect the relative amounts of the four major types of vegetation present. Light grazing actually benefits the grasses in comparison to other types of plants. As long as more than half of the length of a grass stalk is not eaten, grazing causes little damage to the grass plant. Unfortunately, heavier grazing is extremely detrimental. When most of the grass leaf is eaten, it is more difficult for the plant to recover—particularly if the plant is already stressed by dry, hot weather.

Varieties of grass that are easily damaged by even moderate grazing are called **decreasers**. Unfortunately, these are often the ones that grazing animals find most desirable. Given their choice of grasses to feed on, grazers eat those they find most palatable and leave those that are unpalatable. Thus, the more desirable grasses tend to be eaten until they are damaged. In other cases, the decreasers are simply more susceptible to damage from grazing because they are not especially hardy plants.

Other rangeland plants that tend to thrive under heavy grazing are called **increasers**. Many increasers are successful merely because the

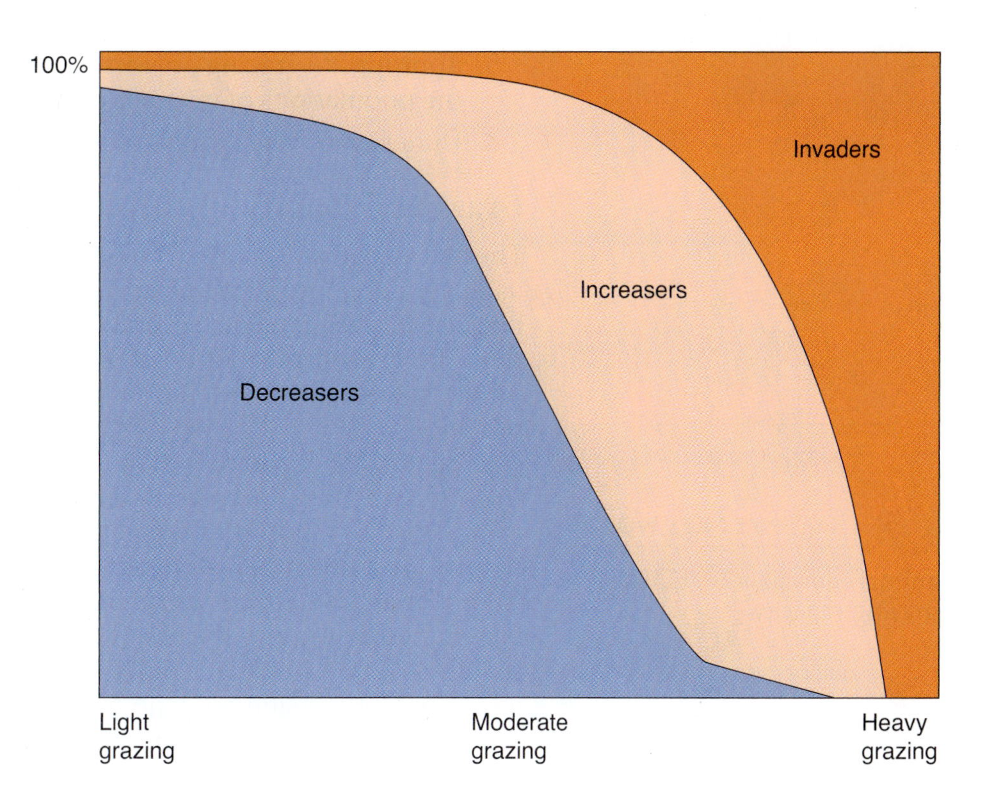

FIGURE 8–8 Relationship between levels of grazing and type of vegetation on rangeland.

grazers find them unpalatable and so avoid eating them. In some cases, the plants are simply better able to get at the limited soil, water, and nutrients.

Plants that move into an area after it has been badly overgrazed are called **invaders**. Oddly enough, badly **undergrazed** areas are also subject to being taken over by such less desirable plants. Prickly cactus and some shrubs are particularly potent invaders (Figure 8–8). Ideal management involves a moderate level of grazing. A rangeland without grazing as well as with overgrazing will produce less total biomass than one with moderate grazing.

Range Management Techniques

Objectives of Range Management

Management implies doing something: decisions and actions. If they are to be consistent and productive, however, those decisions and actions should be guided by objectives. In the case of range management, most range managers generally accept a single major objective: long-term maximization of livestock productivity. That sounds simple enough, but several additional objectives are implied by this:

1. Current grazing must not damage the land's capacity for grass production.
2. Current grasslands that are not in their best condition must be treated so that they will improve. This can mean reseeding with natural grasses and fertilization.
3. Soil conservation techniques must be used to control soil erosion and depletion.
4. Water management must ensure that the groundwater supply is not depleted.

5. Appropriate grazing pressure must be maintained to help reduce the populations of invaders into a managed area.
6. Undesirable vegetation must be controlled.

Grazing Capacity

The first step in any planned range management program must be establishing an area's carrying or **grazing capacity**. This is not a simple task because it must consider several factors, including types of local vegetation, the growth rates of desirable grasses, the effects of grazing rates on the specific local vegetation, climate, soil type, slope, and expected rainfall.

The grazing capacity is then used to determine an acceptable *stocking rate,* which is expressed in terms of **animal equivalent units (AEU)**. An AEU is the amount of forage that is required to feed a 1,000-pound animal for a given period of time. For instance, 1 AEU for one month is known as an *animal unit month* (Figure 8–9). In general, one steer is assumed to require 1 AEU. Five goats or five sheep are assumed to require 1 AEU. A horse or a bull is assumed to require 1.25 AEUs. Wildlife also must be considered. One elk needs approximately 0.67 AEU, and four deer require approximately 1 AEU.

animal equivalent unit, AEU
grazing capacity

Grazing Management

Once the grazing capacity of an area has been determined, grazing rates must be controlled to ensure that the rate is not exceeded. Even with controlled stocking rates, other management techniques also must be used. Undergrazing or overgrazing discourages the growth of desirable grasses and promotes invaders in an area. Left to themselves, animals do not graze uniformly across an entire area. They will tend to overgraze near water sources and near salt blocks. Separating the salt blocks from water sources helps to encourage better grazing patterns. Even then, animals will still overgraze parts of an area and undergraze others.

FIGURE 8–9 This grazing area for domestic cattle is located in the grasslands of the Monument Rock chalk flats of Grove County, Kansas. (Courtesy of the U.S. Department of Agriculture's Natural Resources Conservation Service.)

FIGURE 8–10 An example of a holistic management schedule designed for a 30-day rotation with 25-day recovery time per paddock. The number of days can be varied if the grass-growing conditions change.

Paddock A, 5 days	F, 5 days	E, 5 days
B, 5 days	C, 5 days	D, 5 days

A system that controls grazing distribution so that it is more even across an entire range area is necessary. Three common management systems are used to meet this goal: continuous grazing, deferred-rotation grazing, and holistic management.

With continuous grazing, livestock are kept in the same area year-round and are allowed to graze as they choose. The major disadvantage of this system is uneven grazing. The advantage is that it is less work and less expensive than other systems.

With deferred-rotation grazing, a range area is fenced into two or more separate grazing areas. The livestock are moved periodically to allow the grasses time to recover.

The newer method of holistic management is similar to deferred-rotation grazing. In this system, because healthy grass is not damaged by being bitten off once but may be damaged by being eaten down more than once, the livestock grazing area is divided into multiple paddocks. Animals are moved every few days to a new paddock according to a predetermined schedule or as the grass appears to reach the point where it needs time to recover (Figure 8–10).

Range Restoration

grass, range restoration

The best time to restore a range area is before it is damaged too severely. Once too much damage is done, major restoration efforts may be needed (Figure 8–11). Several management strategies are available, including controlled grazing, reseeding, and elimination of undesirable vegetation.

FIGURE 8–11 This example of severe soil erosion developed as a result of the loss of grass cover from overgrazing. (Courtesy of the U.S. Department of Agriculture's Natural Resources Conservation Service.)

FIGURE 8–12 Extreme overgrazing like this can present major erosion and other types of hazard to the land. (Courtesy of Dr. Jay McKendrick, University of Alaska, Fairbanks.)

Controlled grazing is the first step in range restoration. Because overgrazing or undergrazing probably caused the problem in the beginning, controlled grazing is essential in restoration. If the range must be reseeded, then grazing must be withheld completely until the grasses are reestablished (Figure 8–12).

A range area that has been stripped bare will require reseeding. Although many varieties of "improved" grasses have been tried over the years, generally grasses that are native to an area have been found to be the best long-term reseeding solution. Grasses that livestock will eat and that are less susceptible to overgrazing should be selected.

If an area is overgrown by invaders, then the undesirable vegetation must be eliminated first before reseeding can be done. Regardless, during the reestablishment of grasses in an area, application of fertilizers can make a substantial difference in the speed with which the recovery takes place. In less severe cases, just controlling grazing, light overseeding, and fertilization may be adequate to restore a productive range.

WETLANDS

What Are Wetlands?

Wetlands are a premier, underrated, and overlooked natural resource. They are home to birds, amphibians, waterfowl, small mammals, and various plants and fungi. The variety of life surrounding wetlands should be reason enough to realize their importance. Wetlands, however, have a major ecological role in controlling floods, acting as a filter for pollutants, adding to underground water sources, providing habitat for many species (most notably waterfowl and amphibians), and providing recreational use. According to the Environmental Protection Agency (EPA), more than one-third of all threatened and endangered species live only in wetlands and one-half of threatened and endangered species use wetlands at some point in their lives (Figure 8–13).

Depending on what group or individual you ask, you can get many different definitions of wetland. Even the "official" definition of a wetland varies because different governmental agencies use several differing definitions, as you will read later. Farmers have many different ways of defining wetlands, depending on where they live and on what kinds of crops and livestock they produce. Environmentalists use still another set of definitions. Different individuals' and groups' perspectives and agendas drive what they perceive to be wetland. Still, many types can be found in a variety of habitats across the United States, and two principal definitions are used by federal agencies.

Characteristics

Most experts agree that wetlands exhibit three characteristics. First, there will be a prolonged presence of water. An area whose **hydrology** (water characteristics) includes frequent saturation with free water is probably a wetland. Second, this prolonged presence of water affects how the soil develops and what plants will grow in the water-saturated soil. Plant types that are attracted to these growing conditions are generally called **hydrophytes** (Figure 8–14). A soil that is covered by hydrophytes is probably a wetland.

Figure 8–13 Bicyclists on the Tidelands Trail in the San Francisco Bay National Wildlife Refuge, California. This saltwater wetland area serves as habitat for migratory birds and endangered species, and it offers superb recreational activities. (Photo courtesy of the U.S. Department of Agriculture's Natural Resources Conservation Service.)

Third, because the soil is subjected to frequent saturation and may have a permanent water table close to the surface, it develops into **hydric soils**, ore soils that tend to be saturated with water most of the time. They are usually low in air content, generally have different colors than other soils in the same region, and may have a completely different soil structure. Hydric soils may be mottled with white or gray coloring or may be extremely yellow where a darker coloration might be expected. The structure may be sticky and wet rather than granular. An area with hydric soils is probably a wetland.

INTERNET KEY WORDS

hydrophytes, hydric soils, pictures
jurisdictional wetlands

Two Definitions

The first definition of wetland is the legal definition. The U.S. Army Corps of Engineers defines wetland as an area that has the three characteristics mentioned above: frequent flooding or saturation, coverage by hydrophytes, and inclusion of hydric soils (Figure 8–15). The corps is responsible for these **jurisdictional wetlands**.

From an ecological standpoint, on the other hand, some lands may be wetlands, but they are not jurisdictional wetlands. As noted before, there are many different ecological definitions of wetlands. Here we will use the definition used by the U.S. Fish and Wildlife Service (USFWS) because it is broad and this agency maintains many wetland refuges across the country. Wetlands as defined by the USFWS are lands that are transitional between terrestrial and aquatic systems where water tables are usually at or near the surface or the lands are covered by shallow water. For the purposes of this classification, wetlands must have one or more of the following three attributes: (1) At least periodically, the land supports predominantly hydrophytes; (2) the substrate is predominantly undrained hydric soil; and (3) the substrate is nonsoil and is saturated with water or covered by shallow water at some time during the growing season each year.

Figure 8–14 These green pitcher plants in Little River Canyon, Alabama, are an example of hydrophytes. (Photo courtesy of the U.S. Fish and Wildlife Service.)

FIGURE 8–15 Wetland areas are a natural habitat to many organisms.

As you can see, the ecological definition used by the USFWS is much broader than the legal definition used by the Corps of Engineers. According to the USFWS definition, **ecological wetlands** may or may not have all three characteristics of the jurisdictional wetland definition. As an example, the Corp of Engineers would not consider a mudflat or coral reef as a wetland, but the Fish and Wildlife Service would.

ecological wetland

History of Wetlands in the United States

Until the last half of the nineteenth century, wetland areas in the United States were perceived as harsh and forbidding places, a problem rather than an asset. A perfect example of the new federal government's position on wetlands occurred in the 1780s and 1790s. To build the new District of Columbia as the nation's capital, the federal government determined to drain and fill in a large area of swampland.

In another example, the whole southern part of Louisiana is designated as wetlands, but the citizens of Louisiana felt that building a city on a water crossroads site was important. When New Orleans was founded, it was surrounded by various types of wetlands (swamps, marshes, rivers, and lakes), which has led to constant problems over two centuries with flooding. Over the last few decades, the U.S. Army Corps of Engineers had developed and built a series of levees to help keep the waters of the Mississippi River and the surrounding wetlands from encroaching on the city. Unfortunately, the city suffered devastating flooding when these human-made levees were breached by the flood surge from Hurricane Katrina in 2005.

The very names of some wetlands show early American attitudes toward them—for example, the Great Dismal Swamp of southeastern Virginia and northeastern North Carolina. We would never name a wetland area "The Great Dismal Swamp" today (Figure 8–16). On the other hand, Americans in the 1700s would hardly consider the Everglades a national treasure worthy of safeguarding. It is important to understand that our social values change over time; sometimes they change

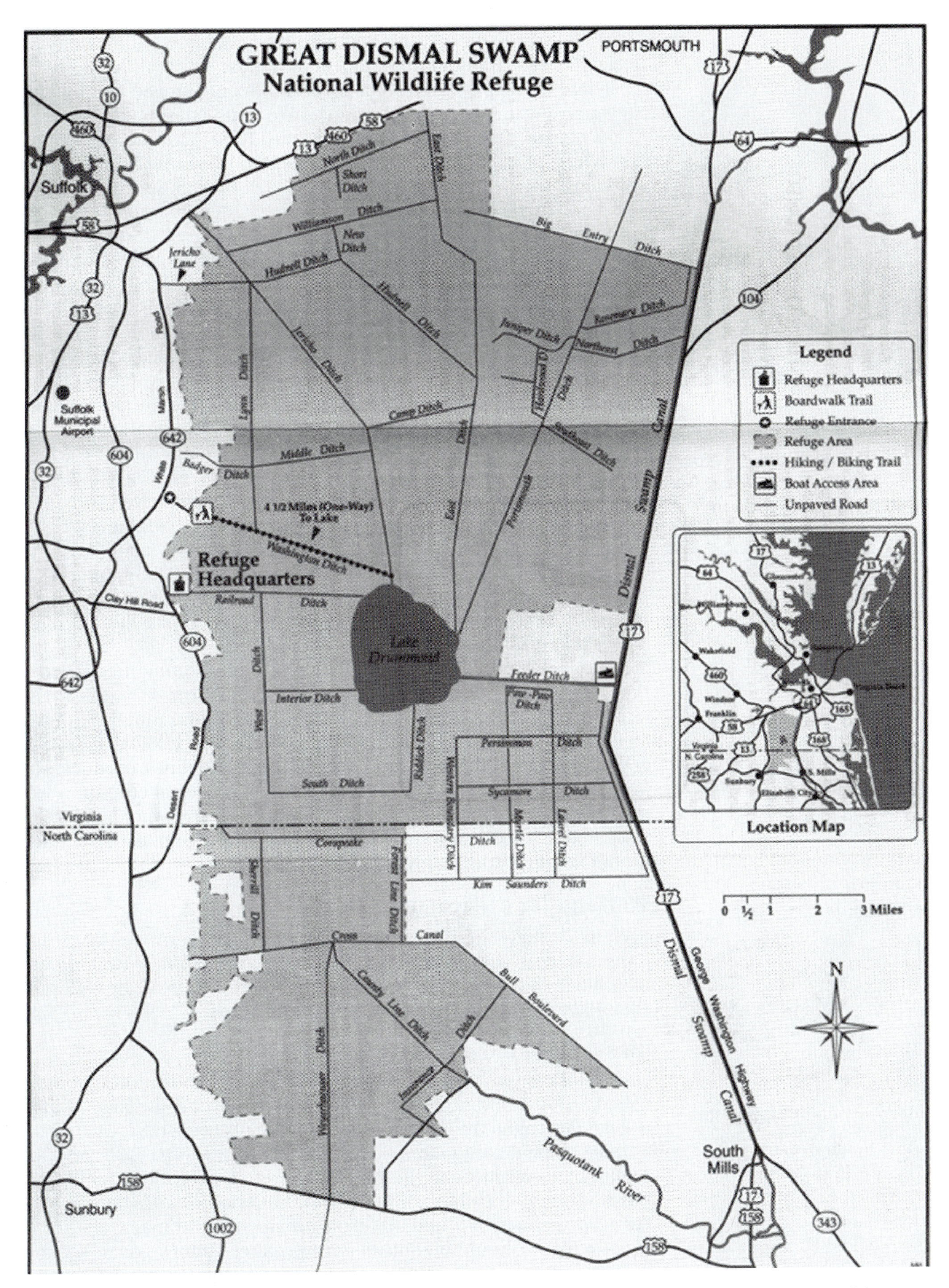

FIGURE 8–16 The Great Dismal Swamp on the border of Virginia and North Carolina (Map courtesy of the U.S. Fish and Wildlife Service.)

drastically. Until recent decades, governmental programs sought to "reclaim" swamps rather than protect wetlands.

Before the 1970s, the federal government encouraged the idea that wetlands were unhealthy and unproductive areas that should be avoided. The governmentally authorized and subsidized draining of wetlands began in 1849 with the passage of the Swamp Lands Act for the state of Louisiana, which was later broadened to cover the entire United States. This act gave states permission to fill in and change areas that "were unfit for cultivation." For more than a century, the process of draining wetlands was thought of as the recovery of swampland for productive use.

stagnant water, insect-borne diseases, examples
West Nile fever, USA

Even now in many metropolitan areas with nearby wetlands, some experts recommend that during "bug season"—that is, the majority of the summer months—people stay away from all areas of standing water such as swamps and small pools. These areas should be avoided because mosquitoes and other nuisance insects breed in stagnant water. Each year, more foreign insect-borne diseases seem to appear in the United States because of increased foreign travel and trade. Traditional insect-borne diseases include malaria and encephalitis, among many others. The newest fear over the last decade has been the spread of West Nile fever, a mosquito-borne illness that that killed a handful of people in New York in the summer of 1999 and has now spread to most states in the country.

In 1972 the federal government changed its policy about wetlands and their importance and took steps to protect and restore wetlands by requiring permits under the Clean Water Act. This not only helped to protect current wetlands from being filled but also allowed for the restoration of degraded wetlands.

Under the Wetland Conservation provision (commonly known as the "Swampbuster") of both the 1985 and 1990 farm bills, farmers are required to protect the wetlands on their farms and ranches in order to be eligible for U.S. Department of Agriculture (USDA) farm program benefits. According to the Swampbuster guidelines, producers are not eligible for benefits if they planted an agricultural crop on a wetland that was converted by drainage, leveling, or any other means after December 23, 1985, or if they converted a wetland to make agricultural production possible after November 28, 1990.

Wetland Identification

Because of the wide variety of legislation that governs wetlands, it is important that each site be correctly determined. How are wetland sites identified? The two most common techniques for wetland identification are *off-site* and *on-site* identification methods.

Off-Site Identification

wetland identification, off-site
wetland identification, on-site

Off-site inspection involves checking maps and wetland inventories maintained by the relevant federal agencies. Although off-site identification will not determine the exact size and location of a wetland, it can be used as a screening device to find possible wetland locations. Three principle resources are available to the landowner who is researching for wetland sites: the National Wetlands Inventory, the National Resource Conservation Service's soil survey list, and U.S. geological topography maps.

The U.S. Fish and Wildlife Service produces the National Wetlands Inventory (NWI). Following the enactment of the Emergency Wetland Resources Act of 1986, the service was required to map wetland areas every 10 years. The NWI is concerned with wetlands and open water in the United States.

The National Resource Conservation Service (NRCS) maintains a list of hydric soils—a soil survey list—in county-by county survey maps. Individual landowners can visit the local NRCS office to view the maps for their land.

U.S. geological topography maps look at vegetative covers, surface characteristics, bogs, and marshes. The maps use specific terminologies. Each map spells out what is growing on a piece of land or if water is standing in the area. By using several map years, a historical record of the area can be determined. This will be helpful if a restoration program is undertaken.

On-Site Identification

The location of the wetland will determine which agency is responsible for the site analysis (Figure 8–17). If the site in question concerns dredging, filling, or discharge into a suspected wetland, then the U.S. Army Corps of Engineers will take charge. If the site is close to a lake or inland water, then the U.S. Fish and Wildlife Service will be called. The USDA's Natural Resources Conservation Service will be needed to identify wetlands on agricultural lands or nonagricultural lands that border on agricultural lands.

The NRCS determines the use of the wetland under the Swampbuster guidelines. It manages how wetlands are used for agricultural crops—as pastureland and for orchards, cranberries, and rice—as well as for aquaculture purposes such as crawfish fields. An individual who knows and understands plant identification and soil science should complete an on-site inspection. That person should determine if hydrophytic vegetation is present and then determine what plant species are present and identify the plants that tolerate wetland conditions. The condition of the plants will also be an indicator of wetland status. Swollen tree bases, trees with multiple trunks having adventitious roots (roots that branch above the ground and help provide support to the plant), trees with shallow

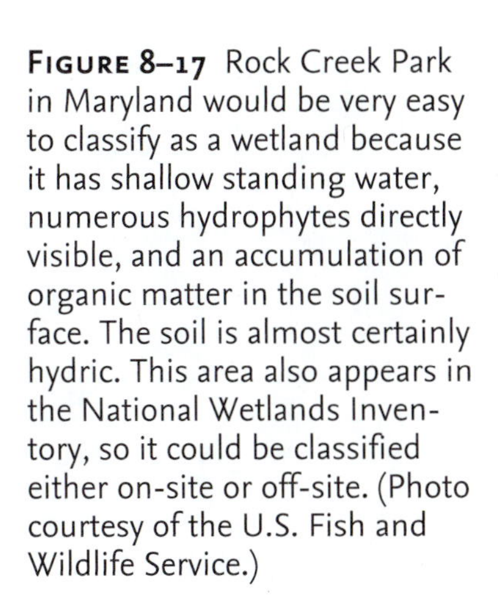

FIGURE 8–17 Rock Creek Park in Maryland would be very easy to classify as a wetland because it has shallow standing water, numerous hydrophytes directly visible, and an accumulation of organic matter in the soil surface. The soil is almost certainly hydric. This area also appears in the National Wetlands Inventory, so it could be classified either on-site or off-site. (Photo courtesy of the U.S. Fish and Wildlife Service.)

roots, and trees with leaves that are of various shapes are all examples of wetland adaptations for plants.

The soil also will need to be examined. Hydric soils are often highly organic soils. A large amount of black humus indicates a less decomposed plant life, which shows that the soil has been subjected to long periods of saturation by water. The soils also could be poorly drained with the water table less than 6 inches from the surface. It is not uncommon for the soil to emit a sulfurous odor and smell like rotten eggs. Closer examination would show rustlike iron pockets with gray to black-looking color spots called *mottling*.

The last part of the on-site inspection would deal with the hydrology or water supply of the site. Standing water is definitely an indicator of a possible wetland location. A closer examination may need to be conducted. The inspection might be conducted during the dry part of the year. Further field investigation could show water-stained leaves, water-borne sediments or plants, watermarks on trees and other fixed objects, or observation of soil saturated areas. By looking at the plants present, the condition and development of the soil, and the water makeup of the area, a technician can determine if the site should be classified as a wetland area.

Types of Wetlands

wetland, marsh

As mentioned earlier, there are many types of wetlands, including marshes, floodplains, lakes and ponds, rivers and streams, swamps, bogs, prairie potholes, and vernal pools. Depending on how a wetland is categorized, there are many more possibilities.

Marshes

Marshes are found throughout the United States. There are freshwater, saltwater, and tidal marshes. The largest area of freshwater marshes is found in the Dakotas and Minnesota (Figure 8–18). The majority of saltwater and tidal marshes are found along the East Coast and in the Gulf of

FIGURE 8–18 There has been a big effort to restore and maintain wetlands in North America. (Courtesy U.S. Fish and Wildlife Service.)

Figure 8–19 This photograph of the Everglades National Park shows dense marsh grasses in a freshwater marsh ecosystem. As you can see from the cranes in the foreground, the water is only inches deep. (Photo courtesy of the U.S. Fish and Wildlife Service.)

Mexico. Marshes do not rely on rainfall for their water supply, and soft-stemmed plants are the dominant plant type. Examples of marsh plants are reeds, sedges, rushes, grasses, and cattails. Other plants that may be found in marshes are arrowhead, lizard's tail, water parsnip, common pipewort, spider and swamp lily, sweetflag, marsh marigold, tall meadow rue, Virginia meadow beauty, thread-leaved sundew, Turk's-cap lily, marsh Saint John's wort, pickerelweed, marsh fern, marsh cinquefoil, arrow arum, water pennywort, and the carnivorous Venus flytrap.

Florida's Everglades is a well-known marsh (Figure 8–19). Urbanization, conversion of wetland to agricultural production, and the use of underground water supplies for cities in south Florida have all combined to cause serious degradation of the Everglades wetland ecosystem.

Floodplains

Floodplains are those areas that border rivers, lakes, and streams and that are flooded periodically. The common trees found in floodplains are American sycamore, paw paw, black willow, American elm, cottonwood, bald cypress, water tupelo, several oak species, and various birch species. One of the more common plants found in floodplains is the Virginia bluebell. Another less common plant is phlox (*Phlox divaricata*). The most well-known floodplain is the Mississippi River floodplain. The Mississippi floods sporadically, therefore leaving some warning for those residents who choose to live in a well-known floodplain.

The last major Mississippi flood occurred in the summer of 1993. That flood was judged to be a 500-year flood—meaning that we would expect such an extreme flood only once every 500 years. Many towns were evacuated permanently; in other places, the former residents chose to build their houses on higher ground. Those who live in a floodplain take a great risk because floodplains exist for a reason, and no matter how many dams and levees the Army Corps of Engineers builds, the floods will inevitably reoccur from time to time (Figure 8–20).

wetland, flood plain
wetland, pond

Figure 8–20 This farm near Winfield Kansas lies in a floodplain that is protected by a levee. In 1998, the levee was breached by floodwater and this damage resulted. (Photo courtesy of the Federal Emergency Management Agency, or FEMA.)

FIGURE 8–21 Herons thrive well in a lentic habitat. (Courtesy of Getty Images.)

Ponds

Not all lakes and ponds should be considered wetlands, although some do have wetlands around their edges. Some ponds may even turn into marshes if they become shallow enough. In this process, cattails will grow around the edges and water lilies and duckweed may take over the water's surface. Lakes and ponds are home to an abundance of life that will usually draw people who seek outdoor recreation. Examples include, but are not limited to, a variety of fish species, frogs, salamanders, aquatic insects, birds, waterfowl, and small mammals such as beaver and muskrat (Figure 8–21). A variety of plant species may also be found along the shores, including various species of alder and birch, blueberry, and winterberry. Lakes and ponds are permanent bodies of water, whereas prairie potholes (discussed below) are not.

Rivers and Streams

Once again, not all rivers and streams are considered wetlands. However, some do move slowly enough that certain types of vegetation start to take over. Sometimes this process leads to a larger area that turns into a marsh, bog, or forested wetland. A beaver may build a dam and cause a small pond, which will also change the characteristics of the river or stream. Rivers and streams also harbor a great variety of plants and animals. To the list of animals found in lakes and ponds may be added various snake species as well as shorebirds. Plants might include watercress, turtlehead, cardinal flower, globe mallow, checker mallow, water hyacinth, monkey flower, wood nettle, and red osier dogwood. The slower and warmer a river or stream becomes, the more prolific the plant life becomes as well.

INTERNET KEY WORDS

wetlands, rivers, streams
wetlands, swamps
wetlands, bogs

Swamps

Now we come to the wetland that is most synonymous with dark, eerie, and scary. Several years ago, a Dr. Pepper television commercial took place in a swamp in which a man jumped out of his boat to save his can of soda and was eaten by an alligator (Figure 8–22). Unfortunately, the

FIGURE 8–22 Many people associate alligators with swamps, such as this one in Mississippi. Alligators are found in wetlands in many parts of the southeastern United States. They can live 30 or more years, grow as long as 14 feet, and weigh as much as 1,000 pounds. At one time, people killed so many alligators for their skins to make purses, boots, and shoes, that they were placed on the U.S. list of Endangered and Threatened Wildlife and Plants. Conservation measure have resulted in the recovery of the alligator to the point that alligators are becoming a problem in many areas. (Photo courtesy of the U.S. Fish and Wildlife Service.)

media has also added to the public's generally negative attitude toward swamps. Movies and television shows such as *Swamp Thing* did not help the swamp's image either.

There are various types of **swamps**: cypress, southern bottomland hardwood, shrub, and northern. Each differs in the type of vegetation found there, but all have the same general characteristic of standing water with trees or shrubs growing in the water. This is extremely stagnant water that tends to be dark and nontranslucent.

Along with the shadow of the trees, the swamp leans toward being dark and eerie. The Great Dismal and the Big Cypress swamps are the most well-known swamps in the United States. These swamps, however, are highly productive areas. They represent small remnants of earlier times when the larger areas surrounding river systems were swamps. Some areas have giant trees that have been untouched by foresters.

There are also animals found in swamps that are rarely found in other areas; these animals remain unchanged because of the lack of human interaction. The cougar, alligator, red wolf, alligator gar, ivory-billed woodpecker, and sirens (similar to salamanders) are examples. Many plant species also thrive in swamps: spicebush, cowslip, swamp buttercup, skunk cabbage, elderberry, swamp dewberry, showy lady's slipper, butterfly orchid, swamp rose, hibiscus, red iris, spotted touch-me-not, great hedge nettle, spotted joe-pye weed, large purple-fringed orchid, blue flag, cinnamon fern, and jack-in-the-pulpit.

Bogs

Bogs are areas that are extremely damp, usually with evergreens present, and with a floor covered with moss and peat. This makes for a spongy walk, if you are lucky enough to not get stuck in the mud. Most bogs are found in the Northeast and in Michigan, Minnesota, and Wisconsin. If you were to take a hike on the Appalachian Trail through the eastern states, many boardwalks or logs have been placed by trail clubs to traverse through these areas. In other areas you may be forced to pick your way through bogs. That is often quite difficult when you step in the wrong place and almost lose your shoe to the suction of the muck. Bogs are known for a few plants that are found nowhere else such as cranberries and carnivorous pitcher plants. Typically, you will find only acid-loving plants in bogs because of the high acidic level produced by the leaching of nutrients and slow fermentation of organic matter. Small mammals such as shrews and voles are typically found in and near bogs. Trumpets, cotton grass, round-leaved sundew, leatherleaf, bog rein orchid, swamp pink, and sheep laurel are among the few plant species found in a bog.

Prairie Potholes

wetlands, prairie potholes
wetlands, vernal pools

Prairie potholes are found mainly in North Dakota, South Dakota, Minnesota, and Nebraska. They rely on periodic rainfall for their water supply and are usually full in the spring and early summer before water levels start to drop off and the potholes start to disappear for the rest of the year. They are best known for providing habitat for migrating waterfowl. Unfortunately, many waterfowl species rely on this critical habitat that declined steadily until recent years. Prairie potholes have always been in the middle of debates between landowners and naturalists.

Recently, many people have come to understand the importance of potholes as critical waterfowl habitat, but not before many potholes were reclaimed for agricultural production and lost forever.

Vernal Pools

Vernal pools are a special type of wetland that may last for only a few months each year. Like prairie potholes, vernal pools rely on periodic rainfall to form in the spring. They disappear in early summer. These pools become a haven for amphibian species because they lack predators. Many species of amphibians rely on these pools for breeding grounds, including the tiger, spotted and marbled salamanders, spring peepers (a frog), and wood frogs to name just a few. The fairy shrimp is one highly interesting animal that also uses these pools for its entire life cycle of approximately two weeks. According to Jim Petranka, a biologist at the University of North Carolina at Asheville, "About one-third of all amphibians in the eastern U.S. are strongly dependent on vernal ponds as breeding sites."

Status of Wetlands in the United States

It has been estimated there were more than 200 million acres of wetlands in the United States in the 1600s. According to the EPA, approximately one-half of that area remains in wetland today. Many states have lost far more than half of their original wetlands. Many inland freshwater wetlands in the southern states were drained for agricultural or urban use.

What are the major causes of the loss of wetlands? The main causes can be attributed to urbanization, industry, agriculture, timber harvesting, and mining operations.

wetland loss, urbanization
wetland loss, industry
wetland loss, agriculture

Urbanization

As our cities grow out toward rural areas, the construction of roads, parking lots, and buildings move and shift rainfall flows. The runoff from these areas moves fertilizers, wastes, organic matter, and road salts. These compounds cause a change in the water quality even if structures are not in wetland areas. Because wetlands were considered low-value land for so long, roads and bridges were constructed through wetlands rather than around them. Road runoff thus became a problem with wetland ecosystem maintenance. Many times non-native plants and animals invaded these areas and disrupted the normal succession of the wetland (Figure 8–23). Construction may not even be in the wetland itself, but in the watershed area draining into the wetland.

Industry

Industry can affect the wetland habitat through its increased use of water, which changes the hydrology of the area. Industrial manufacturing plants may discharge warmed water (thermal pollution) or change the acidity levels of the water. Pollutants from abandoned industry sites also may significantly damage wetland areas.

Agriculture

Agriculture has been a major factor in the loss of wetland areas. For decades, it was common for devices to be installed to "drain" wetland areas, making them available for tillage operations. As discussed earlier,

FIGURE 8–23 Many wetlands lie near urban areas. The John Heinz National Wildlife Refuge at Tinicum, Pennsylvania, near Philadelphia offers trails, fishing, and environmental education. (Photo courtesy of the U.S. Fish and Wildlife Service.)

the U.S. Congress enacted the Swampbuster provision with the 1985 federal farm bill. This program has prevented the conversion of wetland to agricultural production. The act stated that agriculture production could not alter the water hydrology, water quality, and the plant and animal life of an area. Areas that provide habitat for waterfowl were also protected.

Farmers were encouraged to allow prior wetland areas to revert to their original state. Since the Swampbuster program started, duck and migratory bird populations have increased by more than 25 percent. Farmers were also encouraged not to graze wetland areas. Wastes from cattle as well as their physical destruction of wildlife homes were considered damaging to wetland areas. Programs were established that help farmers manage the grazing. It was found that the grazing could coexist with wetlands, therefore benefiting farmers as well as protecting the wetland area.

INTERNET KEY WORDS

wetland loss, timber harvest
amphibians, indicator species

Timber Harvesting

Most wetland loss from timber harvesting is only temporary. The destruction usually lasts for three to five years. During the harvesting, new timber roads are constructed, and the rutting of existing roadways may impair wetland quality. The use of heavy equipment in the wetland area will affect the water quality, soil structure, and development. Through careful management practices, timber harvest can be accomplished with limited damage to wetlands.

Mining

The most common mining practice that affects wetlands is the mining of peat moss. Peat moss is used in the horticultural area as a planting medium for greenhouse production. The mining operation involves removing vegetation, draining the wetland, and extracting the peat. The wetland is totally transformed from a wetland to an open water area. The area surrounding the wetland is also changed, thus affecting the habitat of many species of wildlife.

Wetland Preservation

The lower 48 states have lost more than half their original wetlands. According to the EPA, the annual loss of wetlands is estimated at 70,000 to 90,000 acres on nonfederal lands. In the last 30 years, wetlands have been a focus for environmentalists. Wetlands give us those amphibians that are indicator species of degradation in the environment. Sometimes environmental problems are finally noticed when amphibians are born with extra limbs or missing limbs or die at quicker than normal rates.

In recent years, the preservation of our nation's wetlands has drawn increased national interest. With the federal government's goal of a "no-net-loss" policy, a closer examination of wetlands has taken place. Wetland preservation efforts can be classified into three major areas: (1) protection of our natural wetland areas, (2) construction of new wetland areas to enhance water quality, and (3) restoration of previously altered wetlands.

Natural Wetland Protection

Natural wetlands can be defined as those that have not been constructed by human activities. Most government regulations are intended to prevent the destruction of these wetlands (Figure 8–24). These regulations work to keep construction projects such as roads, bridges, and buildings from invading these areas. Regulations also examine the watersheds around these natural areas, trying to keep them as intact and natural as possible. The constant challenge to preserving the natural wetland is urban growth and all of the construction that accompanies that growth.

Construction of New Wetland Areas to Improve Water Quality

A relatively new approach to wetland management is the construction of natural structures to treat all forms of water pollution. This might include everything from animal wastewater to municipal effluent. Most modern municipal programs use wetlands as secondary treatment after the primary removal of solid waste. Through the use of plants, wetlands

FIGURE 8–24 Wetlands are being improved and maintained along active migration routes.

FIGURE 8–25 The long-billed curlew inhabits marches, mud flats, and open plains.

are one way nature can clean nutrient-rich effluents from wastewater. The cost of wetland construction is considerably less than that of the construction of water-treatment facilities.

Restoration of Previously Altered Wetlands

As we lose wetlands to the pressures previously discussed, researchers look to areas that may have been overlooked as altered wetland areas. Some restoration may be demanded because of requirements that an industry must replace or restore wetlands at locations other than the areas altered by industry. These "new" wetland areas will be welcome additions in helping to implement the goal of no net loss.

The major concern for new wetland areas is the development of a large variety of wildlife. This will include mammals, fish, migratory game birds, and waterfowl (Figure 8–25). A constructed marsh wetland will achieve these habitat goals (Figure 8–26).

The U.S. Forest Service has promoted the redevelopment of cleared forested areas. These areas help to maintain water quality in shallow groundwater. With livestock excluded, these reforested areas are allowed to develop naturally. Urban areas also are included in the restoration efforts. Urban restoration deals mainly with water quality in the landscape. Other concerns include sedimentation, buffer-zone construction, and the development of smaller land areas.

Wetlands Management

Wetlands management is a concern of many government and private agencies, including the Environmental Protection Agency, the U.S. Fish and Wildlife Service, the U.S. Army Corps of Engineers, Ducks Unlimited, and the Nature Conservancy to name a few. Most of the difficulties in management occur when wetlands are found on private land. Such agencies, whether private or part of the local, state, or federal government, cannot force private landowners to manage their wetlands in any particular way. This is why so many conservationists are deeply concerned about the future of wetlands. However,

FIGURE 8–26 The maintenance of existing wetlands and the restoration of former wetlands on private land are both important. This shows a farm before and after a restoration project. The dark area in the first photo shows a wetland area that had been drained for farming, a process that used to be called "reclamation." The second photo shows the same farm after the wetlands area was restored; it is now a farm pond. (Photo courtesy of the U.S. Fish and Wildlife Service.)

the Corps of Engineers is trying to tighten policies and permits regarding wetland development and to help in the conservation of such vital ecosystems.

Since June 2000, the Corps of Engineers has been enforcing a tighter permit program regarding wetlands. This new program allows the destruction (grading, draining, filling, etc.) of one-half acre of wetland without an individual permit; that figure is down from three acres, which was previously allowed. Another change requires landowners to notify the Corps of Engineers of activities that affect any wetland area one-tenth of an acre or larger in size. These changes allow the government to more closely monitor development in wetland areas. The intention is to limit developers from building on wetland sites.

The causes of wetland loss and degradation include drainage, dredging, stream channelization, tilling for crop production, logging, mining, construction, air and water pollution, introduction of non-native species, and grazing by domestic animals. Natural threats that play a part in wetland loss and degradation are erosion, droughts, hurricanes, tornadoes, and other climate and weather conditions.

For the personal management of wetlands, the best thing to do is to leave them alone. The best way to protect a wetland area is to provide a buffer zone around it in which destruction or degradation is not allowed. Waterways or wetlands should not be polluted. If pollution is taking place, it should be reported to the local fisheries and wildlife agency or the Natural Resources Conservation Service. Individuals can also take the initiative and start a local adopt-a-wetland program. Schools and communities can become involved by participating in a wetland awareness program.

Restoration plays a big part in the management of degraded wetlands and may involve the removal of non-native species; planting native trees, shrubs, or other aquatic plants; and creating walkways to minimize or eliminate further degradation. Along with these activities, proper signs should be placed to alert people that an area is being restored, hopefully to the original state.

Private landowners working with local, state, and federal government agencies as well as private companies and agencies can manage and preserve wetlands and ensure the future of these pristine and critical habitats. For more information on your wetlands and what you can do, contact the EPA.

Other Government Programs

1973 Endangered Species Act

In 1899, the Rivers and Harbor Act required approval from the Secretary of War before any construction activities were started that disposed wastes into a navigable river. In 1967, the Fish and Wildlife Coordination Act required the U.S. Army Corps of Engineers to study the ecological effects of all water-related areas. Then in 1969, the Environmental Protection Agency began conducting environmental impact statements that affected wetlands. This legislation led to the Federal Water Pollution Control Act of 1972. This bill is commonly referred to as the Clean Water Act.

Several sections in the Clean Water Act address the regulations of wetland areas. This bill instructs that both the Corps of Engineers and the EPA will regulate the wetlands of the United States. The bill was amended in 1977 to exempt some farming and mining activities. The agencies administer the permits to both companies and individuals when a wetland will be affected by some human-directed activity.

Other legislation affecting wetland management includes:

- the 1973 Endangered Species Act;
- the Flood Disaster Protection Act;
- the Floodplain Management Order;
- the 1985 farm bill;
- the 1986 Emergency Wetland Restoration Act;
- the 1989 North American Wetland Conservation Act;
- the 1990 Coastal Wetlands Planning, Protection, and Restoration Act;
- the 1990 farm bill; and
- the 1990 Water Resources Development Act.

State governmental agencies have also taken the lead in wetland protection. They are also concerned with similar areas as federal regulations but are more aware of individual state needs. Most states direct educational programs that help the public understand the value and function of the wetlands. The states can help fill the gaps that federal programs might not address. Examples could be the establishment of sanctuaries, refuges, or wilderness areas that would coincide with state-owned wetlands. Local governments can also develop local wetland areas. They can help local citizens develop workable wetland-management plans that will be of value to all. Local governments can lend support to private wetland areas on an individual basis.

Looking Back

The world's grasslands are known by many names. Whatever they are called, grasslands have been fundamental in the development of humans and civilization as we know it. They remain critical in feeding our growing human population. The advantage in productivity that grasslands have over forests is that the sun's energy is captured near the ground rather than in the crowns of trees. At ground level, grazing animals can feed in large numbers and move from one feeding site to another.

In what is now the United States, approximately one-fourth of our land area was covered by grasslands before European settlement. The most productive part of that area was the tallgrass prairies. Throughout the world, humans have tended to overuse grasslands, which has resulted in soil damage and reduced carrying capacities. Today we know how to manage grasslands to produce the maximum amount of grazing for animals useful to humans, while protecting the land from damage because of overuse.

Historically, wetlands have been regarded as swamps and bogs that were breeding grounds for unwanted insects and reptiles. The main management procedure was to install drainage devices so that agricultural crops could be grown or structures could be built on the "reclaimed" land. Through careful examination and research, it has been determined that wetland areas provide valuable assets to the ecosystem. The added population of wildlife, their natural water-filtration qualities, and the buffer-zone possibilities make the wetland areas worth the regulation and concern.

Self-Analysis

Essay Questions

1. Why do some parts of the world tend to be naturally covered by trees and other parts by shrubs, grasses, and other small plants?
2. What happens to the vegetation in a range area if there is overgrazing for an extended period of time? Undergrazing?
3. What factors led to the development and eventual end of the open-range cattle industry in this country?
4. What is the Taylor Grazing Act? What has been the effect of this legislation in the United States?
5. What factors affect the grazing capacity of a range area?
6. Define wetland from both the legal and ecological perspectives.
7. What are the three characteristics of a wetland area?
8. What would be included in an on-site evaluation for determining a wetland?
9. What resources are available for off-site wetland identification?
10. Explain how marshes, floodplains, lakes and ponds, rivers and streams, swamps, bogs, prairie potholes, and vernal pools are alike and how they differ.
11. List and explain the major causes for the loss of wetlands.
12. What governmental agencies regulate wetland management?
13. What is the primary federal law that regulates wetlands?

Multiple-Choice Questions

1. Of all the land in the United States, how much is rangeland?
   a. one-fourth
   b. one-fifth
   c. one-third
   d. one-tenth
2. In the southwestern United States, what percent of the land is rangeland?
   a. 15 percent
   b. 75 percent
   c. 35 percent
   d. 85 percent
3. The area between Indiana and Nebraska and southward to Texas was historically known as
   a. tallgrass prairie.
   b. transition prairie.
   c. shortgrass prairie.
   d. tundra prairie.
4. Which class of animals is primarily supported by a grassland habitat?
   a. omnivores
   b. carnivores
   c. herbivores
   d. producers
5. Varieties of grass that are easily damaged by even moderate grazing are called
   a. increasers.
   b. decreasers.
   c. invaders.
   d. diminishers.
6. The amount of forage that is required to feed a 1,000-pound animal for a given period of time is called
   a. an animal equivalent unit.
   b. a capacity ratio.
   c. a rationale.
   d. an average intake unit.

7. With which type of soils are wetlands primarily associated?
   a. organic
   b. sandy
   c. hydric
   d. loamy
8. Which of the following is least likely to be considered a wetland?
   a. a plateau
   b. a prairie pothole
   c. a marsh
   d. a floodplain
9. Approximately how many million acres of wetlands are found in the United States today?
   a. 100
   b. 50
   c. 200
   d. 250
10. Which of the following is most commonly mined in wetland areas?
    a. phosphate
    b. coal
    c. peat moss
    d. lead

Learning Activities

Which learning activities you can do will be determined largely by where you live. You will need to select those activities that are possible in your location.

1. Locate a grassed area that is not being grazed or mowed. Measure and mark off a typical 10-foot by 10-foot area. Estimate the percent of the 100 square feet that is covered by grasses and by shrubs. Next, locate a grassed area that is being grazed. Measure and mark off a typical 10-foot by 10-foot area. Determine the percent of the area that is covered by grasses and by shrubs. How do the two areas compare? What do you think made the difference?
2. Look through the last several years of *National Geographic* magazine. Find as many articles as you can about parts of the world with important grasslands. Bring the editions to class for an open discussion on the grassland ecosystems represented. What problems are they facing? How are people using them?
3. Find a natural range area. Dig up several grass plants and single examples of some of the other plants present there, particularly the woody ones. Try to identify the plants. Clean off the root systems. As you compare the different plants, which ones would be more effective in preventing erosion? Why? Which ones would be more useful to cattle? Why?
4. Contact your local NRCS office for the location of wetlands in your area. Visit one of the sites and observe the components of the wetland area.
5. Design a wetland plan for a previously altered area.
6. Obtain a set of maps and make an off-site evaluation of a possible wetland area in your school district.
7. Start an adopt-a-wetland project for your class or school.

TERMS TO KNOW

- botany
- zoology
- food
- forage
- mast
- simple stomach
- ruminant
- crop
- gizzard
- herbivore
- predator
- omnivore
- mortality
- vegetative cover
- adaptive behavior
- gregarious
- gamete
- fertilization
- zygote
- parasitism
- warm-blooded animal
- mutualism
- predation
- commensalism
- competition
- extinct
- endangered species
- threatened species
- imprinting
- encroachment
- economic value
- ecological resource
- alien species
- nonadaptive behavior
- biotic potential
- fecundity

CHAPTER 9

Wildlife Biology and Management

Objectives

After completing this chapter, you should be able to

- identify the differences between the sciences of botany and zoology
- list the basic needs of wild animals
- explain why each of the basic needs must be met for an animal to survive
- describe the relationship between energy and wildlife survival
- identify some differences in the digestive process between animals with simple stomachs and those that are ruminants
- describe the process by which birds are able to process and digest food
- explain why clean water is needed by wild creatures
- discuss the needs of wildlife for shelter
- explain the cell division processes of mitosis and meiosis and how they relate to growth and reproduction
- describe the differences in the wildlife relationships of parasitism, mutualism, predation, commensalism, and competition
- explain why some natural resources are also known as *ecological resources*
- name the most common causes of extinction among wildlife species
- list some approved practices in wildlife management

Biology is the science that deals with the life processes and characteristics of plants and animals. It includes the study of the origin, history, and habits of the many forms of plant and animal life. Biology is divided into two main branches. **Botany** is the branch of biology that deals with plants, and **zoology** is the branch of biology that deals with animal life.

One principle of zoology is that all animals have basic needs that are required to sustain life. Among these needs are food, water, shelter, and space. Food provides a source of energy for those organisms that lack chlorophyll to capture energy from the sun. Water provides habitat for many animals in addition to being a liquid medium in which food is transported to the cells of the body. Shelter provides protection from heat, cold, and wind and also provides protection from predators. Space is needed so that adequate supplies of food can be produced. It also provides room to grow physically and to reproduce a new generation.

Prioritizing basic wildlife needs is difficult. Food, water, cover, and space—in a suitable arrangement—are of equal importance. Imagine these basic habitat requirements as a chain. If just one link is removed, the chain is broken and becomes worthless. If any of the essential elements in an animal's habitat are removed, that habitat is of less value.

Animal Behaviors and Habits

The life of an animal can be described as the distinct behaviors and habits that make it possible for the animal to find, consume, and obtain energy from food; grow physically; adapt to the environment; and reproduce. The differences that we observe among the many species of animals make it possible for them to live in a wide variety of environments. For example, some kinds of birds eat only small seeds, whereas other kinds of birds eat meat. In most cases, birds that are adapted to eating seeds would be unable to exist if they had to depend on their ability to capture and kill prey. In fact, the seed-eating birds are often killed and eaten by birds of prey.

The manner in which an animal goes about obtaining its basic needs is often distinctive to it and enables the animal to occupy a specific *niche* in an environment. For example, a woodpecker is adapted to finding and capturing insects that live beneath a tree's bark and within its woody tissues. Most other birds are unable to do this, and therefore the woodpecker has a competitive advantage over most other birds in obtaining this source of food (Figure 9–1).

FIGURE 9–1 The woodpecker has a competitive advantage over other birds because it is equipped with a strong bill and skull that allow it to capture insects by drilling holes into the bark and wood of trees.

Some animals are able to obtain food from several different sources, giving them a big advantage over animals that depend on a single source or type of food. If one source of food becomes unavailable, they simply seek food of another kind. Some animals, however, cannot do this and are in grave danger of starving when their food is no longer available. They may be completely unable to adapt to a different diet even though other kinds of food might be abundant.

In some instances, an animal may be completely incapable of digesting a particular kind of food and extracting adequate nourishment from it. For example, a member of the cat family would die of starvation if it had to eat grass and twigs like the deer family. Cats are

FIGURE 9–2 A newborn fawn instinctively hides from danger by lying motionless in an area where a combination of sunlight and shadow allows it to blend with its surroundings.

incapable of digesting large amounts of fiber, but the bacteria in the stomach of a deer are able to break down fiber obtained from roughage. The deer then digests the bacteria, which are composed of high-quality nutrients.

Some behaviors of animals are *instinctive* and others are learned. An instinctive behavior is one that is evident at birth, such as the suckling instinct of a newborn mammal and the tendency of a duckling to go into the water and swim. Many of us have heard the old saying, "He took like a duck to water." An instinctive behavior is a behavior that is natural to an animal and that is characteristic of other animals of the same species (Figure 9–2).

instinct, learned behavior, examples

Many behaviors are also learned by young animals through their life experiences. For example, young animals must learn to find food. They do so by watching their parents and mimicking their behaviors. It is a common sight to see a young robin learning to listen for the movement of a worm hidden under the soil surface. Most young animals must learn to drink water. They also learn to seek safety from their natural enemies. These are learned behaviors.

FOOD

Animals require food to live because energy is required to perform their life functions. **Food** is anything an animal consumes and digests to provide such energy. The growth and repair of living cells requires nutrients from which energy and materials are obtained. Energy is necessary to form chemical bonds that hold the materials together that are used to form cells. Warm-blooded animals also require energy to maintain body temperature. Energy is used each time a muscle flexes or a nerve sends an impulse racing through the body. Without energy, there is no animal life. Without nutrients, energy is not available to an animal. Each animal uses the energy provided by food to meet its daily needs.

Animals require varying amounts of food during different times of the year. An animal's food requirements, for example, generally increase during extreme cold, reproduction, migration, or any other situation that causes the animal stress. Animals use the energy they

FIGURE 9–3 A coyote listens for the sound of rodents scurrying through a snow-covered field. To survive, the coyote must get an adequate supply of food and water. (Courtesy of the U.S. Fish and Wildlife Service.)

get from food just as humans do: to stay warm, to walk and run, and to reproduce (Figure 9–3).

If the quality or quantity of their food is inadequate for some reason, then animals will likely suffer from weakness and malnutrition. An animal that is not getting enough to eat is more likely to succumb to sickness, disease, or predators. If the nutritional deficiency is severe or occurs during a period of increased stress, such as a severe winter, then the animal may starve to death. Lack of adequate nutrition almost always results in decreased reproduction or weak, small offspring that are much more vulnerable to predators, disease, and sickness.

animal malnutrition

The actual amount of food required by an animal depends on its age, sex, size, location, and the season of the year. An animal may not necessarily have its greatest nutritional need during the northern hemisphere's winter (December, January, and February). In the Southwest, for example, August and September can be extremely hard on wildlife. Often the green spring and early summer **forage** dies and dries up before the fall **mast** crop (fruit of forest trees) and plants other than grass (such as forbs) are available. If a drought limits the mast crop and there is an early hard winter, then wildlife can suffer terribly.

Ideally, food is present in a habitat in adequate quantity, quality, and variety to sustain a population of wildlife without it expending valuable energy traveling to and from cover to get it. The less distance an animal has to travel to secure its meal, the less vulnerable the animal is to predators. It can also use the energy not spent searching for food for other needs, such as keeping warm or cool. As a general rule, the less time an animal spends searching for food and the less distance traveled, the better.

Differences exist in the ways in which animals obtain nutrients. Some animals eat plants and break down plant nutrients through digestion. These nutrients are then reassembled to form animal tissue. Some animals eat the flesh of other animals to obtain nutrients. Some animals, such as bears, are able to eat foods obtained from both plants and animals.

Figure 9–4 The squirrel is an example of an animal that is equipped with a simple stomach in which it digests foods, such as nuts, that contain highly concentrated nutrients.

Most mammals are equipped with a **simple stomach** that has a single compartment in which food is stored and where the first steps in digestion occur (Figure 9–4). The simple stomach does not have a large capacity, so foods with a relatively high concentration of nutrients must be eaten. For example, grains and meat contain high concentrations of nutrients, so they are ideal foods for an animal with a simple stomach.

Foods that are high in fiber (such as grass, twigs, and leaves) are generally low in nutrient concentration. For this reason, much greater amounts of food must be eaten by animals with high-fiber diets. Mammals such as deer, sheep, goats, and pronghorns eat such diets, and their digestive systems are quite different from those of animals who have simple stomachs. These mammals are equipped with stomachs with four compartments; such an animal is called a **ruminant** (Figure 9–5).

The largest of the four stomach compartments is called the *rumen.* It stores large amounts of plant materials that ferment as they are broken down by bacteria. As they reproduce, the bacteria assimilate the nutrients from the plant material. The ruminant animal then digests

Figure 9–5 A Bighorn sheep is a ruminant animal that has a stomach with four compartments. The largest of these is called the *rumen*, and it acts as a storage and fermentation vat for grasses, herbs, and other high-fiber foods that make up the sheep's diet.

FIGURE 9–6 The pheasant is equipped with a different digestive system than most other animals. Its unique digestive organs include the crop and the gizzard. (Courtesy of the U.S. Fish and Wildlife Service.)

INTERNET KEY WORDS

ruminant, simple stomach
crop, gizzard

the bacteria as they pass through the other three compartments of the stomach.

A different type of digestive system exists in birds (Figure 9–6). Because they lack teeth, birds cannot chew their food. As food is swallowed, it is stored in an organ called a **crop**, where it absorbs water and is softened. The food then moves to a muscular organ called a **gizzard**. The food is ground by the gizzard into fine particles by small stones or grit that have been swallowed by the bird for this purpose.

An animal that eats food obtained directly from plants is a **herbivore**. Herbivores may eat the foliage, seeds, fruits, or even the roots of plants. In some cases, they may eat only specific plant parts, such as the seeds of a plant. For example, the Douglas squirrel lives in pine forests where it gathers and caches pine cones and obtains the seeds from the cones for its food. Other herbivores are able to eat more than one item. The wild turkey, for example, has been returned to much of its original range, and it has also been introduced into many new areas. This bird eats a varied diet, but when it has the opportunity, it consumes a diet that consists mostly of seeds such as nuts and acorns.

An animal that eats other animals is a **predator**, or *carnivore*, because it kills other creatures and eats meat. The Canada lynx is a good example of a predator that lives in a forest environment (Figure 9–7). Most of these lynxes live in the northern coniferous forest region. Their favorite food are wild hares, but they also eat birds, rodents, foxes, and even small deer. Another example of a well-known predator is the great horned owl. It ranges across most of the North American continent and eats rabbits, hares, small rodents, and birds.

Some animals—called **omnivores**—eat both plants and animals. Bears and raccoons are good examples. Such animals are usually successful in adapting to new environments and are in little danger of becoming extinct because they do not depend on a narrow range of foods, unlike some birds and animals that depend on only single sources of food.

FIGURE 9–7 The Canada lynx preys heavily on hares. (Courtesy of Eyewire.)

water

All wildlife species need water. It is a necessary nutrient that is required by cells in the body. It acts as a solvent to absorb food products such as proteins, fats, and carbohydrates. It is also used to control body temperature, either by submerging the body in it or by removing heat from the body as water evaporates from the body surfaces.

Each type of digestive system is designed to absorb water from the digestive tract along with the dissolved nutrients it contains. The water and nutrients then become part of the blood supply, and the nutrients are transported to the cells of the body by the blood as it passes through the circulatory system. Despite the differences in the anatomy of the digestive tracts among the different animal classes, each system operates efficiently in extracting nutrients and energy from food.

water, digestion
water, wildlife, guzzler

There is tremendous variation in the amount of water used by different animals. Some animals require bodies of standing water, such as lakes and ponds, but many do not. Most large land mammals—such as bear, deer, elk, bison, and antelope—require standing water. They typically drink two to three times per day, usually in the early morning and late evening (Figure 9–8). Other species, such as the white-tailed deer, may drink during the middle of the day. This is particularly common during the summer and early fall in southern and western states, where midday temperatures may reach 100° F (38° C).

Many smaller mammals and birds obtain water from the food they consume. Most songbirds and upland game birds do not require large bodies of standing water to survive. They use it if it is available, but it is not essential. They can drink the dew off vegetation in the mornings or they may drink from puddles during the day.

The wild turkey is a notable exception among upland game birds. It requires standing water and often prefers river and stream bottoms as habitat. These areas provide water, cover, and food. In the Southwest,

FIGURE 9–8 Most large land mammals require fresh water to survive. (Courtesy of Getty Images.)

where standing water can be scarce, many species of animals and birds obtain moisture from native vegetation or from the prey they consume. Although water may not be a serious limiting factor in an area that receives 30 or 40 inches of rainfall per year, it can be a factor in an area that receives only 10 inches.

In areas with limited surface water, wildlife managers often construct artificial watering devices commonly called *guzzlers.* These structures catch and store rainwater that would otherwise run off and be unavailable to wildlife. In many areas of the arid Southwest, these artificial watering devices have been helpful in restoring and maintaining wildlife numbers.

Of course, many species live in or close to water and cannot survive elsewhere. North America's rivers, streams, marshes, swamps, lakes, and potholes once teemed with a wide variety of wildlife. In most cases, wetlands still do. However, for centuries humans have dammed, drained, channeled, diked, polluted, and otherwise altered wetland areas. Only recently have we begun to realize the damage we have been causing. If we do not stop the loss of wetlands, many species of wildlife may be lost as well.

Most animals obtain water by drinking it, but a few are able to absorb enough water through their skin to sustain their needs. A few animals are even able to survive desert conditions where no water is evident by lapping up dew drops that have condensed during the cool night hours.

Whether an animal requires water for drinking or for its living environment, it cannot live long without it. As soon as water becomes too scarce for an animal to access it in the ways it requires, it must either move to new locations where it is available or it will die. An inadequate water supply is never forgiving. An inadequate water supply is never forgiving. It is like a shrinking habitat in that each creature requires a minimal amount of water to sustain life. When water is unavailable in adequate amounts, the population of animals in the area must decline. The strong will always

FIGURE 9–9 A cottontail rabbit blends well with vegetative cover. (Courtesy of Getty Images.)

crowd out the weak until even the strong become weak also. Drought conditions can be deadly for wildlife.

Shelter

All wildlife, regardless of species, have similar basic habitat needs. An animal's habitat is its home; it is the area where it eats, rests, and reproduces. Wildlife requires food, water, cover, and space, all suitably arranged, for survival. Cover or shelter serves several important functions for wildlife. It provides a relatively safe haven from predators and shelter from inclement weather such as rain, snow, or excessive heat. It must also give animals a place to reproduce and raise their young.

Wildlife must have places to rest, feed, and sleep, as well as places to retreat to when predators threaten. Ideally, cover is arranged so that an animal does not overexpose itself to predators or the elements as it goes about its daily business. A wildlife population can suffer unusually high death or **mortality** rates if it is forced to feed, water, or travel too great a distance from its escape cover.

Shelter takes many forms for wildlife. Although it is typically some form of vegetation, it also can be a pile of rocks, a cliff overhang, a hole in the ground, or a simple depression in the earth. **Vegetative cover** comes in countless variations (Figure 9–9). What works or is preferred by one species may not work for another. Some species, such as white-tailed deer, are highly adaptable and can use a wide variety of habitats. White-tailed deer can be found in mature hardwood forests, marshes, farmlands, grasslands, suburban yards, and other habitats. Other species are found only in specific habitats. For example, the northern spotted owl is found in the Pacific Northwest and the red-cockaded woodpecker in the South and Southeast, both in old-growth forests.

Atwater's Prairie Chicken

Many species have never been especially numerous and probably never will be because they cannot adapt to a wide variety of cover or habitat situations. Each species occupies a specific niche or spot in a given habitat for which it is adapted. The more specific or restricted the animal, the more vulnerable it becomes to changes in habitat. These changes may occur naturally or they may be imposed by humans. An animal species can often adapt to natural changes to its environment because such changes usually occur over many years. Human-driven changes usually happen much faster, giving animals little time to change and adapt.

Interest Profile

ATTWATER'S PRAIRIE CHICKEN

Species such as the Attwater's prairie chicken, which inhabits the native coastal prairie region of south Texas, face an uncertain future. Only a few thousand acres of such pasture remain, with the rest paved, plowed, grazed, or otherwise occupied by humans. The Attwater's prairie chicken will have to adapt to changing habitat, survive in a tiny remnant of its former habitat, or perish.

Wildlife managers are faced with the unenviable task of saving many species in much the same situation as the Attwater's prairie-chicken.

FIGURE 9–10 An elk calf's speckled coat helps it blend in with its surroundings and avoid detection by predators. (Courtesy of Robert Pratt.)

Habitat

The need for a favorable living environment is greatest during extreme weather conditions. For example, during winter months when an animal is weakened by an inadequate supply of food, it depends on shelter to conserve its energy. If the animal is able to get out of the wind, it will be able to maintain its body temperature with less food than if it were subjected to chilling conditions. In some climates, shelter is needed for protection from the sun. Animals living in hot climates often require shade to keep their body temperatures regulated.

One period when shelter is greatly needed is during and immediately after the birth of the young. Plenty of cover is needed to hide and shelter baby animals that are born into wild environments (Figure 9–10). Adequate habitat provides a place of shelter from weather and climatic conditions and a place to hide or escape from predators. It also offers a place where young animals can grow to maturity with sufficient food and water to minimize hunger and malnutrition.

Animals learn to use their environment to improve their chances of survival. For instance, a deer may develop the habit of bedding down during the hot part of the day on a hillside where it is not easily seen, but from which it can see anything that approaches its position. Once it has been detected, the same deer might follow a predictable route up the hill and away from the area to escape its enemy. These are learned or **adaptive behaviors** that improve the chances of the deer to avoid predators and hunters.

Space

All animals have a home range—that is, an area in which they live that provides their daily needs. The size of this area varies greatly for different animals. Some animals require many square miles of habitat—for example, wolves and mountain lions. Others may spend their entire lives within a 1- to 2-acre area—cottontail rabbits, for example. As a general

FIGURE 9–11 Most species prefer habitat with some variety. Mixed brush land with grass and forb-filled clearings makes excellent habitat for a wide range of species, from white-tailed deer to songbirds. (Photo courtesy of the U.S. Department of Agriculture's Natural Resources Conservation Service.)

rule, small animals have small territories and larger animals have larger home ranges.

Naturally, different species require different types of food, different types of cover, and varying amounts of water (Figure 9–11). However, each of these basic elements must be available within an accessible distance for it to survive. All of the elements necessary for excellent wildlife habitat may be within the normal range or territory occupied by a species, but—using whitetail deer as an example—if that habitat is split by a four-lane highway, then it becomes more difficult, if not hazardous, for the deer to use it.

Species such as migratory waterfowl utilize a combination system: They have a small home range during the breeding season and a much larger one during the fall and winter. During their migration to their wintering grounds, and to some extent all winter, they have several temporary home ranges. As the available food supply in an area is consumed, the birds move to other areas, usually fairly close by, that still provide food.

gregarious animals

Mule deer are good examples of a species that may have two or more home ranges, depending on the time of year. Mule deer typically spend the summer at higher elevations, where forage tends to stay greener and more lush for longer. When winter arrives and the snow begins to fall, the deer tend to migrate to lower elevations, where forage is usually easier to find. They establish a home range or territory within each of these areas. Thus, home ranges may be temporary during the course of the year or during a migration.

Some animals tolerate other members of their species to a much greater extent than others do. Geese, for example, are quite **gregarious**, gathering in groups and traveling to their wintering grounds in large flocks (Figure 9–12). Once there, they feed and rest together. In the spring, they even nest in close proximity to one another, but each ferociously defends the area immediately around its nest. In contrast, some animals, most notably large predators such as wolf packs and cougars,

FIGURE 9–12 Snow geese are gregarious birds that form large flocks. (Courtesy of the U.S. Fish and Wildlife Service.)

may not tolerate others of their species within an area of several square miles.

In any habitat, it is important that animals not be overcrowded. If a species is too heavily represented in an area, the resulting competition for food can be a serious problem. Wildlife populations can severely damage their habitats if they are overcrowded. The results can be poor reproductive rates, increased death rates because of malnutrition and stress, and generally poorer health for an entire species.

Another serious problem when wildlife become too numerous is disease. For example, in areas of the south Texas coast where hundreds of thousands of geese spend the winter, an outbreak of a disease such as avian cholera can devastate the population. In any situation in which large numbers of a species are in close proximity, disease spreads rapidly. For these reasons, adequate space is an important component of an animal's habitat.

Animal Growth

Each newborn animal experiences a period of rapid growth in size. This growth is necessary before the animal can mature into an adult of its species. Cells divide to form new cells throughout the lifespan of an organism as old cells die and are replaced. Growth occurs when body cells divide at a faster rate than they die.

embryo growth, cell division
animal reproduction

Cell clusters become specialized in an undeveloped embryo to form different kinds of tissues such as heart or lung tissue. In the process of growth, the cell masses expand the size of the tissue from which they originated. As a result, the baby fawn develops into a mature deer, the gosling grows into a mature goose, and the tadpole becomes a mature frog or toad (Figure 9–13).

As growth occurs in young animals, new demands are placed on the habitat. Large animals require more food and space than small animals. A range that was adequate for the parent animals must be expanded if space is available to do so. Later on, the forces of nature will reduce the population to fit the amount of habitat that is available, but usually an abundance of food is available during spring and summer months. As growth of the young occurs, predators also have high demands for food. Gradually, the predator population reduces the food animal populations. Once this has

FIGURE 9–13 The process of mitosis is highly active in a young gosling as it grows to a mature-sized bird during just a few months.

occurred, predatory animals become stressed for food, and the predator population declines in the area because of migration or starvation.

Animal Reproduction

Animal reproduction is the sexual process by which new individual animals are created. The formation of reproductive cells, also called **gametes**, occurs in both male and female animals. **Fertilization** is the process by which the male and female gametes join together in a single cell. This cell is called a **zygote,** and it is the first cell that is formed in the creation of a new individual.

Once the gametes are formed, other aspects of reproduction in animals are quite different among the many species. These differences between animal species are discussed in greater detail in later chapters of this textbook. In summary, some animals reproduce by laying eggs. In some instances, the eggs are fertilized inside the body of the female; in other instances, the eggs are fertilized outside her body. Birds and reptiles are examples of reproduction in which eggs are fertilized inside the body. Frogs, toads, salamanders, and fish are examples of animals whose eggs are fertilized outside the female's body.

Mammals give birth to live young that have developed from eggs fertilized inside the female's body. The sperm from the male is deposited inside the reproductive tract of the female in close proximity to the unfertilized eggs. When the timing and conditions are favorable, the sperm cells migrate to the egg and penetrate it to complete the process of fertilization.

Snakes and sharks are examples of animals in which the fertilization of eggs occurs inside the female's body (Figure 9–14). In some instances, the females lay eggs before they hatch; in other instances, the eggs hatch inside the body of the female and live young are born. In addition to these examples of differences in animal reproduction, there are many other unusual behaviors exhibited by animals during the reproductive process.

FIGURE 9–14 A snake is an example of an animal whose eggs are fertilized internally.

Arrangement

Last, but certainly not of least importance, is the arrangement of habitat components. As stated earlier, all wildlife requires food, water, cover, and suitably arranged space. If these components are present in a habitat, regardless of their arrangement, then that habitat will almost certainly be used by wildlife. However, optimal populations will not be achieved unless the food, water, cover, and space are appropriately arranged. Let's take a covey of bobwhite quail as an example. The birds have food, escape cover, water, and space all within a reasonable distance of one another. However, their primary feeding areas are located across an area of native pasture that is not sufficiently dense or tall to protect them from overhead predators. Such a covey will likely suffer a high death rate because of the number of birds that end up as hawk dinners.

Another example, which is much more likely to occur these days, is a habitat fragmented or split up by malls, road construction, subdivisions, and parking garages. On the outskirts of an imaginary midsize American town we'll call Wildlifeville is a lake of several acres. Two hundred yards away is an old-growth forest of several acres. Numerous species of wildlife use this habitat. The mud around the pond is thick with the tracks of deer, rabbit, and great blue heron, just to name a few, but the majority of the tracks belong to the large population of raccoons that live in the wooded area and come to the pond to feed every night. The raccoons thrive on the small fish, frogs, and shellfish they catch in the pond's shallows.

Wildlifeville is growing, however, and city officials become alarmed at the volume of traffic traveling through their city. A four-lane loop around town is the solution. When it is finished, it splits the lake from the woods. Because their main food source is now located across a four-lane highway, there is an immediate rise in the number of "road-killed" raccoons (Figure 9–15). This example shows how the arrangement of food, water, cover, and space is a crucial part of wildlife habitat.

FIGURE 9–15 Cities and the transportation arteries that link them often fragment what was once good wildlife habitat. (Photo courtesy of the U.S. Department of Agriculture's Natural Resources Conservation Service.)

WILDLIFE RELATIONSHIPS

Every type of wildlife is part of a community of plants and animals in which all individuals are dependent on others. Any attempt to manage wildlife must take into account the naturally existing relationships, because they are constantly changing within the wildlife community, and it is difficult to set standard procedures for their management.

The balance of nature is actually a myth, because wildlife communities are seldom in a state of equilibrium. The numbers of various species of wildlife are constantly increasing and decreasing in response to each other and many external factors such as natural disasters. These include fires, droughts, and disease outbreaks. The interference of humans often upsets sensitive relationships in nature.

Among the most important natural relationships that exist in the wildlife community are parasitism, mutualism, predation, commensalism, and competition.

Parasitism

The relationship between two organisms, either plants or animals, in which one feeds on the other without killing it is called **parasitism**. Parasites may be either internal or external. An example of a parasitic relationship is the wood tick, which lives on almost any species of **warm-blooded animal**—that is, animals that have the ability to regulate their body temperatures.

parasitism, examples
mutualism, examples

Mutualism

In **mutualism**, two types of animals live together for the mutual benefit of both. There are many examples of mutualism in the wildlife community. Tick pickers are African birds that remove and eat ticks from many of the continent's wild animals to their mutual benefit. For the wild animals, the parasites are removed; for the birds, the ticks provide nourishment. Other examples include a moth that lives and feeds only on

FIGURE 9–16 Predatory animals play an important role in nature by keeping populations of rodents, birds, and other animals from expanding beyond the capacity of their environments to provide food and shelter for them. (Courtesy of the U.S. Fish and Wildlife Service. Photo by Pedro Ramirez.)

a certain plant but is the only pollinator of that plant and certain plant seeds that germinate only after passing through the digestive tract of a specific bird or animal.

Predation

When one animal eats another animal, the relationship is called **predation** (Figure 9–16). Predators are often critical in controlling wildlife populations. Foxes, for example, are needed to keep populations of rodents and other small animals under control. Populations of predators and prey tend to fluctuate widely. When predators are abundant, prey becomes scarce because of overfeeding. When prey becomes scarce, predators may starve or move to other areas, permitting the population of the prey species to increase again.

Commensalism

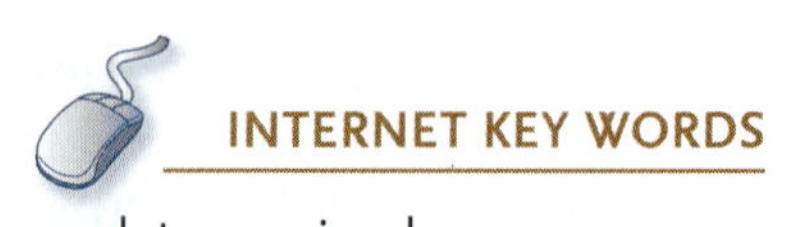

predatory animals
commensalism, examples

In **commensalism**, a plant or animal that is in, on, or with another, sharing its food but neither helping nor harming it. The commensalist species is helped, but the other is unaffected. Vultures waiting to feed on the leftovers from a cougar's kill is an example of commensalism.

Competition

When different species of wildlife compete for the same food supply, cover, nesting sites, or breeding sites, **competition** exists. Competition may exist between two or more species that share the same resources. It also exists among members of the same species, especially when food or shelter is in short supply or during a mating season (Figure 9–17). When

FIGURE 9–17 Competition among wildlife species helps keep animal populations in balance. (Courtesy of the U.S. Fish and Wildlife Service. Photo by John D. Wendler.)

competition exists among species, one species may increase in number, whereas the others decline. Often the numbers of both species decrease as a result of competition. For example, owls and foxes compete for the available supply of rodents and other small animals.

The various relationships that exist among species of wildlife make it necessary to consider more than just one species any time that management is contemplated. Understanding the relationships that exist in the entire wildlife community is essential if wildlife management programs are to be successful.

Preserving and Restoring Wildlife Populations and Habitats

The settlement and colonization of North America by European immigrants opened new frontiers to the European nations. They saw an abundance of natural resources and raw materials in the New World that eventually attracted new industries. Forests were cut to provide homes and fuel, land was cleared, swamps were drained to produce crops, and wild animals were harvested to supply food and clothing.

This pattern continued as settlements moved west. Industries were established to process raw materials, and vast acreages of land were converted from wildlife habitat to farms (Figure 9–18). Many wildlife populations were reduced, and some animals became extinct as land was converted for new uses. Some species were completely eliminated from the ranges that they occupied before human settlement of the land.

Wildlife habitats can never be restored as they were when the Native Americans were the only people who occupied the land. Herds of bison will never roam the Great Plains as they once did, and the rivers will probably never have the abundance of fish they once held. All of this is in our past. Some wildlife habitats and populations can be restored, however, and those that still exist can be preserved.

Figure 9–18 As industries' needs for raw materials increased, so did the conversion of wildlife habitat to farms. (Courtesy of Michael Dzaman.)

The U.S. Endangered Species Act

Congress passed legislation in 1969 that protected animal species whose numbers were declining. The act was expanded in 1973 to require the U.S. Fish and Wildlife Service to identify species of animals and plants that might become **extinct** because of the death of the entire population. The act identifies two classes into which those species that are found to be at risk may be placed. Those in immediate danger of extinction are classed as **endangered species**. These are the plants and animals that have small numbers in their population. In many cases, a species' population also is becoming smaller throughout most or all of the range. Species that are in less danger of extinction, but which are at risk of becoming endangered, are classed as **threatened species**. These species can reasonably be expected to survive if immediate steps are taken to protect the environment in which they live.

The Endangered Species Act protects both the species and its habitat. A *habitat* is defined as the environment in which an organism or living creature makes its home and from which it obtains its food. Species that are protected under this legislation cannot be hunted or killed without heavy legal penalties being assessed. The act also protects the organism's habitat from development or other disruptive uses by humans. Restricted use of the land area where protected species live is considered to be necessary to prevent the delicate balances of nature from being destroyed.

endangered species list

Several proven methods are available to restore fish and wildlife to suitable habitats. They include transplanting birds or other animals from areas where they are abundant to areas where populations have been depleted or no longer exist. Some of the most successful of these programs have moved species such as elk, bighorn sheep, and wild turkeys to locations where they were no longer found or where they never existed.

Many different species of fish are raised in hatcheries for the purpose of transplanting them to streams, rivers, and lakes (Figure 9–19).

FIGURE 9–19 Many species of fish are currently being raised in hatcheries for the purpose of supplementing populations of wild fish.

science connection profile

CALIFORNIA CONDOR RECOVERY EFFORT

One of the most interesting recovery efforts ever undertaken by scientists is the effort to save the California Condor from extinction (Figure 9–20). Condor eggs are removed from the nests of wild breeding pairs, and the young birds are raised in captivity. They are fed by a human using a puppet that resembles a condor parent from the time of hatching until they learn to gather their own food. This practice was adopted to keep them from imprinting on human parents. Several birds have been returned to the wild with mixed results. Some have been recaptured because they were not adapting well to life in the wild. Others appear to be adjusting, but long-term survival may prove difficult for them.

FIGURE 9–20 The endangered California condor has been preserved from extinction, at least for now, through the recovery efforts of science. (Courtesy of the U.S. Fish and Wildlife Service.)

This is done to supplement wild fish populations and to provide adequate numbers of fish for sport fishing. Large numbers of game birds also are raised each year for release into the wild for hunting. Birds and fish that are raised in captivity are released with the intent of harvesting them. This is because they are likely to have a hard time surviving without the skills that wild fish and birds have learned.

Some endangered species of birds are raised in captivity with the intention of adding them to wild flocks. The people who care for young birds must take steps to prevent them from **imprinting** on humans. Normally, imprinting is a learning process by which young animals learn to mimic the behavior of a parent or trusted caregiver to establish a beneficial behavior pattern such as recognizing and being attracted to its own kind. In this case, it is important that birds raised in captivity learn to recognize and be attracted to birds of their own species and not to humans. In other words, if they are to survive in the wild, they must learn to act naturally like their species, not like the people who raised them.

INTERNET KEY WORDS

California condor recovery
wildlife environment

Human Impacts

Humans affect wildlife habitat and therefore wildlife in many ways—more than we can cover in this chapter. However, we discuss here the most obvious human impact on our world. *Habitat destruction* is the single greatest threat facing wildlife today and for the foreseeable future. Habitat destruction results from a variety of human activities, including construction, farming, mining, and timber harvesting. All are related to human encroachment on wildlife habitat. Pollution of the Earth's air, water, and land is certainly a serious threat.

Humans have always used natural resources and wildlife for their own purposes. We developed our food plants from wild plant varieties, and our domestic breeds of livestock and poultry came from wild animals and birds. Therefore, our very existence is possible because of wild plants and animals; most of our food and clothing can be directly linked to them. Wildlife and other natural resources also are valuable to humans for recreational and other purposes. Wild animals bring pleasure to humans, and they add to our enjoyment of the world we share. It is difficult to imagine a world in which the songs of birds did not exist or where wild creatures were gone from the landscapes.

Habitat Loss

Increasing human populations require more and more living space each year. As human communities spread, wildlife habitat is reduced. This **encroachment** has serious effects on wildlife communities. We are converting farmland into subdivisions, supermarkets, office complexes, and shopping centers at an alarming rate. It is estimated that 2 million acres of farmland are lost in the United States each year because of human expansion. Much of this land would be excellent wildlife habitat, so as a result of human encroachment, wildlife often takes a double hit. Faced with this annual loss of productive acreage and an increasing demand for food, farmers are forced to find additional land to bring into production. To meet our needs for food and fiber, farmers then often utilize land once used exclusively by wildlife.

The North America wetlands are some of our most productive wildlife "factories." They support an astounding variety of wildlife and are often the only habitat in which many species are found. The native forests and prairies have also been drained, harvested, and plowed. Wildlife habitat will never be the same as it was before it was converted to serve human needs.

As noted in Chapter 8, *wetlands* refers to a variety of wet environments, including wet meadows, ponds, fens, bogs, coastal and inland marshes, wooded swamps, mudflats, and bottomland hardwood forests. These areas have traditionally been looked on as having little **economic value**. In fact, for more than 200 years, Americans have drained and developed these areas, often directed by laws and programs that were designed to benefit these developments. The results have been devastating to American wetlands (Figure 9–21), with

FIGURE 9–21 Millions of acres of wetlands have been lost to refineries and other forms of coastal development. (Photo courtesy of Texas Parks and Wildlife Development, copyright 2002.)

FIGURE 9–22A This is a healthy prairie pothole wetland area. (Courtesy of the U.S. Fish and Wildlife Service.)

FIGURE 9–22B This area was once similar to the one shown in Figure 9–22A, but this wetland has been ditched and drained for farming purposes. (Courtesy of the U.S. Fish and Wildlife Service.)

more than 100 million acres of wetlands lost, or more than half of all American wetlands.

Perhaps our most valuable wetlands are the prairie potholes of southern Canada, Iowa, Minnesota, Nebraska, North and South Dakota, and northern Montana. Thousands of these shallow depressions were created after the most recent ice age as glaciers withdrew from these areas. These potholes filled with water and became major breeding grounds for North American waterfowl. These inland marshes remain important nesting areas for many waterfowl, especially ducks, and provide food, water, and cover for a great variety of wildlife (Figure 9–22A). Their importance to hundreds of species cannot be overemphasized. Unfortunately, these prairie potholes are spread across much of our most productive farmland and many hundreds, perhaps thousands, have been drained and turned into cropland (Figure 9–22B).

In the past few years, however, people have realized what great value these wetland areas hold. These precious **ecological resources** are extremely important to wildlife, slow the destructive power of floods and storms, purify polluted waters, and provide a variety of recreational opportunities. Over the past 20 years, our change of attitude toward wetlands has resulted in federal and state laws and programs designed to protect and preserve wetlands.

The dominance of humans in the food web has contributed to the struggle that often surrounds the movement to maintain and restore natural environments. Clearly, humans constitute a highly competitive species that tends to exploit natural resources to benefit itself. It is equally evident that using a particular natural resource in ways that interfere with its renewal will eventually destroy it.

In response, extreme positions have been taken on both sides of these issues. Some people contend that human dominance of other species of organisms is a natural process that has evolved since the beginning of human existence, and, as the dominant species, humans have the right to exploit all other living organisms. Other people recognize humans as

science connection profile

ENVIRONMENTAL CLEANUP THROUGH BIOTECHNOLOGY

One of the most remarkable scientific developments in the last 20 years has been bio-engineered bacteria. Such organisms have been developed for a variety of purposes, especially breaking down accidental spills of crude oil. These spills from ships have caused serious damage to marine animals and environments, and the bacteria are important tools in reducing environmental damage. The bacteria ingest the oil and convert it to a form that is more compatible with and less damaging to the environment.

the only one among the living organisms that is capable of changing its behaviors to preserve other species. They believe that humans have the moral obligation to protect all other forms of life.

INTERNET KEY WORDS

biotechnology, environment

The Principle of Stewardship

The principle of stewardship as it applies to natural resources implies that the agencies and people responsible for managing wildlife and other natural resources should perform their duties with a long-term view and commitment to the resources that are being managed. The ability to exercise good stewardship requires the knowledge of proven management practices and the wisdom and skills to implement them before habitats or resources are damaged. It also requires knowledge of the ecology and habitat requirements of an area's plant and animal communities.

Wise stewardship occurs when managers of natural resources make management decisions based on dependable information. This requires them to distinguish facts from fantasies and dependable research from biased research. It also requires that researchers know what questions to ask in order to get the research information needed to make sound management decisions. Wise managers will adjust their management practices to protect natural resources against uses that cause damage to them. Ultimately, the principle of stewardship requires managers to use resources in ways that will preserve them for their children and grandchildren.

Extinction and Its Causes

The extinction of a species of organisms is not something to be taken lightly. The diversity of species is considered to be an indicator of a healthy environment. Humans have become a dominant species because they are able to adapt to nearly any environment and also are predators toward many of the animals and birds with which they share the environment. They frequently disrupt the habitats on which other species depend.

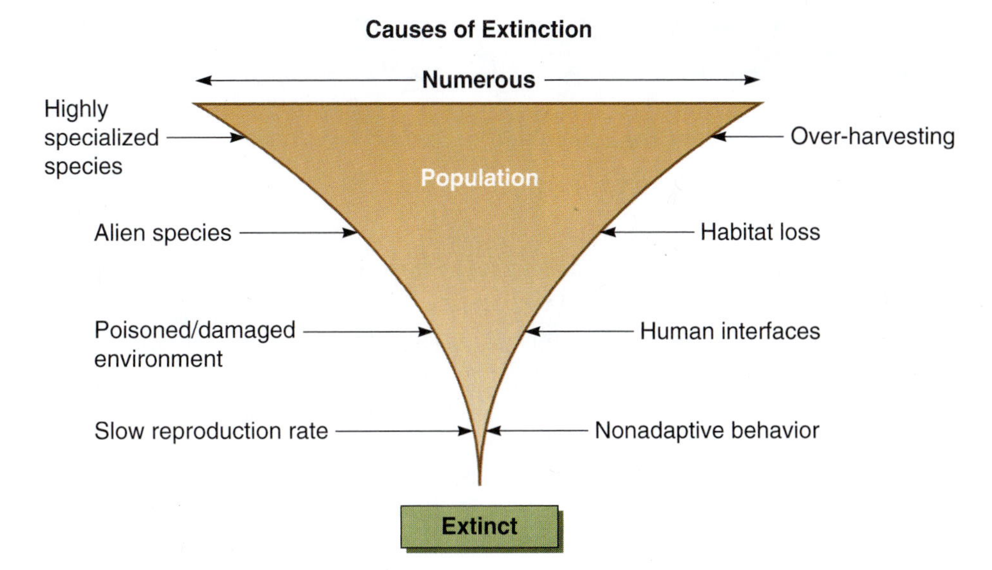

FIGURE 9–23 Many factors can lead to the extinction of species.

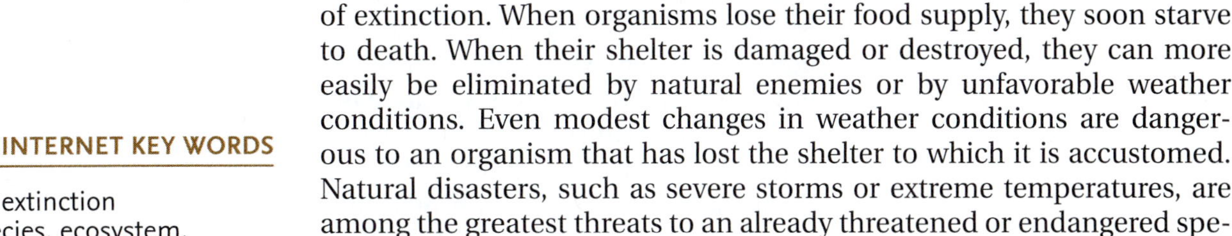

INTERNET KEY WORDS

cause of extinction
alien species, ecosystem, disease

Destruction or modification of a habitat is the greatest single cause of extinction. When organisms lose their food supply, they soon starve to death. When their shelter is damaged or destroyed, they can more easily be eliminated by natural enemies or by unfavorable weather conditions. Even modest changes in weather conditions are dangerous to an organism that has lost the shelter to which it is accustomed. Natural disasters, such as severe storms or extreme temperatures, are among the greatest threats to an already threatened or endangered species of organisms.

Many organisms are unable to adapt quickly to changes in environmental conditions or to heavy losses from predators. They decline in numbers as their environments are modified or as predators increase. A single cause or event can result in the extinction of some plants or animals, but a species is more likely to become extinct by a combination of factors that affect it in a negative way (Figure 9–23).

The introduction of an **alien species** into an ecosystem is one important factor that sometimes leads to the extinction of an established species. The new species may compete with the native species for food and shelter, or it may prey on the native species as a source of food. When this happens, the balance in the ecosystem is upset, and the weaker species tends to decline as the newly introduced species increases in number. Alien species are sometimes sources of new diseases. Entire populations of organisms have been lost because of such diseases.

Still another major factor that contributes to a species' extinction is overhunting by humans. In some instances, this has been done commercially; in other instances, sport hunting has contributed to the extinction of a vulnerable species.

The degree of specialization in a species also affects how vulnerable it is to extinction. Species that cannot adapt their behaviors or their diets to accommodate changes in their environment are at the greatest risk because a highly specialized mammal or bird may depend on a single

Interest Profile

THE EXTINCTION OF THE PASSENGER PIGEON

The passenger pigeon was the most numerous bird species in the world in the 1800s. Nearly 2 billion of these birds were reportedly in a single flock observed by ornithologist Alexander Wilson in Kentucky. A combination of negative factors led to its loss. It was intensely hunted for food, and a large market for passenger pigeon meat developed in population centers in the eastern United States, which was made possible by expansion of the railroads. Destruction of the trees that provided nesting habitats and harvesting of young birds from densely populated nesting areas also played a role in reducing their numbers.

Passenger pigeons lived in large flocks, which made them vulnerable to disease and predation. In addition, a pair of passenger pigeons only produced one egg and raised a single offspring during each nesting period. The last known passenger pigeon died in a zoo in Cincinnati in 1914.

The lesson to be learned from the passenger pigeon is that even the most numerous species can become extinct when their habitats are destroyed and hunting is not restricted.

source of food or shelter. When that particular source of food or shelter is gone, a highly specialized bird or animal will probably be unable to adjust and face extinction. Failure of a species to adapt to a changing environment is called **nonadaptive behavior**.

Most surviving species in the world today are able to adapt to modest and gradual changes in their environments. Abrupt changes in an environment, however, allow no time for living organisms to adjust, and in some cases adjustment is not possible.

A slow rate of reproduction contributes to extinction by reducing the recovery rate of an endangered species. Biologists refer to this problem as low **biotic potential** or **fecundity** (Figure 9–24). Examples of animals and birds that fall into this class include the California condor, which lays only one egg every two years, and the whooping crane, which takes several years to reach breeding age and then lays only two eggs per year. More successful species of birds often lay 8–10 eggs or nest several times per season.

passenger pigeon

FIGURE 9–24 The trumpeter swan is an endangered species partly because of a low biotic potential.

FIGURE 9–25 Nesting boxes are an innovative way to restore shelter in critical management situations.

Managing Endangered and Threatened Species

Humans are the only species that can make conscious decisions to destroy or preserve the other forms of life. They also have a moral responsibility to preserve the other organisms with which they share the environment. People in some parts of the world take this stewardship seriously, but other cultures place little value on preserving threatened and endangered species. Even in societies that accept responsibility for protecting these organisms, there are strong differences of opinion over how much protection should be provided.

One of the most difficult problems facing fish and wildlife biologists is managing the surviving members of a population that is found to be endangered or threatened. They must learn to identify and understand relationships between the organism and its environment. Relationships of this kind are often difficult to define because the species' preferred habitat may no longer exist.

Effective management of endangered species of organisms must be based on reliable research. Those who are responsible for the recovery of endangered or threatened species should explore as many alternative management strategies as they can identify. Innovative ways of restoring acceptable shelter and providing appropriate food sources have been successfully used in critical management situations (Figure 9–25).

management
endangered species

Approved Practices in Wildlife Management

Farm Wildlife

Wildlife management on most farms is usually a by-product of farming or ranching. It is often given little attention by the farmer or rancher, except when wildlife causes crop damage and financial loss.

Much of farm wildlife management involves providing a suitable habitat for living, growth, and reproduction. This may involve leaving some unharvested areas in the corners of fields, planting fence rows with shrubs and grasses that provide winter feed and cover, or leaving brush piles when harvesting wood lots.

The timing of various farming operations is also important in a farm wildlife management program. Crop residues should be left standing over the winter to provide food and cover. Planting crops attractive to wildlife on areas that are less desirable as cropland is an excellent practice. Providing water supplies for wildlife during dry periods is often necessary to maximize the numbers of farm wildlife on the area being managed.

Extensive research has shown that harvesting farm and ranch wildlife by hunting has little impact on spring breeding populations. Excess populations of farm and ranch wildlife that are not harvested by humans usually die during the winter. Even heavy hunting pressures seldom result in severe damage to wildlife populations. The sale of hunting rights to hunters is a way to increase the income of many farms and ranches. In addition, it often means the difference between profit and loss for a farming enterprise.

Wildlife management on game preserves or farms set up specifically for hunting often differs drastically from other management programs. Species of animals and birds that are not native to the area are sometimes raised and released on the preserve. Native wildlife species may also be raised in pens and released to the farm or preserve expressly for harvest by hunters.

Forest Wildlife

wildlife population balance

The types and numbers of forest wildlife in any specific woodland depend on many factors, including the type and age of the trees in the forest, density of the trees, natural forest openings, types of vegetation on the forest floor, and the presence of natural predators.

Forest wildlife management usually aims to increase the numbers of desired species. If such desired populations of wildlife are present, then the management goal is usually to maintain them. Sometimes, however, the numbers of certain forest wildlife species increase to the point where habitat destruction occurs. When this happens, control measures may have to be instituted to restore proper balance. First, the steps in developing a forest wildlife management plan should include taking an inventory of the types and numbers of wildlife living in the managed forest area. Second, goals for the use of the forest and the wildlife living in it need to be developed. Third, the management plan must determine the types and populations of wildlife that the forest area can support and how best to manage the forest so that required habitat is provided (Figure 9–26).

Forest wildlife requires food, water, and cover, and these must be readily available to the desired species at all times. Management practices that meet these requirements include making clearings in forests so that new growth produces twigs for deer to feed on, selectively harvesting trees of various ages to make a more suitable habitat for squirrels and many other species of forest wildlife, leaving piles of brush for food and cover, and preventing the contamination of existing supplies of water.

Deer, grouse, squirrels, and rabbits are the forest wildlife species that are usually targeted for management because they are valuable

FIGURE 9–26 Managing wildlife in forest environments is often difficult because many species of plants and animals occupy the living environment. Interactions among all of the organisms in the environment must be considered in developing the management plan. (Courtesy of the U.S. Department of Agriculture.)

FIGURE 9–27 Deer management is a key activity of fish and game agencies. Management units are studied carefully to determine the size of the deer population and to set hunting season regulations. (Courtesy of Getty Images.)

for recreational purposes, especially hunting (Figure 9–27). They may also be managed to prevent the destruction of valuable forest trees and other products.

Notably, during times of overpopulation of forest animals, especially deer, it is seldom a good idea to provide supplemental food. Natural losses should nearly always be allowed to occur, including starvation of excess animals or allowance of heavier-than-normal hunting pressures. Artificial feeding of wildlife populations usually results in further population increases and an extension of the problem.

Wetlands Wildlife

No area of U.S. land is more important to wildlife than wetlands, which include any land that is poorly drained: swamps, bogs, marshes, and even shallow areas of standing water (Figure 9–28). Wetlands constantly change as wet areas fill in with mud and decaying vegetation. They eventually become dry land that contains forests.

Wetlands provide food, nesting sites, and cover for many species of wildlife. Ducks and geese are probably the most economically important type of wildlife that depends on the wetlands for survival (Figure 9–29). Other types of wetland wildlife include woodcock, pheasants, deer, bears, mink, muskrats, raccoons, and many other lesser known species.

The management of wetlands for wildlife may include impounding water. Open water areas should occupy approximately one-third of the wetlands for optimal wildlife use. The depth of the standing water should not be more than 18 inches.

The management of the plant life in the wetlands is also important. This may include cutting trees to open up the wetland area. Many species of wildlife require large, open areas to thrive. Care must be taken to leave hollow trees; these are used as nesting sites for some wildlife species. Wetland areas can also be opened up by killing excess trees rather than cutting them. This provides resting areas for many types of wetlands wildlife.

wetlands, wildlife

Establishing open grassy areas around wetlands and planting millet, wild rice, and other aquatic plants in the wetlands also helps attract many types of wildlife to the area.

FIGURE 9–28 No area of U.S. land is more important to wildlife than wetlands. (Courtesy of the Chesapeake Bay Foundation.)

FIGURE 9–29 Wetlands are important nesting areas for millions of ducks and geese in the United States. (Courtesy of the U.S. Fish and Wildlife Service.)

Pollution presents a serious hazard to wetlands wildlife. Polluted water flowing into the wetlands area may come from agriculture, industry, or domestic wastes. Because pollutants are trapped in the mud and silt of the wetlands, the effects of pollutants are often long-term.

In areas that lack natural nesting sites, populations of some wildlife species can be greatly increased by providing artificial nesting sites. Wooden duck boxes, old tires, and islands surrounded by open water provide safe nesting sites for many species of wetlands wildlife (Figure 9–30).

Raising certain species of ducks in captivity and later releasing them in wetlands areas has helped in maintaining viable duck populations. This has been important as many natural duck nesting areas have been destroyed to meet human needs.

Stream Wildlife

Stream wildlife can be divided into two general categories: warm water and cold water. These categories are based on the water temperatures at

FIGURE 9–30 Artificial nesting sites are beneficial to many kinds of birds. Wood ducks would benefit from the artificial nesting site shown here. (Courtesy of Cameron Waite.)

FIGURE 9–31 Riparian areas, which include stream banks, can be protected by fencing livestock out. Such practices protect stream environments and the animal and plant life found there.

which the wildlife, primarily fish, can best grow and thrive. There is little or no difference in the management practices between the two categories of wildlife. In general, fish are the type of stream wildlife for which management plans are developed, although many other types of wildlife also depend on streams for their existence.

As land is developed, forests are harvested, and civilization is expanded, streams and their wildlife populations come under increasing pressure. Because we cannot build new streams, it is essential that existing ones be managed properly.

Management practices for streams include preventing stream banks from being overgrazed by livestock. Fencing the stream to limit access by livestock is also wise to reduce pollution and the destruction of stream banks (Figure 9–31).

Good erosion-control practices on lands surrounding streams are important to help maintain clear, clean water. They are also important in preventing silt or chemical pollutants from entering streams. Maintaining stream-side forestation is important in regulating stream temperatures during warm summer months because some fish species stop feeding and may even die when stream temperatures become too high. The amount of dissolved oxygen in warm water is also much lower than it is in cold water. Without adequate oxygen, aquatic wildlife dies.

Anything that impedes the flow of a stream also serves to change it—to the detriment of many species of stream wildlife and to the benefit of other species. Trout thrive only in swiftly moving cool water, whereas catfish are adapted to sluggish streams. Care must be taken to maintain desirable species of wildlife in the stream. Introducing new species in a stream may reduce native wildlife.

fish population control

Maintaining the population levels of stream wildlife that are in balance with the available food supply is important. Too many fish for the available food supply normally results in stunted fish that are of no value to fishermen. This situation provides an increased food supply for some types of birds and animals that use streams for their food supply.

Overfishing of predatory species of fish, such as bass or northern pike, may allow perch or sunfish to overpopulate the stream and become stunted. Often the only way to restore streams to a desired mix of fish species is to remove the unwanted species. This is accomplished by netting, poisoning, or electric shocking. These techniques are legal only for authorized officials and should be done only by specially trained personnel.

The artificial rearing and stocking of desired species of stream wildlife is a management practice that is important in many streams. Typically, game species of fish are stocked in streams for fishermen to catch and remove. Often few or no fish survive to reproduce, and stocking must take place each year.

The regulation of sport fishing is often necessary to maintain desirable populations of game fish. This may include closed seasons, minimum size limits, creel limits, and restricted methods of catching fish.

Lakes and Ponds Wildlife

The management practices for wildlife in lakes and ponds are usually similar to those for managing stream wildlife (Figure 9–32). Pollution must be controlled. Wildlife populations must be managed to maintain a desired mixes of species. Harvest and use must also be controlled to ensure wildlife for the future.

There are, however, some differences between managing wildlife in streams and managing wildlife in ponds and lakes. Because the water in lakes and ponds is normally standing, the amount of oxygen available for aquatic life sometimes becomes critical during the hot summer months. In small ponds, incorporating oxygen into the water artificially may prevent fish deaths.

Water temperatures in lakes and ponds are more variable than they are in streams, which means that different species of fish are usually dominant in ponds and lakes. In many ponds and lakes, fish populations are predominantly large-mouth bass and sunfish.

When it is necessary or desirable to rid a pond or lake of unwanted species of fish such as carp, it is often much easier to do so because the

FIGURE 9–32 Ponds and lakes can accommodate fish and wildlife as well as provide family recreation. (Courtesy of the Chesapeake Bay Foundation.)

water is contained. Sometimes it is possible to drain a body of water to remove the unwanted species. More often, however, the body of water is simply poisoned so that all fish and other species of pond wildlife are killed. The pond is then restocked with desirable wildlife species.

Managing wildlife is an imprecise business. Some species of wildlife are often managed to the detriment of others. It is reasonably clear that any human interference in the wildlife community results in changes that are not always to the benefit of much of that community.

Looking Back

Some of the principles of zoology is that all animals have basic needs that are required to sustain life. Some of these needs are food, water, shelter, and space. The differences that we observe in the many species of animals make it possible for them to live in a wide variety of environments. An instinctive behavior is one that is evident at birth, such as the tendency of a duckling to go into the water and swim. All wildlife species need water. Water acts as a solvent to absorb food particles and transport the nutrients to the cells of the body as the blood circulates. The need for a favorable living environment is greatest during extreme conditions in the weather.

Wildlife requires food, water, cover, and space, all suitably arranged, for their survival. The numbers of various species of wildlife are constantly increasing and decreasing in response to each other and external factors such as fires, drought, and disease. Some of the natural relationships that exist in the wildlife community include parasitism, mutualism, predation, commensalism, and competition.

The U.S. Fish and Wildlife Service is responsible for identifying species of animals and plants that might become threatened, endangered, or extinct. Effective management of endangered species of organisms must be based on reliable research. Much of the management of wildlife involves providing a suitable habitat for living, growth, and reproduction.

No area of U.S. land is more important to wildlife than the wetlands. Wetlands provide food, nesting sites, and cover for many species. The maintenance of population levels of wildlife that do not exceed the available food supply is important.

Self-Analysis

Essay Questions

1. How is the science of biology different from zoology?
2. What are the basic needs of all wildlife species?
3. Distinguish between an instinctive behavior and a learned behavior.
4. Describe the anatomical differences in the digestive tracts of animals having a simple stomach, a rumen, and a gizzard.
5. What uses do wild creatures make of water?
6. How does shelter contribute to an animal's ability to survive?
7. What are two of the most critical times when wild animals benefit most from shelter?

8. Explain how growth occurs in a young animal.
9. In what ways do the arrangements of basic need components within a habitat affect an animal's ability to survive?
10. What are the similarities and differences between the wildlife relationships known as parasitism, mutualism, predation, commensalism, and competition?
11. Identify some factors that are known to contribute to the extinction of wild creatures.
12. Explain how the U.S. Endangered Species Act might be used to protect species of organisms that are in danger of becoming extinct.
13. Speculate on the importance of human ethics in preventing abuses to wildlife and other natural resources.

Multiple-Choice Questions

1. The branch of biology that studies animal life is
   a. limnology.
   b. zoology.
   c. silviculture.
   d. botany.
2. The suckling ability of a newborn mammal is an example of which animal behavior?
   a. learned
   b. experiential
   c. instinctive
   d. manipulative
3. An animal that has canine teeth is most likely to eat a diet of
   a. grasses and herbs.
   b. meat.
   c. insects.
   d. worms and grubs.
4. A ruminant animal is most likely to eat a diet of
   a. grasses and herbs.
   b. meat.
   c. insects.
   d. worms and grubs.
5. A wildlife population that is forced to feed, water, or travel too great a distance from its escape cover is likely to encounter a high rate of
   a. mortality.
   b. growth.
   c. reproduction.
   d. survival.
6. A learned behavior that improves the chances for a wild animal to survive is also known as
   a. manipulative behavior.
   b. nonadaptive behavior.
   c. erratic behavior.
   d. adaptive behavior.
7. When two species of wildlife live together for the benefit of both, the relationship is called
   a. mutualism.
   b. parasitism.
   c. commensalism.
   d. competition.
8. Landowners, managers, and users are responsible not only to their own generation but also to all future generations for the care and management of the land and water resources. This is an example of
   a. nonadaptive behavior.
   b. low biotic potential.
   c. good stewardship.
   d. the Endangered Species Act.
9. The greatest single cause of extinction is
   a. the destruction or modification of habitat.
   b. overharvesting.
   c. a nonadaptive behavior.
   d. an alien species.

10. The U.S. Endangered Species Act defines an *endangered species* as
    a. organisms that are declining in population but can reasonably be expected to survive if immediate steps are taken to protect their habitat.
    b. a population whose members have all died.
    c. a species that is at risk of becoming extinct because of small numbers in the population in all or most of its range.
    d. a species that threatens the safety of humans.
11. Low biotic potential is defined as
    a. the failure to adapt well to a changing environment.
    b. a slow reproductive rate.
    c. a high susceptibility to poisons in the environment.
    d. the production of large numbers of offspring each year.
12. Which of the following is a characteristic of an endangered species?
    a. low biotic potential
    b. ability to eat more than one kind of food
    c. easy adaptation to the presence of humans
    d. easy adaptation to new habitats

Learning Activities

1. With your class divided into two groups, debate the issues that arise when the Endangered Species Act is invoked. One group should debate in favor of restricting the use of resources that are part of the environment of an endangered species. The second group should defend the rights of people who depend on those resources to earn a living.
2. With the United States (or the world) divided into geographic regions, form teams of students to research the species in their regions that are considered to be threatened or endangered. Obtain a list of endangered and threatened species found in your state or region from the U.S. Fish and Wildlife Service site on the Internet at <http://endangered.fws.gov/wildlife.html>. Describe the factors that contribute to the problem and offer solutions for restoring the populations that are in danger of extinction. Each team should report back to the class.
3. Invite a professional from a government agency who works with endangered and threatened species to talk to the class about the criteria for listing or removing a species on the endangered or threatened species list. Request that the presentation address ways that the criteria are assessed. Be sure to select presenters who will present both sides of the issues.
4. Each class member should write a code of ethics for some limited number of outdoor sports (two or three). Examples might include "A Code of Ethics for Deer Hunters" or "A Code of Ethics for Off-Road Vehicles." Write a single paragraph of three to four sentences then combine the papers together for each sport. Use the ideas from the papers to write a single code of ethics for each outdoor sport.
5. Divide your class into two groups. One group will play the part of the U.S. Senate. The other group will play the role of the U.S. House of Representatives. Each group will write a bill to regulate the use of motorized vehicles on land managed by the U.S. Forest Service. When each group has passed a bill, two or three students from each group will meet to create a compromise bill that is acceptable to both houses of Congress. Print a final copy of this bill for each class member.

section Three

The Human Impact

Chapter 10
Agriculture and Sustainability

Chapter 11
Integrated Pest Management

Chapter 12
Population Ecology

Chapter 13
Waste Management

Chapter 14
Fossil Fuels

Chapter 15
Energy and Alternative Fuels

Chapter 16
Toxic and Hazardous Substances

Chapter 17
Careers and Environmental Science

TERMS TO KNOW

- domestic
- stewards
- toxic waste
- sustainable agriculture
- economy of scale
- stewardship

CHAPTER 10

Agriculture and Sustainability

Objectives

After completing this chapter, you should be able to

- discuss relationships between wild and domestic plants and animals
- explain why domestic plants and animals have developed physical traits that are different than those of their wild ancestors
- identify some agricultural practices that sometimes conflict with maintaining healthy living environments
- describe how the conversion of wildlife habitat to farmland has affected wild animal populations
- designate how leftover agricultural chemicals should be disposed of to prevent damage to the environment
- suggest ways in which range improvements for domestic livestock contribute to improved wildlife environments
- identify problems that affect wildlife when domestic livestock are not properly managed on rangeland
- define sustainable agriculture practices
- discuss the effects of human population growth on the environment
- suggest ways in which agricultural policies could be modified to encourage the agricultural industry to adopt practices favoring the environment
- discuss ways in which intensive farming practices contribute to maintaining natural environments

Foundation for Agriculture

Agriculture would not exist without wildlife, plants, and other natural resources. The first **domestic** plants and animals were obtained from wild populations. They were tamed or adapted to cultivation by human ancestors and used to provide dependable sources of food and clothing. Later, draft animals were trained to pull farm implements. Wild plants were eventually raised near the homes of the people who used them. This concentration of food plants in the area made it easier to harvest them.

The domesticated plants and animals used by farmers and ranchers today are quite different from their early ancestors (Figure 10–1). Meat animals were selectively bred to express the genetic traits that are required to produce desirable meat in a short time. Dairy goats and cattle were selected for their ability to produce large amounts of milk, and different genetic lines of poultry were selected to produce eggs and meat.

Selective breeding, however, paid less attention to vigor—that is, resistance to disease and general overall health—in domestic animals because they were protected from severe conditions. The vigor of their wild relatives—animals and plants—results mostly from their ability to survive in the environments in which they live. As a result, wild animals and birds do not produce as much meat, eggs, or milk as domestic breeds; however, they have the genetic ability to withstand severe conditions in their natural environments.

Scientists are examining the genetic makeup of many wild species of animals and plants in their search for natural resistance to diseases and parasites. Wild species have considerable potential genetic value for domestic species of plants and animals. Genetic engineers are

FIGURE 10–1 Through selective breeding, the shape and genetic makeup of meat animals is quite different from their early ancestors. (Courtesy of Michael Dzaman.)

FIGURE 10–2 Domestic species of animals, such as the young chicken that is being examined in this photo, have all descended from wild ancestors. (Courtesy of the U.S. Department of Agriculture's Agricultural Research Service.)

likely to find many important traits among wild species of animals and plants that can be transferred to domesticated species (Figure 10–2).

Farmers and Ranchers

Although a public perception exists that modern agricultural practices are not compatible with a healthy environment, the perception is often false (Figure 10–3). Some conflict does exist between natural environments and agriculture, but most farmers and ranchers provide some protection for the wildlife on their property. Agricultural practices that bring them into conflict with wildlife include tillage practices that lead to soil damage, improper uses of pesticides and other agricultural

FIGURE 10–3 Spraying of pesticides and other agricultural chemicals must be properly managed to avoid damage to public lands and wildlife habitats. (Courtesy of the U.S. Department of Agriculture.)

FIGURE 10–4 Extreme overgrazing such as the pasture on the left imposes major erosion hazards on the land.

INTERNET KEY WORDS

agricultural chemicals, wildlife
soil erosion, habitat, wildlife

INTERNET KEY WORDS

soil conservation
water conservation
pesticides, health effects

chemicals, poor grazing practices on public lands, and conversion of wildlife habitats to fields (Figure 10–4).

Erosion was discussed earlier as a destructive process that occurs when land is unprotected against the forces of flowing water or strong winds. Soil erosion is the leading source of water pollution in North America. It damages wildlife populations by polluting water supplies, killing young fish and aquatic animals, and filling in reservoirs and lakes. It also destroys terrestrial habitats and hinders plant growth, which ultimately has negative effects on the food chain. Soil conservation is the practice of protecting soil from the destructive forces of wind and water (Figure 10–5). Such practices are important to farmers and ranchers because abused land eventually loses the ability to produce profitable yields of crops.

FIGURE 10–5 No-till corn coming up through stubble in Maryland. (Courtesy of the U.S. Department of Agriculture's Soil Conservation Service.)

FIGURE 10–6 It seems unreasonable to believe that farmers would recklessly endanger their own water supplies by being careless with agricultural chemicals. Their own children would be the first to suffer from such abuses. (Courtesy of the U.S. Department of Agriculture.)

Farmers and ranchers are **stewards** of most of the privately owned land and much of the public land in North America—that is, they are administrators and supervisors who manage land resources. Most of the potential farmland in North America is now tilled, and it has been converted from wildlife habitat to the production of crops and livestock. As these changes in land use have occurred, animals such as waterfowl and bison have been displaced, while other wild species such as songbirds have expanded into these regions. Fields of growing crops, however, still provide excellent habitats for many birds and some small animals.

Most agricultural producers are sensitive to the need to manage their property in a responsible manner because they are likely to lose more than anyone else if the soil is damaged or lost because of poor management practices. Farmers operate farm businesses that are just as sensitive to economic losses as businesses in towns and cities.

Agricultural Chemicals

The improper use of agricultural chemicals poses the greatest danger to farm families. Abuses of toxic materials are carefully avoided by most farmers. They live on the land with their families and drink the water from their own farm wells. Their children would be the first to suffer from chemical abuses because they work and play in the fields (Figure 10–6). Many farm families still eat meats, fruits, and vegetables that are homegrown, and the improper use of chemicals would surely affect these foods.

INTERNET KEY WORDS

agricultural chemicals, safe use

There is no great conspiracy by farmers and ranchers to enrich themselves at the expense of wildlife and other human beings by boosting crop yields through the use of poisonous chemicals. The chemicals that are used in agriculture have been tested carefully, and they can be safely applied when used according to the manufacturer's directions (Figure 10–7). Agricultural chemicals that are left over from a job are considered to be **toxic waste** because they are poisonous to living organisms. Laws prescribe how they should be properly handled and disposed of. These materials should be safely stored until they can be delivered to a toxic waste treatment center, or until they can be properly degraded. Chemical abuses must be corrected for the safety of living plants, animals, and people.

INTERNET KEY WORDS

land stewardship

It would be naive to believe that farmers and ranchers never abuse land and water resources, but it would be just as wrong to accuse them of destroying the environment with wanton disregard for their own families, neighbors, customers, and resident wildlife populations. The real issue is to find solutions to our environmental problems.

Educational programs and government regulations that are designed to ensure the safe use of agricultural chemicals are being implemented in most regions of the United States and Canada. There is an increased awareness of the need to follow chemical labels carefully. Empty chemical containers are still a threat to the environment, however, so major efforts have been made to educate those who use agricultural chemicals about the best ways to dispose of used containers.

FIGURE 10–7 When applying chemicals, one must carefully follow the manufacturer's directions to ensure that they are used safely.

Land Use

Those who use land resources are responsible for ensuring that it thrives wherever and whenever possible. For that reason, ranchers who graze livestock on public lands are required to make improvements to the

FIGURE 10–8 Livestock owners who lease government-owned rangelands are required to make improvements such as developing water sources for livestock and wildlife.

range. In this way, water sources have been developed for use by livestock and wildlife populations (Figure 10–8) and new varieties of grasses have been seeded on many ranges. These grasses produce much more forage than the native species they replaced, and they are equally nutritious to wild and domestic ruminant animals (Figure 10–9).

Public lands are managed for both wildlife and domestic cattle and sheep. Scientific research has demonstrated that these uses are compatible with one another. The key to proper rangeland management is to graze an area quickly and then remove the animals from the area to allow the plants to build up their food reserves. Plants that are not allowed to build reserves in their roots are more easily killed during conditions

FIGURE 10–9 Reseeded rangelands are capable of higher levels of production than some of the grasses they have replaced.

of extreme drought or cold. Range management also must consider seasonal variations in plant growth and the different effects of grazing by cows versus sheep versus wild species. Overgrazing is a condition in which domestic livestock or wild animals destroy the plants in an area by harvesting them beyond their ability to recover.

With the exception of drought years, rangelands that are harvested in the summer by livestock will generally produce new growth in the late summer and fall. This forage is then available as winter feed for deer and elk, and it is usually of high quality. When spring growth is not harvested, it tends to become coarse, unpalatable, and of low nutrient value.

Farming practices that damage wildlife populations include the tillage of every acre of land and the drainage of swamps and other wetland areas. Windbreaks, wooded areas, farm ponds, ditches, and fencerows provide habitat for birds and other small animals. Undisturbed grasses and weeds in these areas provide shelter and food for wildlife. As fields have become larger, however, these areas have disappeared on many farms. For this reason, wildlife agencies and organizations are encouraging farmers to skip over small areas as they harvest their grain fields and to leave some areas untilled over the winter season. These small refuges provide winter cover for birds and other animals, helping them to survive on agricultural lands.

Those agricultural producers who fail to use wise management practices that are friendly to the environment and wildlife populations are placing the entire agricultural industry at risk. A single farmer who applies chemicals improperly, thereby causing injuries or death to wildlife populations, is likely to create public sentiment against all farmers, including those who use chemicals safely. One rancher who is lax in his or her stewardship over a grazing allotment on public land can place all ranchers in jeopardy of losing their grazing privileges because publicity tends to focus on problems instead of successes. Agricultural organizations should censure their own members when they are negligent to avoid the abuses of public trust that lead to the loss of privileges.

sustainable agriculture

SUSTAINABLE AGRICULTURE

The term **sustainable agriculture** refers to agricultural practices that will not deplete or damage the land resources on which this and future generations depend for food and fiber. The concept of sustainable agriculture has developed from the policies, beliefs, and practices of people and organizations who represent many walks of life. It is a shared vision for maintaining the ability of agriculture to supply the needs of a growing population of humans while recognizing that we must maintain an environment that is safe for people and other living organisms.

FIGURE 10–10 Horses and other draft animals have long been used to provide power for humans' machines. (Courtesy of the Utah Agricultural Experiment Station.)

Sustainable agriculture is based on compromise and regard for the needs of all life forms. For most of recorded history, agriculture remained much the same. Most humans were engaged in producing their own food and fiber using draft animals to pull their farm implements (Figure 10–10). Much of the work of raising and harvesting crops and caring for domesticated livestock was performed by human

FIGURE 10–11 Manual labor was once required in almost every agricultural job. (Courtesy of Utah Agricultural Experiment Station.)

labor (Figure 10–11). History records little progress in agriculture for thousands of years. Then, starting in the early 1800s, the use of iron spurred inventions that revolutionized agriculture in the United States, the British Isles, and northern Europe (Figure 10–12). Following World War II, new technologies brought a new round of changes to agriculture. As a result, farm production dramatically increased, and the amount of human labor for each unit of production declined.

Some of the changes in agriculture have affected the Earth and its environments in positive ways; others have been negative. On the positive side, the people of the United States enjoy an abundant and relatively safe food supply. A negative impact, however, is that groundwater in many areas has become contaminated by nitrates and phosphates that have leached into the water supply from excess fertilizers. The topsoil continues to be eroded, partly because of insufficient plant cover during some seasons of the year. Social issues include loss of family farms and a cherished way of life from economic pressures to increase farm size to gain greater production efficiency, a tendency toward creating what is known as an **economy of scale**. Much of the hand labor on farms is provided by migratory or immigrant farm laborers, and one social concern that arises is the substandard quality of housing and the poor living conditions these workers and their families endure.

Principles of Sustainable Agriculture

Environmental Health

Healthy environments are vital to the well-being of the human race, as well as to every other living organism. If we permit our living environments to be degraded to the point where they no longer provide the resources needed to sustain life, it will not be just the plants and animals that will suffer. Humans will also find it extremely difficult to survive in damaged environments. Among the environmental issues that must be resolved is how to provide food and shelter to the growing human population as the land base for food production declines. Cities and towns are building on the very farmland that is needed to provide food and fiber products for their citizens.

The world's human population today is approximately 6.5 billion people (Figure 10–13). How long will it take for the population to reach 7 billion? Somewhere, a new child will have the distinction of being the

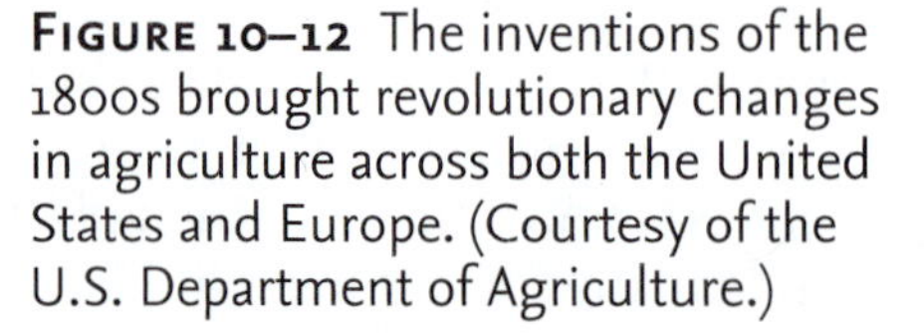

FIGURE 10–12 The inventions of the 1800s brought revolutionary changes in agriculture across both the United States and Europe. (Courtesy of the U.S. Department of Agriculture.)

FIGURE 10–13 The population of the world is expected to approach 7 billion by 2010. (Courtesy of the D.C. Committee to Promote Washington.)

FIGURE 10–14 What kind of life on planet Earth will "Baby 7 Billion" have? (Courtesy of the National Aeronautics and Space Administration.)

7 billionth human being living on planet Earth (Figure 10–14). What will the home and community be like where "Baby 7 Billion" is born? Will that child be warm, but not too warm or too cold? Will there be sufficient food? Will the child be kept free from serious illness? Will his or her family have a house or good living space they call home? Will they have clothing to permit them to live and work outside the home in relative comfort? What will be the quality of life of others around the 7 billionth human being? Will that child survive, and will he or she go on to live a happy life? Positive answers to these questions would indicate a good environment for a person.

All members of the plant and animal kingdoms, including humans, must share the living environments that are available on Earth. Some living environments are more friendly to their inhabitants than others, and plant and animal life tends to be concentrated in the warm, temperate regions of the world. Other environments, such as the Arctic and Antarctic regions, do not support the variety of plants and animals that are found in other places. The same is true of the high altitude, mountainous areas. One thing is certain: The living environments of the Earth will never grow any larger. In fact, the habitable regions may actually decline for many plant and animal species as humans pave the Earth and create cities. Only a few species are able to adapt to the less favorable environments as they are crowded from their natural ranges.

INTERNET KEY WORDS

human encroachment, natural environments

As natural environments become smaller because of human encroachment, it becomes important that those that remain are healthy living environments that are capable of supporting healthy populations of plants and animals. They need to be protected from pollutants and other influences that are known to be hazardous to the health of living things. Such influences as highway noise or human traffic in an area may make the environment unfit for the life forms that live there.

Unhealthy conditions in living environments are followed by the loss or reduction of populations of native species. Healthy environments do not just happen: They are the result of conscious efforts by the people who are responsible for preserving them for the benefit of future generations. One principle of sustainable agriculture is that agricultural production must be maintained, or even increased, without contributing to the decline of the living environments that are shared by all living organisms.

Economic Profitability

The concept of sustainable agriculture is a major shift in thinking from the years when farmers were encouraged by government policies to plant every available acre from fence to fence in an effort to maximize productivity. Policies that reward farmers and ranchers for making environmental improvements on the lands under their control can potentially accelerate the rate of environmental improvement.

genetic engineering, crops, immunity to pests

Among the ways in which environmental conditions can be advanced is by developing farming methods and new crop varieties that require the use of fewer chemical fertilizers and pesticides while crop yields are maintained. Examples include the use of satellite imagery and global positioning systems to deliver fertilizers and pesticides to the sectors of a field where they are needed and in the exact amounts required. This is accomplished by viewing field images to determine plant densities and field sectors that exhibit stress from weeds, insects, or drought. Chemical inputs are applied only where they are needed and in amounts that are used up by the crop. Another example is the development of genetically engineered crops to impart natural immunity to specific insects that feed on the plant. Such crops can be produced without the use of pesticides.

Such an environmentally friendly farming system reduces costs to the farmer while maintaining yields and improving profitability. The secondary benefit is to the living environment for wild plants and animals. Fewer chemicals are used, and the materials that are added are consumed by the crop, leaving little or minimal residue. The technology that supports these farming practices are now in use. It is possible to increase profitability while improving environmental health.

Social and Economic Equity

Sustainable agriculture includes changes in public policy and societal values. For example, government and institutional policies often define the agricultural practices that are used on farms and ranches. Refocusing the requirements for farm-commodity support payments to include the adoption of environmentally friendly production methods would have an immediate impact on the adoption of new production strategies.

environment, corporate farms

One of the biggest changes in agriculture since World War II has been reductions in the numbers of family farms and increasing numbers of corporate farms. Corporations often tend to focus less on environmental issues and more on the efficiency of production. Such operations are often controlled by absentee owners who do not live in the farm environment and who are unaffected by the environmental issues to which intensive farming practices sometimes contribute. As a result, decisions are often based more on economics than on environmental needs. Incentives in the form of tax breaks or commodity payments for improvements to farm environments would encourage more progress toward healthy environments.

Seasonal labor is often required to harvest and care for crops. A social issue that has developed in the current agricultural structure is the use of substandard housing for these workers. Although this is not often discussed in environmental science terms, it has become an issue in sustainable agriculture discussions, affecting the way people view natural living environments. Furthermore, substandard housing usually has a negative impact on the immediate environment.

FIGURE 10–15 Responsible stewardship must be practiced to maintain habitat requirements of plant and animal communities. (Courtesy of the U.S. Department of Agriculture's Agricultural Research Service.)

Environmental Stewardship

The understanding that we must meet current and future needs for products obtained from natural resources such as soil, water, and forests leads us to the concept of **stewardship** for the environment—that is, the responsible management or care of land, property, or resources—should be practiced by all those who use them. In the context of environmental science, each person who uses natural resources or spends time in a natural environment also has an obligation to preserve it for future generations. A land steward is an administrator or supervisor who manages land resources. Federal, state, and provincial governments own large tracts of land in North America, and they depend on professional managers to oversee these lands in a responsible manner (Figure 10–15).

Farmers and ranchers are stewards of much of the privately owned land and large tracts of public land in North America. Most of the potential farmland in North America is now tilled, converted over two centuries from forests and wildlife habitat to the production of crops and livestock. As these changes in land use occurred, animals such as waterfowl and bison were displaced and other wild species such as songbirds expanded into these regions. Most agricultural producers are sensitive to the need to be good environmental stewards. After all, they are likely to lose more than anyone else if lands or water resources are damaged or lost because of poor management practices.

Agriscience in a Growing World

The keys to a prosperous future—indeed, the bottom line for survival of the world's population—can be found in agriscience, or the science of food production, processing, and distribution. It is the system that supplies fiber for building and construction materials, fabrics for clothing and industry, and medicines. It provides the grasses and ornamental trees and shrubs that beautify our landscapes, protect the soil, filter out dust and sound, and supply oxygen to the air.

By 2005, the average American farmer was able to produce enough food and fiber for approximately 140 people. Agriscience will become even more important in the next 100 years. As the world's population increases, it will require a highly sophisticated agriscience industry to meet all the needs of the world's billions of people. Americans will have to participate more in the international arena, as more countries become highly competitive in agriscience and trade barriers are removed. Research and development will continue to play a dominant role as they lead the way in agriscience expansion in the future.

Agriscience accounts for 20 percent of jobs in the United States. It is the mechanism that permits the United States and other developed countries of the world to enjoy high standards of living. It is the system that nondeveloped countries are using in their efforts to feed and clothe their bulging populations. People look to agriscience for the necessary technology to compete on a par with other nations in the twenty-first century and maintain and even improve our quality of life.

The United States is a major world supplier of food. It is also a major supplier of fiber for clothing and of trees for lumber, posts, piling, paper, and wood products. The use of ornamental plants and acreage devoted to recreation has never been greater in our nation's history.

Impact of Agriscience

impact of agriscience

Agriscience and its attendant technology have brought about rapid and substantial improvements in the yields of farm products. For most of the world, however, progress has been much slower. In some nations, government leaders are slow to implement agriscience because massive unemployment would initially follow as machines displaced human labor.

Improving Plant and Animal Performance

Humans have improved on nature's support of plant and animal growth since they discovered that loosening soil and planting seeds could result in new and better plants. Even before that discovery, they probably aided plant growth by keeping animals away from them until fruits or other plant parts edible to humans were harvested.

The human touch has permitted plants and animals to increase production and performance to the point where fewer numbers of people are needed to produce the American food supply. Surplus food is exported to many other nations.

One of the remarkable occurrences of the twentieth century was the mechanization of agriculture (Figure 10–16). The many technologies

FIGURE 10–16 A conventional seedbed is smooth and free of crop residues. Although it is excellent for seed germination, it is prone to water and wind erosion. (Courtesy of the John Deere Company.)

Interest Profile

PRESERVING THE ENVIRONMENT WITH INTENSIVE FARMING PRACTICES

When the first colonists arrived in North America from Europe, there were abundant land resources available for food production, and trees were removed to make way for the farms. When the world's population approaches 10 billion people midway through the twenty-first century, the forests of the world will be further endangered unless we can increase the food production of current farmland to produce an adequate food supply for a growing world population. To do so, we will need to increase the efficiency of existing farmlands through the responsible use of intensive farming practices (Figure 10–17). If we fail to increase the food production of our land, forests will probably be converted to farms.

Agricultural chemicals contribute to high production of food by controlling weeds and insects. Fertilizers also have been a proven method for sustaining high levels of production. However, good judgment must always be exercised in the application of fertilizers and chemicals to ensure that they are used safely and do not pollute the environment. It is an interesting paradox to consider that farm fertilizers and pesticides may be our best hope for preserving the forests and other natural environments in the world. As many foreign governments can attest, preserving the environment is of low priority to people who are starving.

FIGURE 10–17 An uneven-aged population of trees is the result of selecting only the most mature trees in a forest for harvesting.

that were developed for the agricultural industry led to farms becoming larger. Many people, however, have been displaced from family farms because they did not adopt the new technologies and farming practices that were needed to make their farms more efficient in a timely manner. Many of these people have learned trades other than farming and have become productive citizens in other industries.

Without the farming revolution that occurred after World War II, Americans would not have been free to pursue other occupations. The efficiency of U.S. farms has contributed to our freedom to engage in many new and exciting occupations, including the development of computers and other technologies that have resulted in the current information age.

Looking Back

Agriculture would not exist without wildlife, plants, and other natural resources from which domestic plants and animals were derived. When tillage practices lead to soil damage or when fertilizers and pesticides are applied improperly or in excessive amounts, agricultural practices conflict with the goal of maintaining healthy environments. Erosion is another serious threat to the living environments of aquatic animals. Some animals and birds are able to thrive in farm environments, but less-adaptable animals have been pushed out of their original habitats by farms, towns, and cities.

Those who use natural resources need to become good stewards by considering how agricultural practices might affect the environment. *Sustainable* agriculture refers to agricultural practices that will not deplete or damage the land resources on which this and future generations depend for food and fiber. The growing human population continues to encroach on farmland and natural environments, yet the shrinking agricultural land base must continue to provide adequate yields of food and fiber to meet the needs of the nation and the world. New advances in agricultural production methods will help to provide these commodities. Farm fertilizers and pesticides may actually be our best hope for preserving the forests and other natural environments in the world.

Self-Evaluation

Essay Questions

1. How did agriculture evolve from the natural environments and the living organisms that were found in them?
2. What impacts do modern farming methods have on ecosystems in North America?
3. What are some agricultural practices that are sometimes in conflict with healthy living environments?
4. Explain how the development of land for agriculture purposes has affected the native species of plants and animals.
5. Why are leftover agricultural chemicals considered to be toxic waste?
6. What are some ways in which range improvement projects have benefited wild animals?
7. Identify some agricultural practices that are used to improve habitat for wild animals and birds.
8. What is sustainable agriculture and how can practicing its concepts lead to improvement in natural environments?
9. Explain why healthy living environments are as important to humans as they are for wild animals and native plants.
10. Define stewardship and explain how it relates to greater compatibility between modern farming and healthy living environments for wild creatures.

11. How has the farming revolution of the past 60 years affected the growth of other occupations in North America?
12. Explain how intensive farming practices may be our best hope for preserving forests and other natural environments of the world.

Multiple-Choice Questions

1. The term that describes a farm animal in contrast to a wild animal is
   a. feral.
   b. primitive.
   c. ferrous.
   d. domestic.
2. A practice that protects soil from the destructive forces of wind and flowing water is
   a. soil conservation.
   b. erosion.
   c. siltation.
   d. sedimentation.
3. Over the next 20 years, the productive farmland near population centers in the United States is expected to
   a. increase in acreage.
   b. decrease in acreage.
   c. stabilize in acreage.
   d. decline in value.
4. Which of the following is not an approved way to deal with agricultural chemicals?
   a. Take leftover chemicals to a hazardous waste station for disposal.
   b. Dilute leftover materials with water before pouring them out.
   c. Carefully follow the directions for use that are printed on the container.
   d. Avoid spraying chemicals when windy conditions prevail.
5. Where do the greatest concentrations of plants and animals exist on Earth?
   a. polar zone
   b. temperate zone
   c. arid zone
   d. mountainous zone
6. The average American farmer produces enough food and fiber each year for
   a. 63 people.
   b. 140 people.
   c. 92 people.
   d. 211 people.

Learning Activities

1. Invite a resource specialist to instruct the class on the correct procedure for conducting an environmental impact study. (The names of such people can be obtained from government agencies such as the Environmental Protection Agency, Bureau of Land Management, Forest Service, and so forth.) Choose an area near your school where your class can conduct a limited study of the environmental impacts that might be expected if a subdivision or other development were to be constructed there. Use this exercise to see how developing land affects the suitability of the area as wildlife habitat.
2. Contact the local Soil Conservation Service district office and request the help of its professionals in locating an area with severe soil erosion. Take a field trip to the area and observe the damaged site. Measure the depth of the soil layers and compare your findings to a soil map of the area. Discuss the ways in which the soil might be managed to prevent further topsoil losses.

TERMS TO KNOW

integrated pest management (IPM)
element
compound
inorganic compound
disease
vector
insect
arachnid
defoliate
weed
pathogen
annual weed
biennial weed
perennial weed
rhizome
node
stolon
meristematic tissue
noxious weed
exoskeleton
entomophagous
plant disease
causal agent
disease triangle
drift
vapor drift
cultural control
clean culture
trap crop
biological control
cultivar
chemical control
pesticide resistance
pest resurgence
pheromone
organic compound
quarantine
targeted pest
eradication
key pest
pest population equilibrium
economic threshold level
monitoring

CHAPTER 11

Integrated Pest Management

Objectives

After completing this chapter, you should be able to

- describe the multipronged approach to pest control known as integrated pest management (IPM)
- identify the role of natural enemy species in the IPM approach to pest control
- explain why pest-management strategies are needed
- list the types of pests that afflict agriculture and identify a class of pesticide that is commonly used to control each
- explain how the major pest groups adversely affect agricultural activities
- describe both the beneficial and detrimental roles that insects play
- explain the role of chemical pesticides in an IPM program
- identify the three factors involved in the disease triangle
- list the natural resources that are most vulnerable to pollution that results from improper pesticide use
- describe some pest-control alternatives that have been proven to be effective tools
- identify the key elements of successful IPM programs.

integrated pest management

A revived approach to controlling insects and other pests such as weeds or rodents is gaining acceptance in the agriculture and natural resource industries. **Integrated pest management (IPM)** is a multipronged attack on harmful insects and other pests that has proven to be a practical approach to pest control (Figure 11–1). IPM is a method of controlling harmful pests and providing protection to useful organisms such as honeybees. It is proving to be a more practical approach to insect and pest control than separately applied cultural, chemical, genetic, legal, and biological control methods. The IPM approach is effective because it uses a variety of strategies to control insects and other pests, but it also allows for the use of chemical pesticides when the targeted pest population begins to get out of control.

Integrated pest management is not a new idea. It was widely used before the introduction of modern pesticides, and it has emerged in recent years as the best alternative to a complete reliance on pesticides. Integrated pest management is an ecosystem approach to controlling insect problems, taking into account the effects that a particular form of pest control might have on other living things found in an ecosystem.

An IPM program depends on the use of natural enemies and other forms of control to reduce harmful pest populations. Whether the pests consist of insects, weeds, or rodents, among others, the approach to controlling the pest is the same. Total destruction of harmful insects is

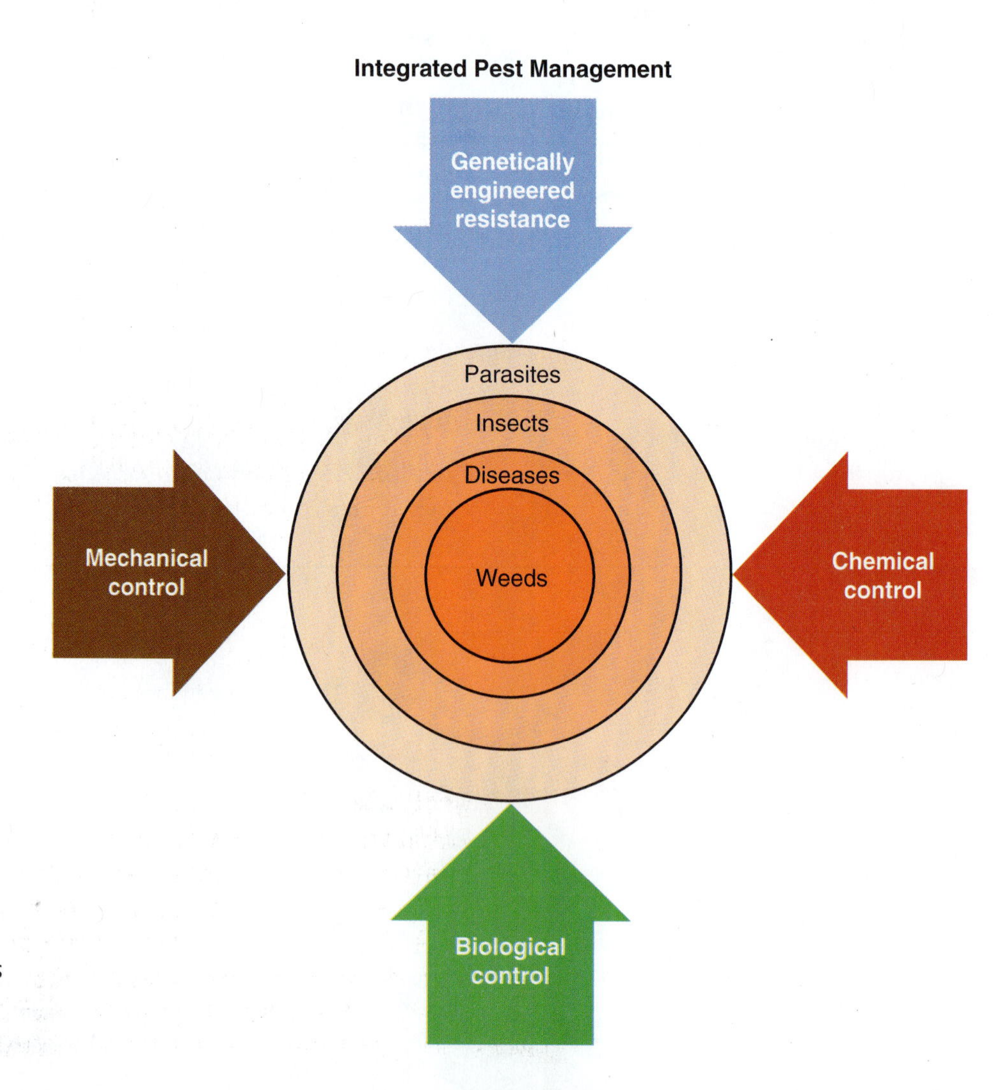

FIGURE 11–1 The most effective form of pest management uses a variety of pest-control methods and is the most acceptable form because it does minimal damage to nontarget species.

not the objective of IPM. Instead, this kind of control seeks to keep some of the pests alive as food sources for their natural enemies. An IPM program strives to establish a natural balance between harmful pests and their natural enemies at population levels that are low enough to minimize insect damage.

Since the 1970s, great strides have been made in the development and implementation of IPM programs. The end result has been to reduce dependency on chemical use while still achieving acceptable pest control.

History of Pest Management

The use of chemicals to control pests is not new. Elements such as sulfur and arsenic were among the first chemicals used for this purpose. An **element** is a uniform substance that cannot be further decomposed by ordinary chemical means. In approximately 1000 B.C., Homer wrote that sulfur had pest-control ability. Around A.D. 900, the Chinese discovered the insecticidal properties of arsenic sulfide, a chemical **compound** or substance that is composed of more than one element.

By the early 1900s, entomologists (scientists who study insects) had developed an array of cultural and natural controls for the boll weevil and other insect pests. Until the late 1930s, however, pest-control chemicals or pesticides were mainly limited to **inorganic compounds** (that is, compounds that do not contain carbon). Examples of other inorganic pesticides are mercury and Bordeaux mixture. Used for plant disease control, Bordeaux mixture is a combination of copper sulfate and lime.

During the period from 1940 to 1972, our approach to pest management moved to a major reliance on chemical pesticides. Alternate control strategies were deemphasized because chemical control gave excellent results at a low cost. In the middle 1960s, however, American biologist Rachel Carson shocked the world with *Silent Spring*, one of the first books to provide convincing evidence of the environmental damage being done by pesticides. After 1962, heavy reliance on chemical pest management began to be questioned. Carson's book created a public awareness of the environmental pollution that can result from the overuse of pesticides. Adverse effects from pesticide misuse or overuse were beginning to occur as well. These effects included pest resurgence, resistance to pesticides, and concern over human health from exposure to pesticides. Not until 1972 did a major change in policy occur in the United States to encourage other pest-control strategies. Biological and cultural control programs began to be introduced as alternatives to chemical pest control.

DDT, effects
natural pesticides

In 1972, dichlorodiphenyltrichloroethane (DDT) was banned in the United States because of its damaging effects on the environment. This insecticide had been used to control mosquitoes, which carried the dreaded malaria organism, among other diseases. DDT was also a highly effective chemical used against flies, and it enjoyed widespread use in homes, on farms and ranches, and wherever flies were a problem. Because it was determined that DDT was responsible for interfering with the reproduction of birds by weakening their egg shells, it had to be discontinued; safer substitutes have since been found.

FIGURE 11–2 This opened cotton boll shows a pick bollworm and the damage it has done. (Courtesy of the U.S. Department of Agriculture.)

FIGURE 11–3 These pear fruits yellowed and shriveled when the first blight disease cut off the flow of nutrients from the tree to the fruit. (Courtesy of the U.S. Department of Agriculture's Agricultural Research Service.)

The Need for Pest Management

The ability to control pests by chemical, cultural, or natural control methods has afforded people in the United States an unprecedented standard of living. We often take for granted an unlimited food supply, good health, a stable economy, and an aesthetically pleasing environment. Without effective pest-control strategies, however, our standard of living would decrease.

INTERNET KEY WORDS

vector, disease, examples
potato blight disease, Ireland

Good pest-management practices have resulted in dramatic increases in yield for every major crop. A single U.S. farmer in 1850 could only support himself and four people; now a farmer can provide food and fiber for approximately 140 people. The ability to control plant and animal **diseases** or disorders *vectored* by insects and arachnids has reduced the incidence of malaria, typhus, and Rocky Mountain spotted fever. A **vector** is a living organism that transmits or carries a disease organism. An **insect** is a six-legged animal with three body segments. An **arachnid** is an eight-legged animal, such as a spider or a mite.

The impact of pest management in maintaining a stable economy can be seen regionally and nationally. The regional Southern economy suffered shortly after the cotton boll weevil's introduction into the United States in 1892, devastating much of the cotton crop in the early 1900s (Figure 11–2). Similarly, the potato blight disease in Ireland caused famine and mass emigration of Irish citizens to other parts of the world in 1845. Blights remain serious threats to our crops (Figure 11–3).

Types of Pests

The word *pest* is a general name for any organism that may adversely affect human activities. We may think of an agricultural pest as one that competes with crops for nutrients and water, tends to **defoliate** plants (that is, strips a plant of its leaves), or transmits plant or animal diseases. The major agricultural pests are weeds, insects, and plant diseases, but other types of pests also exist. Examples and the classes of pesticides or chemicals used for killing them are listed below.

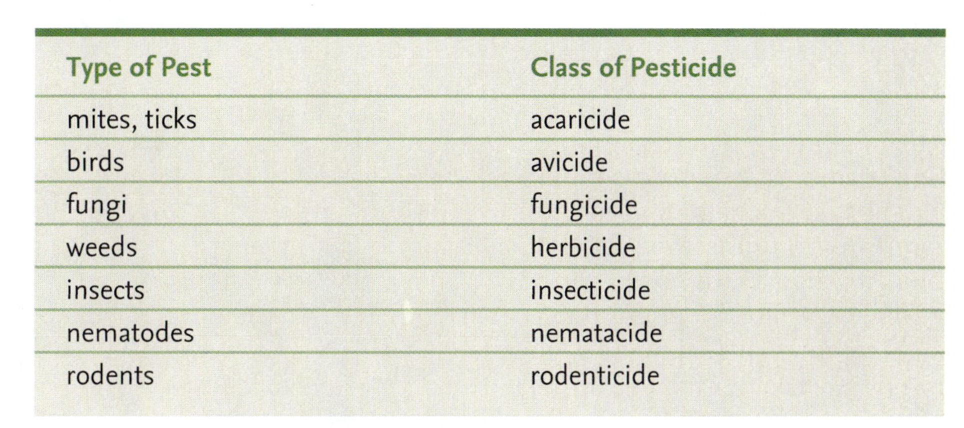

Type of Pest	Class of Pesticide
mites, ticks	acaricide
birds	avicide
fungi	fungicide
weeds	herbicide
insects	insecticide
nematodes	nematacide
rodents	rodenticide

INTERNET KEY WORDS

weed
weeds, annual, biennial, perennial
noxious weed

Damage by pests to agricultural crops in the United States has been estimated to be one-third of the total crop production potential. An understanding of the major pest groups and their biology is thus required to ensure success in reducing crop losses caused by pests.

Weeds

Weeds are plants that are considered to be growing out of place (Figure 11–4). Such plants are undesirable because they interfere with plants grown for crops. The definition of a weed is therefore a relative term. Corn plants growing in a soybean field or white clover growing in a field of turfgrass are examples of weeds, just as crabgrass is considered to be a weed when it grows in a lawn or garden.

FIGURE 11–4 Different types of plants considered to be weeds. (Courtesy of Maryland Cooperative Extension Service.)

Weeds can be considered undesirable for any of the following reasons:

- They compete for water, nutrients, light, and space, resulting in reduced crop yields.
- They decrease crop quality.
- They reduce aesthetic value.
- They interfere with maintenance along rights-of-way.
- They harbor insects and disease **pathogens** (organisms that cause disease).

Weeds can be divided into three categories—annual, biennial, and perennial—based on their life spans and their periods of vegetative and reproductive growth.

Annual Weeds

An **annual weed** is a plant that completes its life cycle within one year (Figure 11–5). Two types of annual weeds occur, depending on the time of year in which they germinate. A *winter annual* germinates in the fall and actively grows until late spring. It then produces seed and dies during periods of heat and drought stress. Examples of winter annuals are chickweed, henbit, and yellow rocket.

A *summer annual* germinates in the late spring and grows vigorously during the summer months. Seeds are produced by late summer, and the plant will die during periods of low temperatures and frost. Examples of summer annuals are crabgrass, spotted spurge, and fall panicum.

FIGURE 11–5 An annual weed is a plant that completes its life cycle within one year. Yellow mustard is a common example.

FIGURE 11–6 The bull thistle is a widespread example of a biennial weed.

INTERNET KEY WORDS

rhizome stem, picture
stolon stem, picture

Biennial Weeds

A **biennial weed** is a plant that will live for two years (Figure 11–6). In the first year, the plant produces only *vegetative growth:* as leaves, stems, and root tissue. By the end of the second year, the plant will produce flowers and seeds. This is referred to as *reproductive growth.* After the seed is produced, the plant will die. Only a few plants are considered biennials. Examples are bull thistle, burdock, and wild carrot.

Perennial Weeds

A **perennial weed** can live for more than two years and may reproduce by seed or vegetative growth (Figure 11–7). By producing rhizomes, stolons, and an extensive rootstock, perennial plants reproduce vegetatively. A **rhizome** is a stem that runs underground and gives rise to new plants at each joint or **node**. A **stolon** is a stem that runs on the surface and gives rise to new plants at each node. These plant parts have **meristematic tissue** (that is, tissue that is capable of starting new plant growth). Examples of perennial weeds are dandelion, Bermuda grass, Canada thistle, and nutsedge.

FIGURE 11–7 An example of a perennial weed is hoary cress or white top.

Noxious Weeds

A **noxious weed** is a plant that causes great harm to other organisms by weakening those around it. Most states have developed lists of noxious weeds, and great effort is made to control or eradicate them. Most noxious weeds are difficult to control, and they require extended periods of treatment followed by close monitoring. Noxious weeds should be handled carefully to avoid spreading seeds to unaffected areas. Noxious weeds are often spread when seeds become airborne, fall into flowing streams, become attached to the hair of an animal or to human clothing, or are eaten and distributed by birds. Examples are purple loosestrife, orange hawkweed, leafy spurge, and tansy ragwort.

cotton boll weevil,
insects, beneficial, useful,
harmful, pests, destructive

Insects

Insects have successfully adapted to nearly every environment on Earth. In fact, there are more species of insects than any other class of organism. Part of their success is because of the large numbers of offspring they are capable of producing and the short time they require to reach physical maturity. The human race depends on insects in many ways, and they perform great service to us. Some of the most beneficial insects are the ladybug, preying mantis, parasitic wasps, and honeybees. Insects also cause great losses to crops, livestock, and people by damaging or injuring them.

Insect Pests

When compared with the total number of insect species, relatively few species cause economic loss. Still, it has been estimated that annual crop losses caused by such insects total more than $4 billion.

Insects can cause economic loss by feeding on forests, cultivated crops, and stored products (Figure 11–8). They can also vector plant and animal diseases, inflict painful stings or bites, or act as nuisance pests.

Insect Anatomy

Insects are considered to be one of the most successful groups of animals on Earth. Their success in numbers and species is attributed to several characteristics, including their anatomy, reproductive potential, and developmental diversity.

Insects are in the class Insecta and are characterized by the following similarities (Figure 11–9):

- Each insect has an **exoskeleton**, which is the body wall of the insect. It provides protection and support for the insect.

FIGURE 11–8 Insects called *defoliators* injure or kill trees by feeding on the leaves and needles. (Courtesy of Boise National Forest.)

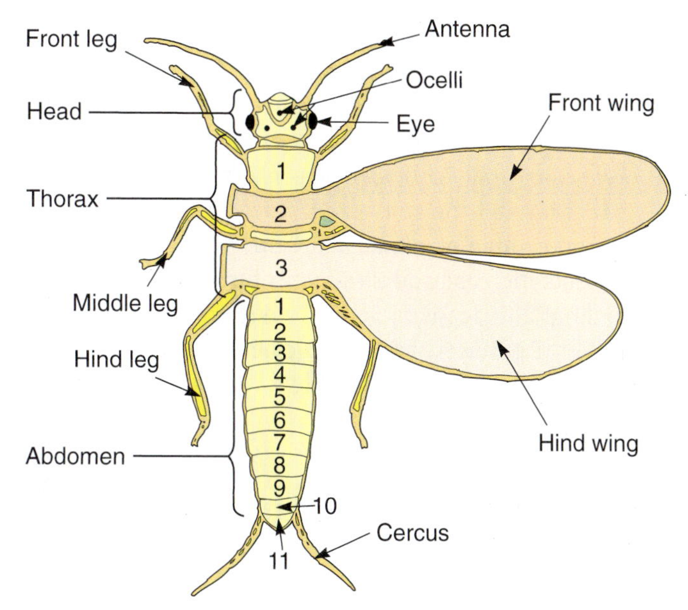

FIGURE 11–9 Diagram of an adult insect.

- The exoskeleton is divided into three regions: head, thorax, and abdomen.
- There are segmented appendages on the head called *antennae* that act as sensory organs.
- Three pairs of legs are attached to the thorax of the body.
- Wings are present (one or two pairs) in the majority of species. This permits mobility and greater use of habitat.

INTERNET KEY WORDS

insects, feeding damage

Feeding Damage

Insects have either chewing or sucking mouthparts. The types of damage caused by chewing insects are leaf defoliation, leaf mining, and stem boring; chewing insects also engage in root feeding. Insects with sucking mouthparts produce distorted plant growth, leaf stippling, and leaf burn (Figure 11–10).

INTERNET KEY WORDS

insect metamorphosis

Rodents

Within their food web, rodents—small mammals such as rats, mice, voles, lemmings, and marmots—fill the role of food animals (Figure 11–11). These animals are equipped with four large incisor teeth. The teeth never stop growing and rodents keep them worn down by gnawing. These animals

science connection profile

BENEFICIAL INSECTS

Scientists estimate that more than 1 million species of insects inhabit the Earth. A majority of them are beneficial or helpful to humans. For example, insects are necessary for plant pollination. In the United States, bees pollinate more than $1 billion worth of fruit, vegetable, and legume crops per year. Honey, beeswax, shellac, silk, and dyes are just a few of the commercial products produced by insects. Many other insects are **entomophagous**—that is, they feed on and naturally control other insect species. Insects that inhabit the soil, act as scavengers, or feed on undesirable plants all play important roles. These insects increase soil tilth, contribute to nutrient recycling, and act as biological weed control agents. Because insects are at the lower levels of the food chain, they support higher life forms such as fish, birds, animals, and humans.

By pollinating plants, honeybees provide an essential and invaluable service to humans and animals.

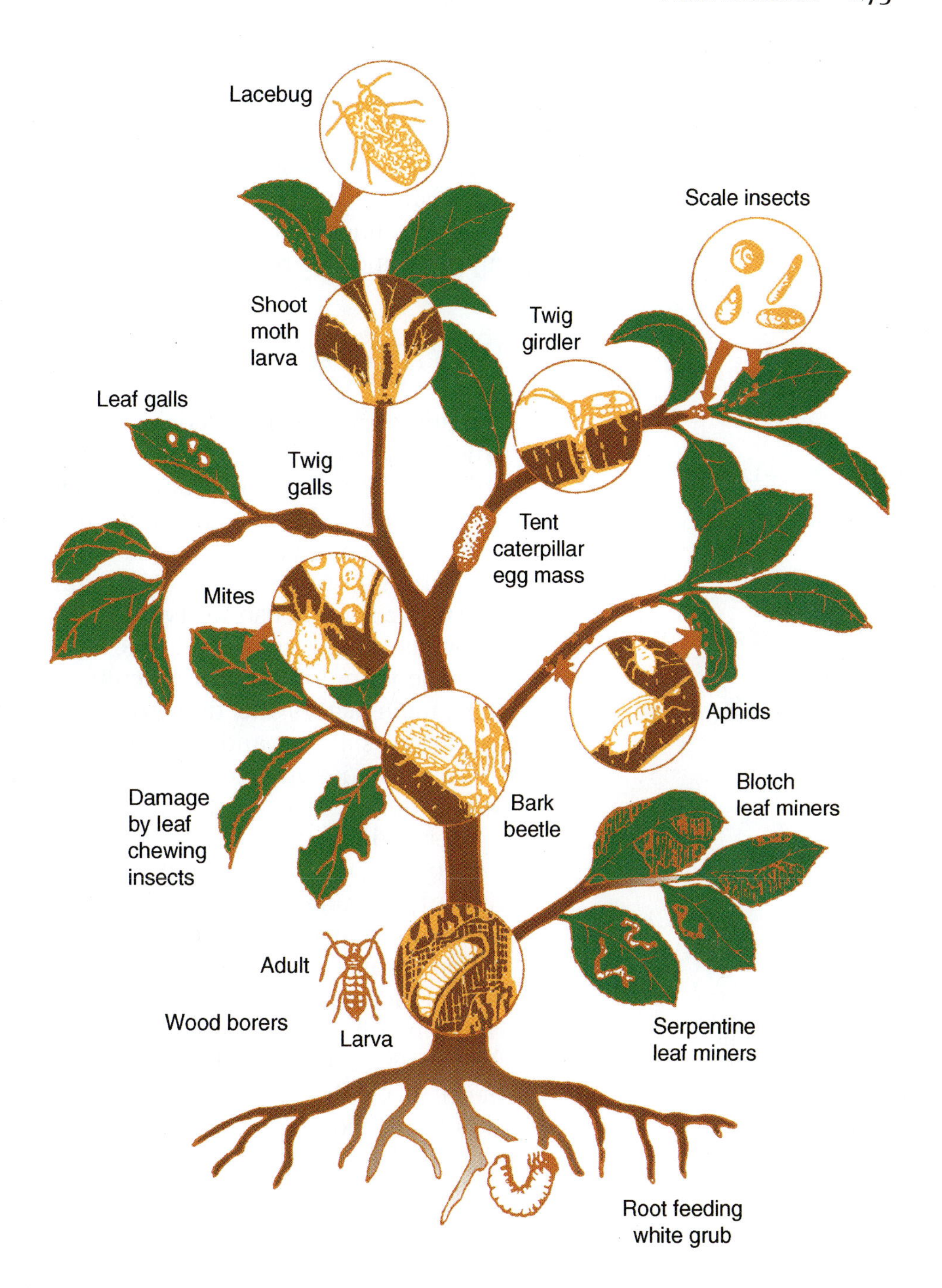

FIGURE 11–10 The different types of insect damage. (Courtesy of the Maryland Cooperative Extension Service.)

FIGURE 11–11 A Norway rat. (Courtesy of Leonard Lee Rue III.)

reproduce at rapid rates and can easily overwhelm their living areas. They are generally considered to be pests because of the amount of damage they inflict on agricultural crops and livestock feed supplies.

PLANT DISEASES

A **plant disease** is any abnormal plant growth. The occurrence and severity of plant disease is based on three factors: (1) A susceptible plant or host must be present, (2) the pathogenic organism or **causal agent** (an organism that produces a disease) must be present, and (3) environmental conditions must be conducive to support the causal agent.

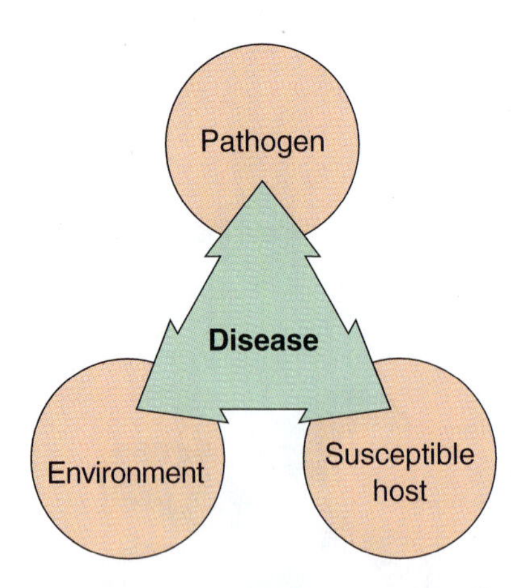

FIGURE 11–12 Components of the disease triangle

plant diseases, pictures
pesticides, environment

The relationship of these three factors is known as the **disease triangle** (Figure 11–12). Disease-control programs are designed to affect one or more of these three factors. For example, if crop irrigation is decreased, then a less favorable environment will exist for the disease organism. Breeding programs have introduced disease resistance into new plant lines for many different crops. Pesticides may also be used to suppress and control the disease organism.

Environmental concerns

Sometimes pesticides drift or leach into unintended areas (Figure 11–13). When this happens, the pesticide is often considered an environmental pollutant. The movement of a pesticide from the designated area may occur in several ways, including through drift, soil leaching, runoff, improper disposal and storage, and improper application.

Natural resources that can be contaminated are groundwater, surface water, soil, air, fish, and wildlife. Surveys have shown that more than 50 percent of the counties in the United States have potential groundwater contamination from agricultural chemicals, or agrichemicals. The three main factors affecting groundwater contamination by agrichemicals are (1) soil type and other geological characteristics, (2) the chemicals' persistence and mobility within the soil, and (3) the production and application methods of the chemicals' users.

Pesticide **drift** is a major cause of soil and air contamination when a pesticide moves through the air to nontarget sites. It will occur at the time of pesticide application when small spray particles are moved by air currents. In addition, pesticide **vapor drift** may occur after an application because of chemical volatilization of the product.

The adverse effects of pesticides on fish and wildlife may directly result in animal death. Pesticides may also indirectly influence animal feeding or reproduction. Pesticide labeling will indicate any potential harm to wildlife, and this information and application instructions should be followed to minimize risk. Fish, birds, bees, and other animals will be affected when pesticides reach them or their habitats.

Environmental contamination by agrichemicals can be decreased through several management practices. Following an IPM program will reduce pesticide use, but if pesticides are used, then proper mixing, application, storage, and disposal must be performed. These practices will decrease the potential for adverse environmental effects. It is also important to minimize any effects that temperature, soil type, rainfall, and wind patterns may have on a pesticide becoming an environmental pollutant.

Pest-control Alternatives

Cultural Control

Cultural control is the attempt to alter the crop environment to prevent or reduce pest damage. It may include such agricultural practices as soil tillage, crop rotation, adjustment of harvest or planting dates, habitat destruction, pest trapping, and irrigation schemes. Other practices that are considered cultural control are clean culture and trap crops.

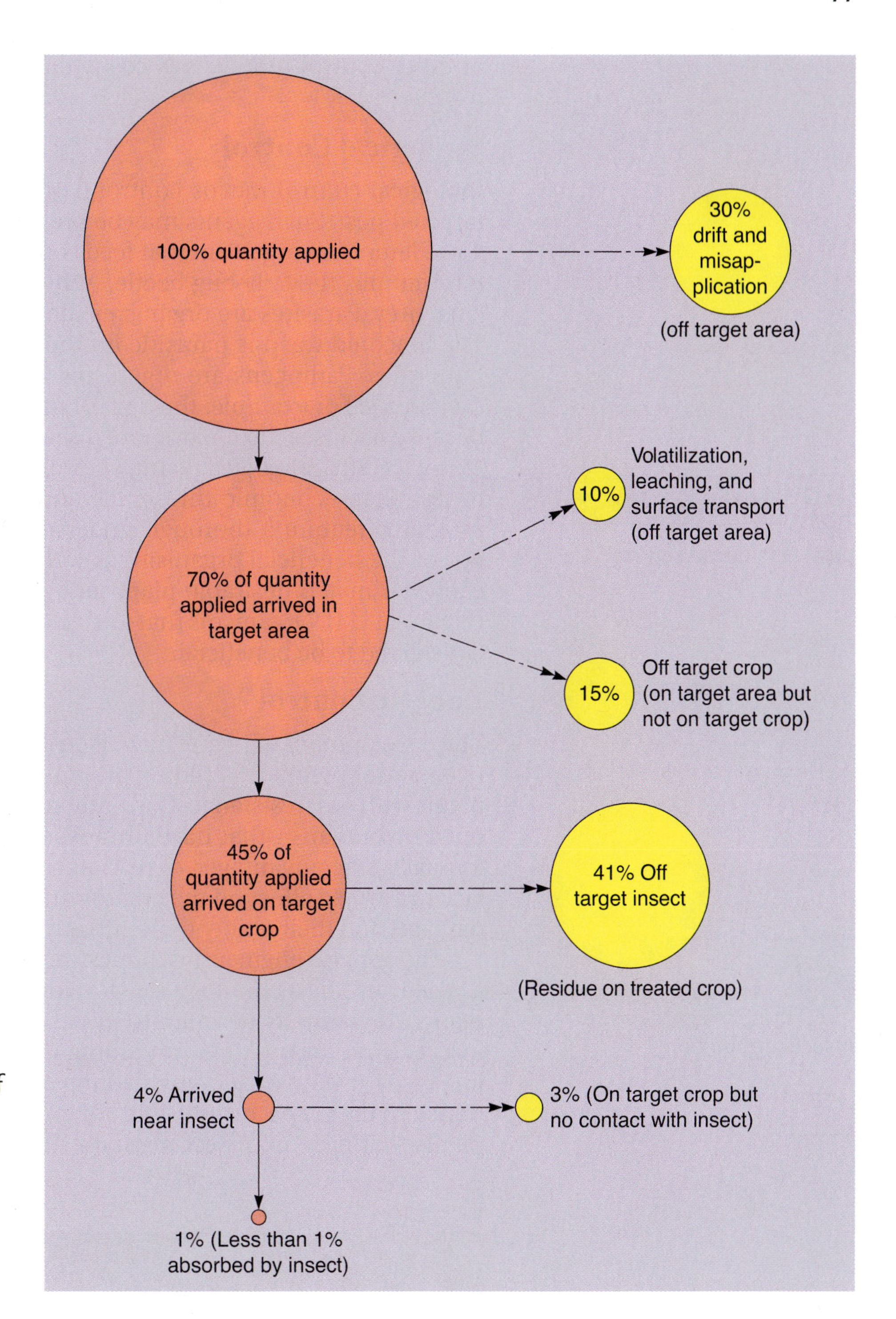

FIGURE 11–13 The movement of a pesticide after discharge from an aerial spray plane (Adapted with permission of M. L. Flint and R. Van Den Boschy, from *Introduction to Integrated Pest Management*, New York: Plenum Press, 1981, 240 pp.)

Clean culture refers to any practice that removes pests' breeding or overwintering sites. This may include removing crop leaves and stems, destroying alternate hosts, or pruning infested parts. A **trap crop** is a pest-susceptible crop that is planted to attract a pest to a localized area. The trap crop is then either destroyed or treated with a pesticide.

pests, trap crop

Both physical and mechanical control programs are examples of cultural control methods. They use direct measures to destroy pests. Examples of such practices are insect-proof containers, steam sterilization, hand removal, cold storage, and light traps. Implementation of these control practices is costly and provides varying pest-control results.

Biological Control

Biological control means control by natural agents or enemies of the targeted pest. Such agents may be predators, parasites, and pathogens. A predator is an organism that feeds on a smaller or weaker organism—for example, the lady bug beetle. Aphids are the lady bug beetle's principal prey. Parasites are organisms that live in or on another organism. The braconid wasp is parasitic on the caterpillars of many moths and butterflies. Pathogens are organisms that will produce disease within their hosts. For example, the bacterium *Bacillus popilliae* is a pathogen because it causes the milky spore disease in Japanese beetle grubs.

pests, biological control

Successful biological control programs reduce pest populations to less than economic thresholds and keep the pests in check. Such programs require a thorough understanding of the biology and ecology of the beneficial organism, as well as of the pest. Careful research can even match desirable plant pathogens against undesirable weeds (Figure 11–14). Organisms that are used to control insects and weeds are considered to be beneficial.

Genetic Control

The development of plants having pest resistance is an extremely effective control practice. Breeding programs attempt to identify and select plants with pest resistance. Currently, new plant **cultivars**—plants developed by humans—that have improved resistance to pests are released annually. The advantages of resistant varieties are their low cost, their lack of adverse effect on the environment, significant reductions in pest damage, and their ability to fit into any IPM program.

insects, irradiation

The genetic engineering of pest resistance within a plant's genes is a recent useful technology for developing pest-resistant plants. Some plants are naturally resistant to insects, perhaps giving off an odor that insects avoid or perhaps containing natural insecticides in their plant juices. By using genetic-engineering techniques, it is now possible to transfer the genes responsible for producing natural insecticides to plants that have no insect resistant traits. The result is that only those

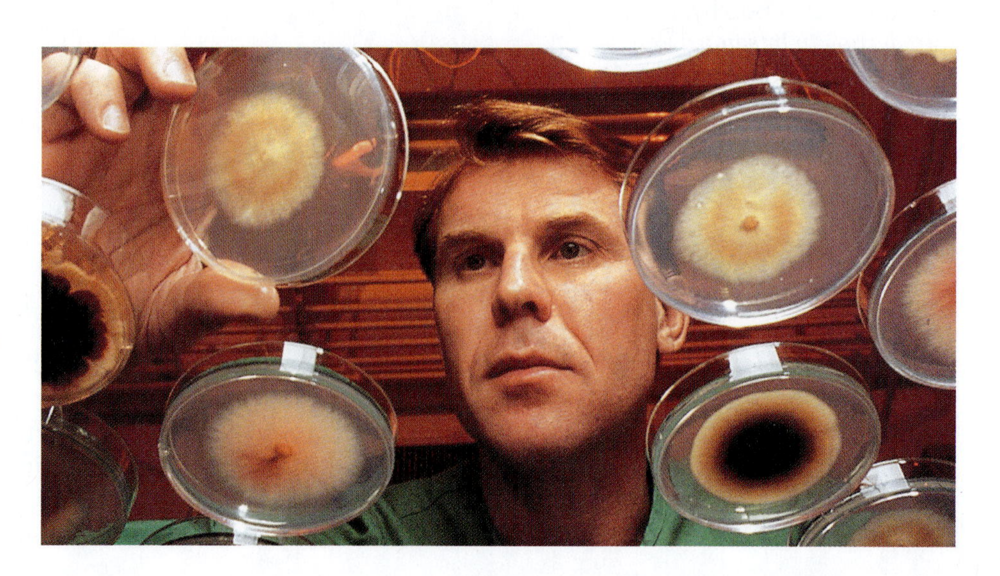

FIGURE 11–14 Plant pathologist Rick Bennett examines fungi that may be used for biological control of weeds. (Courtesy of the U.S. Department of Agriculture's Agricultural Research Service.)

Interest Profile

INSECT CONTROL USING STERILE MALE INSECTS

Imagine a laboratory that raises millions of insects that are harmful to agricultural crops. Once the insects are mature, they are exposed to radiation, which eliminates their ability to reproduce. The treated insects are then released in huge numbers into areas where the same species of insect pest is found. Because the treated insects vastly outnumber the untreated insects, the odds of treated insects mating with untreated insects is high, but no off-spring are produced from these matings. By repeating the releases of sterile insects over a two- or three-year period, the original population of harmful insects may be completely eradicated.

insects that attempt to eat these plants are killed. Pollinating insects and natural insect enemies are not subjected to insecticides, so they survive to help control the damaging insect species.

Chemical Control

pesticide resistance

Chemical control is the use of pesticides to reduce pest populations. Although chemical control programs have become cost-effective, various problems occur if this practice is misused or overused. Problems that can develop are environmental pollution, pesticide resistance, and pest resurgence. **Pesticide resistance** is the ability of an organism to tolerate a lethal level of a pesticide. **Pest resurgence** is a pest species' ability to repopulate after control measures have been eliminated or reduced.

Chemicals can have both helpful and harmful effects on living environments (Figure 11–15). Biological and cultural controls are favored when they are effective. Sometimes, however, it is extremely difficult to control certain pests without using chemical pesticides. Under such circumstances, chemical pesticides must be used safely.

Insect Pheromones

Pheromones are natural hormones found in insects and other organisms that give off a distinctive odor. Insects of each species follow their

FIGURE 11–15 Chemicals are needed in our modern society, but when abused or misused they may threaten the health of animals and people. (Courtesy of Getty Images.)

Pesticide type	Targeted pests
Acaricide	Mites, ticks
Algaecide	Algae
Attractant	Insects, birds, other vertebrates
Avicide	Birds
Bactericide	Bacteria
Defoliant	Unwanted plant leaves
Desiccant	Unwanted plant tops
Fungicide	Fungi
Growth regulator	Insect and plant growth
Herbicide	Weeds
Insecticide	Insects
Miticide	Mites
Molluscicide	Snails, slugs
Nematicide	Nematodes
Piscicide	Fish
Predacide	Vertebrates
Repellents	Insects, birds, other vertebrates
Rodenticide	Rodents
Silvicide	Trees and woody vegetation
Slimicide	Slime molds
Sterilants	Insects, vertebrates

FIGURE 11–16 Classification of pesticides based on the target pests (Adapted from material from *The New Pesticide User's Guide.*)

own odor to find a mate. When a specific pheromone for a harmful insect is released in agricultural fields, it confuses the target insect as it seeks a mate and interrupts its reproductive process. Pheromone applications do not harm any other organisms in the area because they are only attracted to the pheromones that are specific to their species. Pheromones are also used to bait insect traps, attracting the harmful insects to selected locations where they can be destroyed.

INTERNET KEY WORDS

pest control, pheromones

Biological Controls

One form of insect control is achieved using toxins or poisons that are produced by living organisms such as bacteria. Other toxins are obtained from such plant parts as seeds. Such toxins can be as effective as any insecticide and, in fact, are naturally occurring insecticides. Other biological controls are attained by introducing nematodes or other microorganisms into living environment to control specific insects that damage crops or injure animals.

Pesticides

A majority of pesticides currently being used are synthetically produced **organic compounds** (compounds that contain carbon). The organic chemistry involved in pesticide production is often complex and extremely diverse. Pesticide users can consult a classification system for based on the type of pest being controlled (Figure 11–16). The major pesticide groups are herbicides, insecticides, and fungicides.

In the United States in 1999, pesticide use totaled 500,000 tons of 600 different pesticides at a cost of $7.6 billion. Currently, the EPA has registered more than 600 chemicals that are formulated into some 30,000 products for pest control. Of the three major pesticide categories, the largest volume was for herbicides followed by insecticides and fungicides.

FIGURE 11–17 Petroleum or chemical spills in water environments severely damage populations of wild animals and plants. (Courtesy of Getty Images.)

More than 10,000 different pesticides are reportedly registered for use in the states that surround one of our major coastal bays. Just as oil spills and industrial chemical discharges have caused serious problems in oceans, lakes, rivers, and streams (Figure 11–17), chemical pesticides continue to threaten wildlife, fish, shellfish, beneficial insects, microscopic organisms, plants, animals, and humans. Needless to say, careful management and control of so many different chemicals is absolutely essential. It requires the utmost care to avoid unacceptable damage to our environment.

Regulatory Control

Federal and state governments have created laws that prevent the entry or spread of known pests into uninfested areas. Regulatory agencies also attempt to contain or eradicate certain types of pest infestations. The Plant Quarantine Act of 1912 provides for inspection at all ports of entry. Plant or animal **quarantines**—the isolation of pest-infested

HOT TOPIC

A BALANCING ACT

Honeybees are a major insect species of importance to U.S. food, fiber, flower, and ornamentals production industries. We rely on them as the only method of pollinating certain plants and count on them to do some of the pollination of nearly all species of plants.

Bees enter flowers of plants to gather nectar and pollen for their own food and nourishment of their young. Their service to humans and animals in pollinating plants, and thereby producing seeds and fruit, is a service we cannot do without. Most plants could not reproduce and survive without producing seeds. Their safety from pesticides used to control harmful insects is always at the top of the agenda for entomologists. Although they can sting if threatened, honeybees in the United States are of the European type and are predictable. Except for the inexperienced person approaching a beehive, honeybee stings are generally a single sting by a single bee. Except for the relatively few individuals who have life-threatening allergic reactions to bee stings, honeybees pose little threat, because they act individually when they are aggravated enough to sting.

Enter the Africanized honeybee. This hybrid cousin of the domestic pollinator is famous for its aggressiveness and has thus been dubbed the "killer bee." Africanized honeybees resulted in 1957 when bees were imported from Africa to Brazil by a Brazilian scientist. The plan was to experiment by crossbreeding them with the domestic European bees prevalent in the Americas to develop a better strain of honeybees for the tropics. Unfortunately, some African bees were inadvertently released in the countryside and promptly interbred with the domestic bees. These new hybrids and their descendents have migrated as far south as Argentina and as far north as the United States.

On October 15, 1990, the first Africanized honeybee swarm to migrate naturally to the United States was identified by entomologists near Brownsville, Texas. The swarm was promptly destroyed, according to standard procedure. Unfortunately, Africanized honeybees have different dispositions than the domestic bees of the United States. They tend to defend their colonies more vigorously, stinging in greater numbers and with less provocation. One bee is likely to inflict many stings. Therefore, there is greater danger in an encounter with the Africanized bees. U.S. agriculture and the beekeeping industry fear that domestic bees interbred with Africanized bees may become harder to manage as pollinators of crops and may not be as efficient at producing honey.

The challenge of observing, detecting, and stopping the northward migration of Africanized honeybees will be a top priority of government inspectors, entomologists, beekeepers, farmers, and citizens at large. At the same time, animal behaviorists will study the bees' habits and look for ways to manage them. Geneticists will study the bees' genes and look for ways to genetically engineer future bees to decrease their objectionable habits and enhance their abilities as pollinators and honey makers.

material—are implemented if shipments are infested with **targeted pests**, or pests that pose a major economic threat if introduced into an area.

killer bees

If a targeted pest becomes established, an eradication program will be started. **Eradication** means total removal or destruction of a pest. This type of pest control is extremely difficult and expensive to administer. In California, the Mediterranean fruit fly was eradicated at a cost of $100 million in 1982. The insect has recurred since then, and the cost is high each time it is eradicated. This program relies on chemical spraying, sanitation, sterile male releases, and pheromone traps to ensure complete eradication.

Principles and Concepts of Integrated Pest Management

The following concepts or principles are important in understanding how IPM programs are designed to function.

Key Pests

A **key pest** is one that occurs on a regular basis for a given crop (Figure 11–18). It is important to be able to identify key pests and to know their biological characteristics (Figures 11–19 and 11–20). The weak link in the biology of each pest must be identified if management of the pest is to be successful.

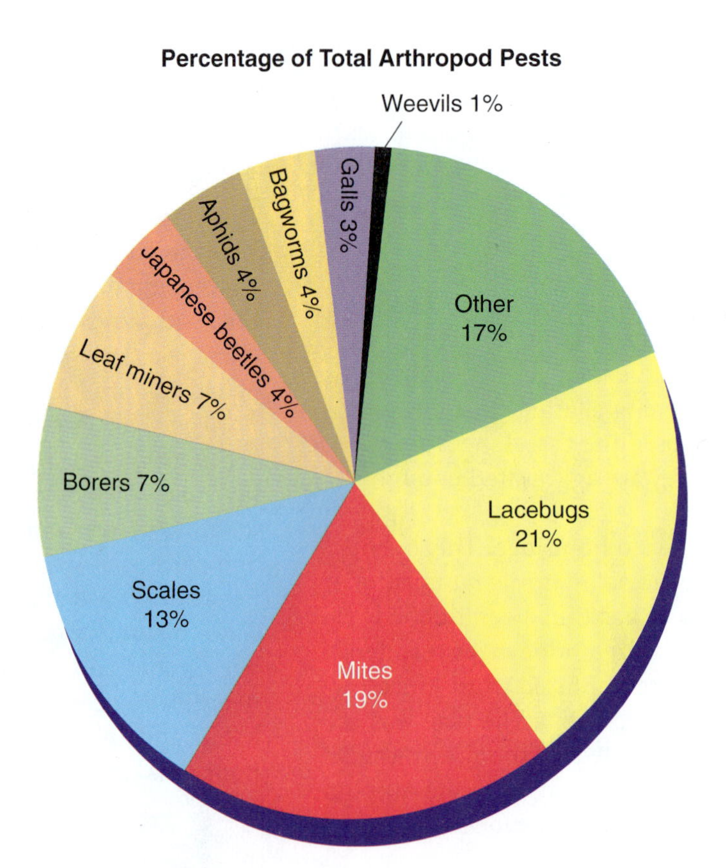

FIGURE 11–18 Arthropod pests and their percentages in six Maryland communities (Adapted from material from the Department of Entomology at the University of Maryland.)

FIGURE 11–19 Grasshoppers, locusts, and crickets are capable of immense crop destruction. The resulting famines have afflicted people throughout recorded history.

FIGURE 11–20 Vast forests suffer severe insect damage throughout North America. Insects account for the death of many trees every year and contribute to the devastating fires that occur in unhealthy forests.

The following insects are examples of pests that cause serious damage in North America:

Damage to Crops	
aphids	beetles
cockroaches	crickets
grasshoppers	leafhoppers
locusts	moths
nematodes	weevils

Damage to Trees	
aphids	carpenter worms
cone beetles	cone borers
cone maggots	cone worms
leafminers	mites
moths	needleminers
psyllids	sawflies
seed bugs	termites

Damage to Humans and Other Animals	
fleas	flies
keds	lice
mites	mosquitoes
screwworms	ticks

Crop and Biology Ecosystem

INTERNET KEY WORDS

integrated pest management

The integrated pest manager must learn the biology of the crop and its ecosystem, which consists of the biotic and abiotic influences in the living environment of the crop. The biotic components of the ecosystem are the living organisms such as plants and animals. The abiotic components are nonliving factors such as soil and water. Examples of human-managed ecosystems are a field of soybeans, a turfgrass area, or a poultry production operation.

Ecosystem Manipulation

With IPM, an attempt is made to understand the influence of ecosystem manipulation on reducing pest populations (Figure 11–21). To illustrate this concept, the manager must ask, "What would happen to the pest population equilibrium if a disease-resistant plant were introduced?" **Pest population equilibrium** occurs when the number of pests stabilizes or remains steady. The introduction of disease-resistant plants

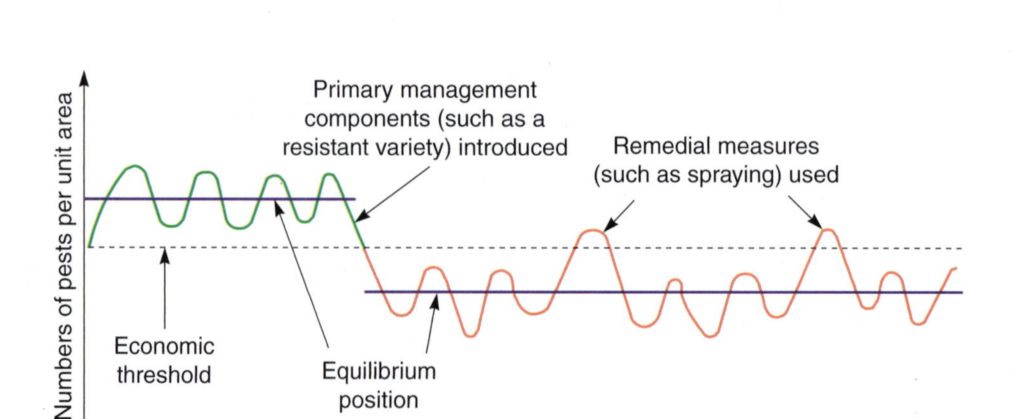

FIGURE 11–21 The effect of lowering the equilibrium position of a pest (Adapted from material from the Department of Entomology at the University of Maryland.)

should decrease the pest population to less than the **economic threshold level**, or the point where pest damage is great enough to justify the cost of additional pest-control measures. Until the pest population increases to a high enough level that the cost of controlling the pest is less than the cost of the losses that the pest causes, no control actions will be taken.

Threshold Levels

The level of a pest population is important. The mere presence of a pest may not warrant any control measures, but at some point the damage created may be great enough to warrant control measures. Various threshold levels are developed to determine if and when a control measure should be implemented to prevent excessive economic loss of plants to pest damage (Figure 11–22).

Economic threshold levels are determined by first developing a pest-damage index (Figure 11–23). It is crucial in the decision-making process to know the level of pest infestation that will cause a given yield reduction. Pest populations are measured in several different ways. They can be counted in number of pests per plant or plant part, number of pests per crop row, or number of pests per sweep with a net above the crop.

Economic Injury Threshold for Alfalfa Weevil; Number of Larvae From 30-Stem Sample

How to use table below:

1. Use plant height category that fits the field.
2. Estimate the value of crop in dollars per ton of hay equivalent and the cost to spray an acre.
3. From monitoring the field, find the number of alfalfa weevil larvae from a sample of 30 stems.
4. The number in each small box indicates the number of larvae per 30-stem sample that is required before a spray application would be profitable under these conditions.

EXAMPLE:

Plants in the field are 20 inches high (use Category II), hay is valued at $80 per ton, cost to spray is $8.00 per acre, and you collected 40 larvae from the sample of 30 stems. The number in the box common to $80 and $8 is 75. This means that under these conditions, 75 larvae are needed before a spray would be profitable. Since you collected only 40 larvae, a spray at this time will not be profitable.

Category I plant height 12 to 18 inches

Value of hay per ton	$8	10	12	14	16	20
$60	91	114	137	160	183	225
$80	68	85	102	119	136	171
$100	54	68	81	95	108	137
$120	45	57	68	79	91	114
$140	39	49	59	68	77	99
$160	34	43	51	60	68	86

Category II plant height 18 to 24 inches

Value of hay per ton	$8	10	12	14	16	20
$60	99	124	149	174	199	240
$80	75	94	113	131	150	186
$100	62	75	90	105	120	149
$120	50	62	75	87	100	124
$140	43	54	64	75	86	107
$160	37	47	56	65	75	93

Category III plant height 24 to 30 inches

Value of hay per ton	$8	10	12	14	16	20
$60	104	130	156	182	209	260
$80	78	97	117	137	157	195
$100	63	78	94	110	126	156
$120	52	65	78	91	105	130
$140	45	56	67	78	90	112
$160	39	49	58	68	79	98

Cost of insecticide application per acre

FIGURE 11–22 Chart to determine economic threshold level for the alfalfa weevil (Adapted from material from Pennsylvania State University.)

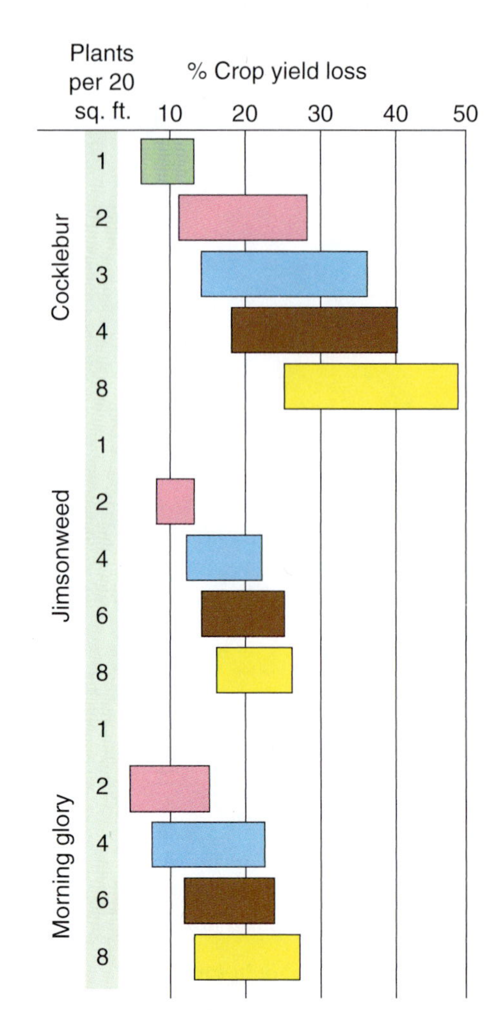

FIGURE 11–23 The pest-damage index for several weeds found in soy beans fields

Monitoring

For IPM to be successful, a **monitoring** (checking) or scouting procedure must be performed. Different sampling procedures have been developed for various crops and pest problems (Figure 11–24). The presence or absence of the pest, amount of damage, and stage of development of the pest are several visual estimates a scout must make. The method used must be speedy and accurate.

Scouts are people who monitor fields to determine pest activity. They must be well trained in entomology, pathology, agronomy, and horticulture.

FIGURE 11–24 Random sampling of plant stems and leaves to determine pest populations and damage (Courtesy of the Maryland Cooperative Extension.)

Integrated Pest Management Program Components

Key elements of successful IPM programs have been identified as these programs have evolved: planning, setting action thresholds, site monitoring, pest identification, and plan implementation. It is important to include each step in the process as a new IPM program is implemented.

Planning

Historical information about a troublesome pest is important in developing a plan to control it. Used in conjunction with current information, a judgment can be made concerning the severity of the pest problem. Planning should be completed well before a problem emerges. This allows managers to act quickly and deliberately in addressing a pest problem instead of reacting to it after it is too late to use all of their options. The key is to see the big picture regarding the pest, both with respect to the pest population and its location.

Setting Action Thresholds

It is important to be able to make an accurate assessment of the true cost of damage from a particular pest infestation. It is also important to know what the cost of treatment will be in order to determine when to implement the treatment plan. It is usually unrealistic to implement a zero-tolerance plan for an insect or pest population, but there is a threshold at which it becomes less expensive to implement the IPM plan than to tolerate the losses. This threshold at which the IPM plan is to be implemented should be determined during the planning phase.

Site Monitoring

Pest numbers and locations must be carefully monitored as a pest infestation develops. It is important to detect a pest problem early and to accurately determine the size of the pest population. Regular monitoring of multiple sites will help in predicting when or if the action threshold has been reached.

Pest Identification

Accuracy in correctly identifying a pest is critical, and mistakes can be costly. Many insects take on different characteristics as they progress through their life cycles, and mistaken identity is all too common, which means managers must look for expert advice when necessary. Mistaken identity leads to expensive and ineffective pest-control efforts.

Plan Implementation

The action to be taken and the threshold at which the IPM plan is implemented is determined by pest numbers and locations. It is also important to apply treatments to attack the pest during a phase in its life cycle when it is most vulnerable. Accurate timing will increase the effectiveness of the treatments that are applied. Implementation of the plan should be almost automatic once the action threshold has been met or exceeded.

Two major factors have accelerated the adoption of IPM. First, some consumers are resistant to the use of chemical insecticides near their homes or to treat pest infestations on the plants or animals they use for food. Second, insects and other pests are able to develop genetic tolerances for toward chemicals. Clearly, an effective alternative for insect control is needed rather than a complete reliance on chemicals. The multipronged approach of IPM for controlling pests appears to be the best alternative to traditional methods.

Looking Back

Integrated pest management is a concept for controlling harmful insects or other pests while providing some protection for organisms that are useful to humans. It not only involves the use of some chemical pesticides, but it also relies on natural insect enemies and other biological control strategies to control harmful pests. Pests may include insects, weeds, rodents, and any other organism that feeds on crops or preys on livestock.

When pesticides are the only source of control, they kill both harmful and useful insects. As a pest-control program, IPM does not attempt to kill all harmful pests since pest control of this kind also kills the natural enemies. For natural enemies of pests to survive, they must have a small population of the target species on which they can prey to obtain a food source. The key to a successful IPM program is to establish a balance between pests and predators that keeps the pest population small enough that it can be tolerated.

Self-Analysis

Essay Questions

1. What kinds of pest control have been most universally adopted during the past 40 years?
2. Name some of the pests that cause the greatest damage to agricultural crops and livestock.
3. What led to the ban of DDT from use in the United States in 1972?
4. How have crop yields been affected by the use of chemical pesticides?
5. What are the differences among annual, biennial, and perennial weeds?
6. Compared to other organisms, what characteristics of insects have contributed to their success?
7. What are the three factors that influence the occurrence and severity of plant diseases?
8. List some of the pest-control alternatives currently used in the agricultural industry.
9. Explain how the use of sterile male insects is effective in reducing populations of harmful insect pests.
10. What are some ways in which pheromones are used to control insect pest populations?
11. What regulatory controls are available to reduce insect damage to crops?
12. List the key elements of successful IPM programs.

Multiple-Choice Questions

1. A management system for controlling harmful insects or other pests while providing some protection for organisms that are useful is called
   a. PHC.
   b. GPS.
   c. GIS.
   d. IPM.
2. The widely used pesticide that was banned in the United States in 1972 because of its environmentally damaging effects was
   a. DDT.
   b. rotenone.
   c. Roundup.
   d. PHC.
3. A living organism that carries or transmits a disease is a
   a. cultivar.
   b. rhizome.
   c. vector.
   d. pheromone.

4. An organism with six legs and three body segments is
   a. an insect.
   b. an arachnid.
   c. a vector.
   d. a nematode.
5. A biennial weed will live for
   a. one year.
   b. two years.
   c. three years.
   d. more than three years.
6. A plant that causes great harm to other organisms by weakening those around it is
   a. a perennial.
   b. an annual.
   c. a biennial.
   d. noxious.
7. The insect-control practice that relies on the introduction of parasites and predators is
   a. cultural.
   b. chemical.
   c. biological.
   d. regulatory.
8. Altering the crop environment to prevent or reduce pest populations is an example of which kind of pest control?
   a. cultural
   b. chemical
   c. biological
   d. regulatory
9. The development of plants having an internal form of pest resistance is
   a. biological control.
   b. genetic control.
   c. biological control.
   d. chemical control.
10. An insect hormone that is used to attract harmful insects is
    a. a pheromone.
    b. an aphlatoxin.
    c. a pesticide.
    d. a dendrotoxin.

Learning Activities

1. Using the Internet, library, magazines, or other sources, prepare a two-minute class presentation on a local pest. Be sure to verify the kind of harm that the pest is responsible for causing along with the integrated pest management techniques that can or may be used to control it.
2. Select a major crop that is produced near you and discuss each of the key pests that afflict the crop. Also discuss the measures that are recommended for keeping the pest under control.
3. Collect and display a collection of weeds or harmful insects that are found in your local community and properly label each sample. Use your collection and those of your classmates to memorize the characteristics of harmful weeds and insects found in your community.

TERMS TO KNOW

population ecology
population
density
distribution
natality
age structure
monogamy
polygamy
polygyny
promiscuity
semelparous species
iterparous species
mortality
sex ratio
fecundity
fertility
production
emigration
immigration
carrying capacity
evolution
biome
freshwater biome
plankton
phytoplankton
zooplankton
turbid
lotic habitat
lentic habitat
thermal stratification
wetland
marine biome
intertidal zone
continental shelf
neritic zone
oceanic zone
estuary
terrestrial biome
desert biome
tundra biome
grassland biome
temperate forest biome
deciduous forest
strata
canopy
understory
shrub layer
herb layer
forest floor
coniferous forest biome
conifer

CHAPTER 12

Population Ecology

Objectives

After completing this chapter, you should be able to

- define population ecology and explain its importance
- identify some characteristics that are useful in describing a population
- explain why species become endangered
- identify some management practices that are used to help endangered populations recover
- explain how events in one ecosystem affect events in a neighboring ecosystem
- describe the relationship between an ecosystem and a biome
- list the distinguishing characteristics of a freshwater biome
- identify similarities and differences between freshwater and marine biomes
- discuss ways in which wetland habitats function to cleanse the environment
- name the terrestrial biomes that are found in North America and describe their similarities and differences
- design a map of the North American continent to illustrate the locations of the major biomes

Population ecology is concerned with the total number of individuals of a species and their relationships with the factors within their environment. It is a study of organisms at the population level rather than as individuals. A **population** includes all of the plants, animals, and other organisms that live in a defined area. The population of each species is considered separately in population ecology studies.

Decisions about managing plants and animals are based on what is happening to the total population, not on individual organisms. This is because a more accurate picture of the environment can be obtained from populations than from individuals. Evaluation of management efforts also addresses effects on populations rather than individuals.

POPULATION CHARACTERISTICS

A measurement that is often used to describe a population is population **density**, or the number of individual organisms living within a defined area. For example, the number of prairie dogs per square mile is a measurement of population density. This kind of measurement takes on added importance when data are available for the population over a period of several years. Such data can be interpreted to know whether a population is growing, is declining, or has stabilized.

Population **distribution** is a characteristic of a population that describes the spacing or dispersion of individuals. Distribution describes how far apart the members of the population live. Some animals, such as prairie dogs, prefer to live close to others of their own kind. Other species, such as the mountain lion, prefer to live alone in large territories. Distribution of a population is seldom uniform because the population tends to congregate near the most easily accessed resources.

animal natality
semelparous species, examples

Birthrates and death rates are important characteristics of a population because they are indicators of the overall health of the population. Birthrate or **natality** is the number of births in comparison with the number of individuals per year. For example, 72 births per 100 females per year is a reasonable number in a herd of elk but an extremely low number among pygmy rabbits.

A population's **age structure** describes its makeup in terms of maturity. Birthrates are affected by a population's age structure. A population of deer that experiences heavy hunting pressure for trophy bucks, for example, is likely to have a high proportion of yearling bucks in the male segment of the population. This can reduce the birthrate when the number of mature males in the population is low. A full description of age structure includes the number or proportion of animals that belong to each age group. Age structure and sex ratio are sometimes combined to identify the number of males and females in distinct groupings by age (Figure 12–1).

Age Structure by Gender
Mule Deer Population: Clearwater Management District II
Winter - 2007

Age	Bucks	Does
Immature	2.76	2.74
2 years	2.13	2.27
3 years	1.57	2.07
4 years	1.12	1.57
5 years	.52	1.21
6 years	.24	1.03
7 years	.13	.86
8+ years	.03	.64

Legend: 1=100

FIGURE 12–1 Age structure by gender of a population of mule deer

Birthrates are also affected by the kind of mating system used by the population. Many species of birds engage in **monogamy**, meaning that they pair up and have only one mate at a time. In other populations, both males and females have more than one mate. This mating system is called **polygamy**. A system in which males attract and mate with several females is called **polygyny**. When males and females mate with numerous members of the opposite sex, the system is known as **promiscuity**.

Some animals reproduce only once in their lifetime. Such species are classified as **semelparous**. Examples include annual plants that

FIGURE 12–2 Trumpeter swans are an iterparous species, producing multiple broods of offspring in a lifetime. (Courtesy of the U.S. Fish and Wildlife Service.)

reproduce only once and the salmon, which uses all of its energy as it migrates to the waters of its birth where it mates and dies. These plants and animals tend to be vulnerable to dramatic declines in population numbers, particularly when catastrophic events occur. In contrast, **iterparous species** reproduce numerous times during their lifetimes (Figure 12–2).

A measurement of equal importance to the birthrate is **mortality**, or the death rate. It is expressed as the number of deaths per year per number of individuals, such as deaths per 100 or per 1,000. These types of data have been assembled to compare the populations of affluent nations with those of the world's poorest nations.

Another characteristic of a population is the **sex ratio**, or the number of males compared to the number of females. For example, the closer the sex ratio comes to 1:1 in whitetail deer, the more likely it is that most of the does will give birth to fawns in the spring. Because the breeding season is relatively short, it is important that enough mature bucks are with the herd during the breeding season.

fecundity

Other population characteristics include the number of eggs produced per female. This is also known as **fecundity**. **Fertility** is a measurement of the percentage of eggs that are viable or fertile. **Production** is the number of offspring of a population.

Population Growth Factors

Several factors affect the growth of a population. The birthrate compared with the death rate is an obvious factor. When population birthrate exceeds death rate, growth occurs. When death rate exceeds birthrate, the population declines. Those species that experience high birthrates often experience wide fluctuations in population numbers.

The age at which the breeding population reaches physical maturity is a factor associated with population growth rates. For example, early maturity tends to promote population growth in comparison with organisms that exhibit late reproductive maturity. The length of gestation

Interest Profile

DEVELOPED NATIONS VERSUS UNDEVELOPED NATIONS

The United Nations has reported that more children than ever before are surviving to adulthood and adults are living longer (Figure 12–3). Together these trends mean more population growth and more pressure on the environment. Advancements in medical science and services have made good health and longer lives a reality, but only for those who can afford good nutrition and modern health services. Similarly, through agriscience, we have made substantial gains in providing food, fiber, and shelter for the world (Figure 12–4). At the same time, the environment has stayed reasonably clean, considering the effects of bulging populations.

In the past, individuals younger than 25 years old constituted the world's largest population group because children were valued for the help they provided in making the family living. Children and young adults were engaged in a nation's labor force and also provided the manpower for armies. In most countries, the young respected their elders and provided for the needs of the elderly within the family.

The age profiles of people in developed countries are quite different from those of developing countries. Honduras has the traditional population pattern, with its largest number of citizens younger than 5 years of age. The number per age group then decreases to the smallest number, which occurs in the age group older than 80 years. When the Honduras population groups are displayed by sex in a bar graph, the graph takes the shape of a pyramid (Figure 12–5). Canada's pattern is slightly different. Its greatest population group is around the 20-year mark and resembles a Christmas tree

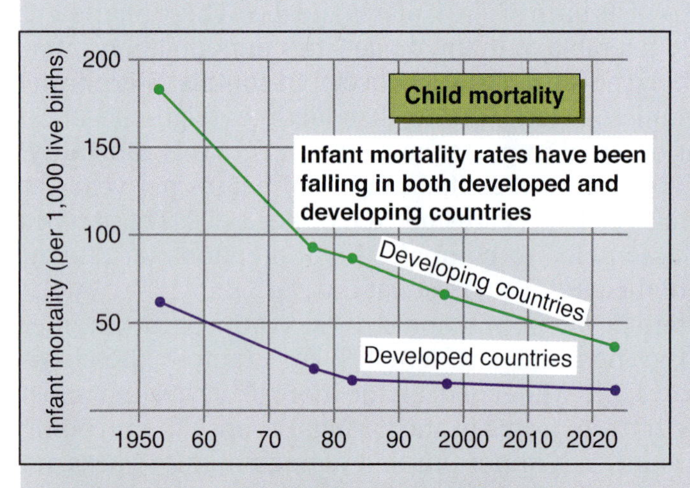

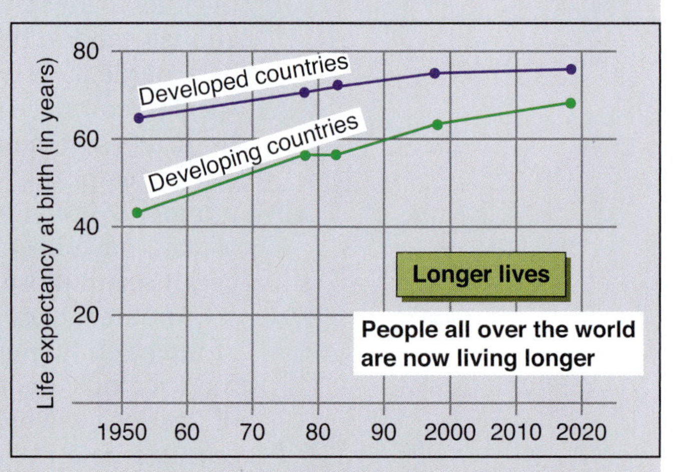

FIGURE 12–3 Worldwide child mortality rates and life expectancy estimates (Courtesy of the United Nations Fund for Population Activities.)

#1. Labor Required to Produce Wheat, Corn, and Cotton (In Hours)

	1800	1935–39	1955–59	1980–84	1990 or later
Wheat (100 bu.)	373	67	17	7	7
Corn (100 bu.)	344	108	20	3	2.88
Cotton (1 bale)	601	209	74	5	5

#2. Yields per Acre of Wheat, Corn, and Cotton

	1800	1940	1960	1985–86	1990 (prelim)
Wheat (bu.)	15	15	20	34	39.5
Corn (bu.)	25	29	55	118	118.5
Cotton (lb.)	154	253	446	630	640.0

FIGURE 12–4 Changes in agriscience productivity

Pyramids and Pillars

The age profiles of developed and developing countries are very different. The diagrams show what percentage of the population of each sex falls within each age band. But tendency all over the world is to move to a more even 'pillar' pattern.

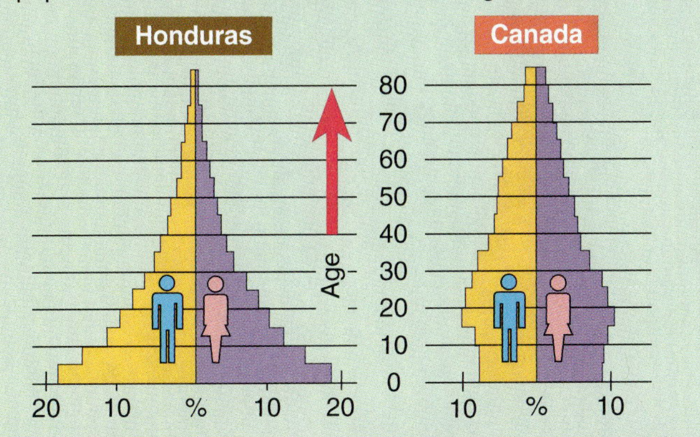

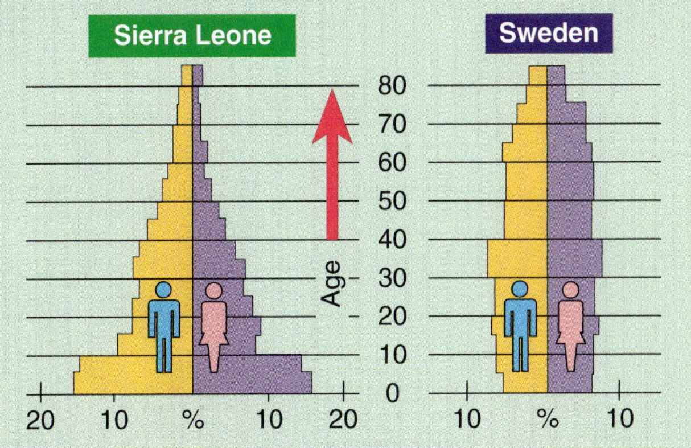

FIGURE 12–5 Age profiles and population patterns for developing and developed countries. (Courtesy of the United Nations Fund for Population Activities.)

with its narrow bottom and cone appearance. Sweden, a country known for its excellent health services and high survival rate, has age brackets that are about equal. The population graph for that country resembles a column.

China has approximately one-fifth of the world's population, yet it has been reasonably successful at feeding its population by keeping some 70 percent of its workforce on farms. In contrast, less than 2 percent of the workforce in the United States is needed to operate the nation's farms. In the mid-1970s, China implemented a policy whereby each couple was limited to one child. This was called the 4–2–1 policy, meaning that extended families consisted of four grandparents, two parents, and one child. What would be the outcome if such a policy were strictly enforced for several generations? The pyramidal shape of China's population graph would be expected to change to the shape of a Christmas tree, and, in time, to an upside-down pyramid! What would be the implications of feeding a nation with a population of mostly elderly people?

affects population growth in the same way. A short gestation period creates the opportunity for additional births, particularly when multiple broods are hatched or multiple pregnancies occur in a single season. Consider the contrast between the mouse and the moose. Several generations of mice are born in a single year and multiple offspring are produced from each pregnancy. A moose, on the other hand, usually matures at two years of age and gives birth to one or two calves per year. The potential for population growth is greatest for the mouse.

The ability to adapt to changing conditions within an environment is a great asset for some species. The coyote population, for example, has benefited from a high level of adaptability (Figure 12–6). This animal is able to eat many different kinds of food, and it is well adapted to living in nearly any environment. It has adapted well to living near humans, and its population as a species continues to increase in many areas.

Some wild animals are quite mobile, especially young males that are driven out by more mature males. They are sometimes unable to establish themselves within the territory of a mature male and are forced to move out of the area. Movement of an animal out of a population is known as **emigration**. Movement of an animal into a new

FIGURE 12–6 The coyote has a high birthrate and adapts well to new living conditions. (Courtesy of the U.S. Fish and Wildlife Service.)

habitat is called **immigration**. Although movement of animals from one habitat to another affects the number of animals within a regional environment, it does not affect a species' population.

Resources within any living environment have limitations. Population growth is related to the availability of resources within the living environment. Increasing the population in an environment requires more food and shelter. **Carrying capacity** is the maximum number of individual organisms that the environment is capable of sustaining. The carrying capacity may vary from one season to another and from year to year. When climatic conditions are favorable for plant growth, the carrying capacity may increase. In unfavorable times, the carrying capacity is likely to decline. A population that exceeds the carrying capacity will experience a high mortality rate until the availability of resources equals or exceeds the demand.

Extinction and Its Causes

causes of extinction

Loss of a species is an extreme example of a crisis in population ecology. The main reason a species declines is the destruction of its habitat. An animal's habitat is its home; without that habitat, the animal cannot survive. There are, however, many factors in addition to habitat destruction that can cause a species to decline (Figure 12–7). Habitat degradation can be as detrimental to a species as habitat destruction. For example, the habitat of a species may become so polluted that it poisons the animals. A habitat may also become so fragmented that there is interference with normal activities such as breeding and raising offspring.

The introduction of non-native species has caused the demise of more than one species. The English sparrow, the European starling, and the Norway rat, for example, compete with native wildlife species

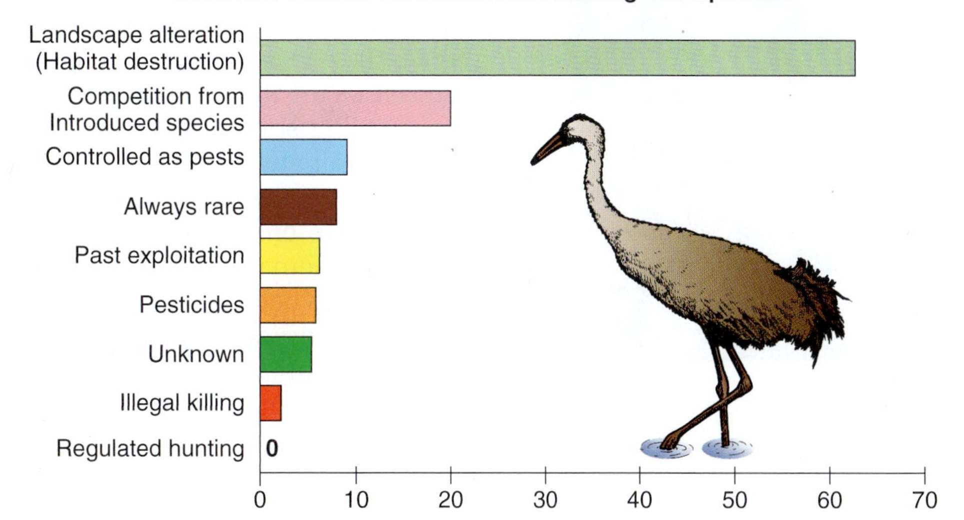

FIGURE 12–7 This graph shows the relative influence of the factors that have endangered wildlife species from most to least influential. (Courtesy of Kevin Deal.)

for habitat. Often native species lose out to more aggressive non-native ones. Such losses to introduced species are even more severe in fragile environments, such as the Hawaiian Islands, where dozens of native species of plants and animals have lost to immigrant competitors.

DDT, effects
black-footed ferret

Some species now listed as endangered were once controlled as pests. Species such as the gray wolf and the grizzly bear were serious predators of domestic livestock during the human expansion across North America. Domestic livestock is generally easier to capture and kill than native wildlife. With severe decreases in the numbers of natural prey such as bison, large predators turned to domestic livestock. Seeking to protect their livelihoods, farmers and ranchers fought back, often with help from the government. As is always the case when wildlife and humans collide, the wildlife lost.

Another factor that contributes to extinction is the fact that some species have always been rare. Some species are highly specialized in their needs or have extremely low reproductive rates. These factors combine to keep their numbers low. Past exploitation has been a factor in several species' fortunes. Market hunting drove the passenger pigeon to extinction and nearly did the same for the bison. The whooping crane was driven to the brink of extinction because of demand for its feathers to decorate women's hats during the early part of the twentieth century.

Pesticides, in particular DDT, caused the decline of several species. The peregrine falcon and bald eagle were both driven to the edge of extinction primarily by DDT (Figure 12–8), which does not break down readily and thus becomes concentrated in lakes and streams after being washed out of fields. Until DDT was banned from use, runoff caused fish and waterfowl to carry major concentrations of the chemical in their bodies. Because the bald eagle and peregrine falcon preyed on the fish and waterfowl, they collected even greater concentrations of the pesticide in their own systems. This caused numerous problems, the most serious of which was the thinning of their egg shells. It became increasingly difficult for them to raise their young because their eggs would crack and break when they tried to incubate them.

FIGURE 12–8 The number of bald eagles in the population has rebounded since the ban on DDT went into effect.

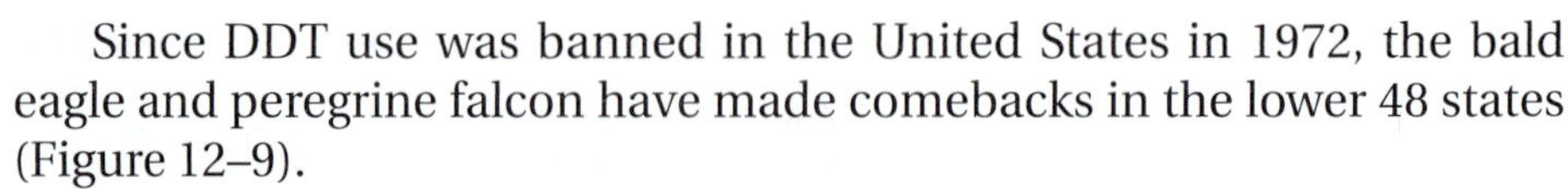

FIGURE 12–9 The peregrine falcon population has rebounded since the use of DDT was banned. (Courtesy of the U.S. Fish and Wildlife Service.)

Since DDT use was banned in the United States in 1972, the bald eagle and peregrine falcon have made comebacks in the lower 48 states (Figure 12–9).

The last factor that is known to have contributed to the endangerment of wildlife species is illegal killing. Sadly, when money can be made, there will always be someone willing to break the law to get it. It has been estimated that illegal killing has been a factor in less than 3 percent of the endangered species in the United States, but it is a significant problem in developing countries around the world.

Some species become endangered for reasons we cannot determine. Extinction is a natural consequence of the process of **evolution**, which has been going on since life first appeared on Earth. Species that are highly specialized or cannot adapt to a changing environment become endangered or extinct.

Some plants and animals may flourish for hundreds or even thousands of years, whereas other species may come and go in a relatively short period. On the North American continent, many great mammals such as camels, mammoths, and saber-toothed tigers flourished until the end of the last ice age. Also present during this period were such species as mule deer, pronghorn antelope, and grizzly bears. For a variety of reasons that no one completely understands, only the latter species survived into our modern era. However, most species that survive through the years have one common trait: adaptability.

An animal or plant species that is capable of changing as its environment changes has a better chance of survival. For example, species such as the white-tailed deer, cottontail rabbit, and common field mouse are highly adaptable. They adjust to a wide variety of habitats and habitat situations, and therefore they are able to survive drastic changes in their environments.

The opposite of highly adaptable species are species that are specialized. As a species becomes more specialized, it tends to lose the ability to change or adapt to changes in its environment. These species are vulnerable to extinction. The everglade kite is a good example of a highly specialized species (Figure 12–10). It feeds exclusively on the apple snail. Therefore, as the snail population decreases, so does the everglade kite. Another example is the black-footed ferret of the Great Plains, which feeds almost entirely on the prairie dog. With the demise of most of the great prairie dog colonies, the black-footed ferret population has plummeted.

FIGURE 12–10 Everglade kites such as this one are an increasingly rare sight over much of their former habitat. (Courtesy of the U.S. Fish and Wildlife Service.)

Plant and animal species get into serious trouble when humans alter their environment at a faster rate than the plant or animal can adapt. Most species are capable of adapting to changes in their habitats if those changes take place over an extended period of time. However, with bulldozers and other modern construction equipment, changes occur much faster than the rate at which most species can adapt. Of course, if we pave, plow, or pollute a species' entire habitat, then it has little chance of survival, regardless of how quickly it can adapt.

If extinction is a natural process, then why should we worry about species becoming extinct during our time? For one thing, the rate at which species are going extinct is far from natural. When the Pilgrims first arrived in the New World, approximately 500 more species and subspecies of plants and animals were on the continent than live here today. In other words, some 500 species, subspecies, and varieties of our nation's plants and animals have been lost forever. In contrast, all of North America lost about three species every 100 years during the 3,000 years of the Pleistocene Ice Age. Most species are lost as a result of such human activities as habitat destruction, habitat degradation, introduction of non-native species, environmental pollution, and exploitation. Each species of plant, wildlife, and fish is of scientific, ecological, recreational, aesthetic, and educational value to our country. They can also be of economic value.

No matter how small and unknown a species is, it could be of great value to us one day. For example, penicillin was discovered in a fungus, and since its discovery has saved millions of lives. The anticancer drug Taxol™ comes from trees, as does rubber. Plants provide us with sources for the drugs quinine, morphine, and curare. Although there are some 80,000 species of edible plants, 90 percent of the world's food comes from fewer than 20 species. This means that there are many species that may prove useful in the future. Imagine what we stand to lose if these species are destroyed before we have a chance to study them. This is exactly what is happening around the world. Species are becoming extinct before we can study them—in some cases, before we have even identified them. Thus, there are many good reasons to save our great diversity of plants and animals.

The Endangered Species Act has saved dozens of species from almost certain extinction. When professional wildlife scientists identify a species as threatened, they can use the act to protect the species and its critical habitat. Through captive breeding programs, animals such as the red wolf, California condor, and black-footed ferret have a chance at being restored to their former habitats. In the lower 48 states, the gray wolf, bald eagle, grizzly bear, and peregrine falcon have increased in numbers and expanded their ranges primarily because of the protections offered by the Endangered Species Act. The habitat protection provisions of the act ensure that these animals will have habitat to return to.

When a species is listed as endangered, a recovery plan is written by an expert or team of experts. The recovery plan may call for captive breeding and reintroduction, habitat protection or improvement, new research, or special wildlife and habitat management techniques (Figure 12–11). A plan may use one or more of these practices.

Figure 12–11 Captive breeding programs are the only hope for many endangered species like this red wolf pup. (Courtesy of the Texas Parks and Wildlife Department. Copyright 2002.)

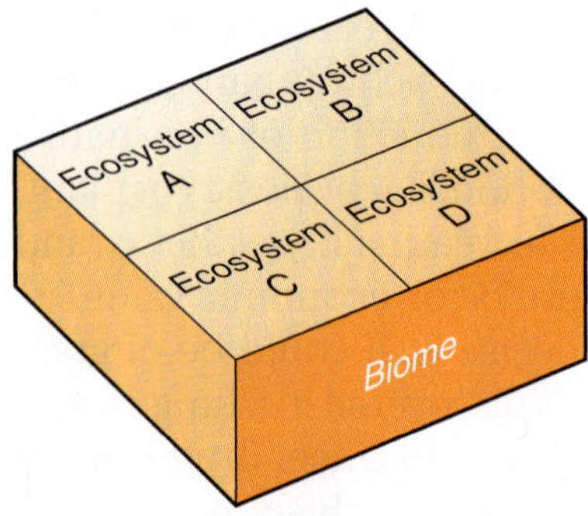

FIGURE 12–12 The boundaries between ecosystems frequently overlap, and they are porous enough that organisms and materials pass through in both directions.

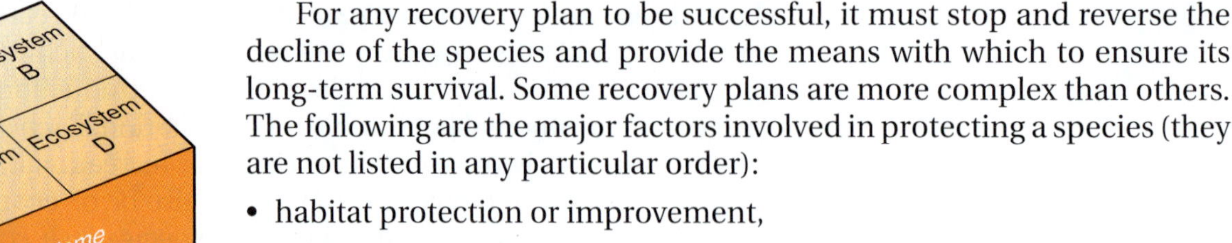

For any recovery plan to be successful, it must stop and reverse the decline of the species and provide the means with which to ensure its long-term survival. Some recovery plans are more complex than others. The following are the major factors involved in protecting a species (they are not listed in any particular order):

- habitat protection or improvement,
- scientific research,
- captive breeding,
- reintroduction into the wild, and
- special habitat or wildlife management techniques.

Biomes

Many different kinds of environments support life. In each instance, the organisms that occupy an environment have adapted to its unique characteristics. Ecosystems that are located at similar latitudes and elevations often have many of the same characteristics. The temperature and the amount of precipitation usually does not vary significantly. A **biome** is a group of ecosystems within a region that have similar types of vegetation and similar climatic conditions (Figure 12–12).

INTERNET KEY WORDS

biomes, North America

The climate of a region is in part determined by its latitude or distance from the equator. Temperature declines the farther the distance from the equator. Climate also is determined by altitude (Figure 12–13). Conditions are even different according to altitude within the same mountain range. Temperature decreases as altitude increases. Ecosystems support different kinds of organisms as we move north or south from the equator and as we move from lower to higher elevations.

Freshwater Biomes

INTERNET ADDRESS

http://mbgnet.mobot.org/fresh/wetlands/

A **freshwater biome** is composed of plants, animals, and microscopic forms of life that are adapted to living in or near water that is not salt (Figure 12–14). The term *freshwater* can be misleading, however, because some freshwater environments depend on water that may be stagnant,

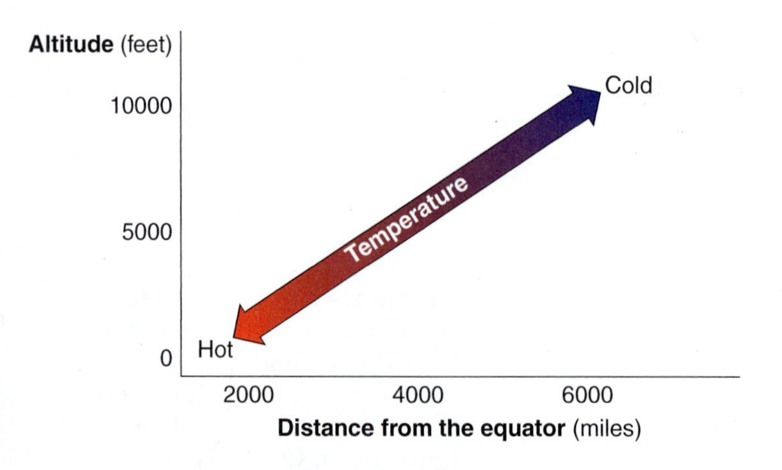

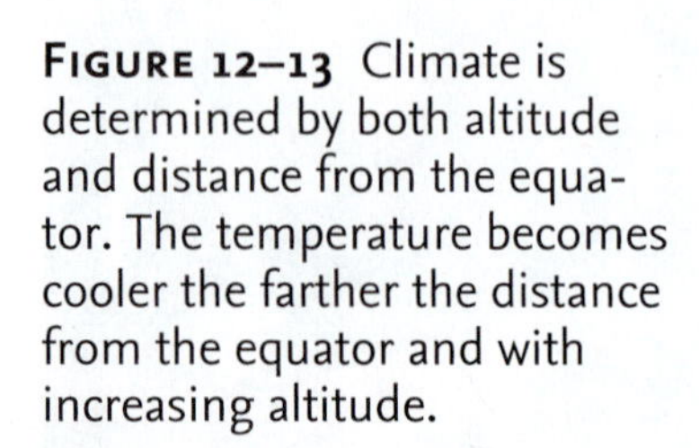

FIGURE 12–13 Climate is determined by both altitude and distance from the equator. The temperature becomes cooler the farther the distance from the equator and with increasing altitude.

Figure 12–14 An example of a freshwater biome

muddy, or heavily polluted. Such water may not appear to be "fresh" at all.

INTERNET KEY WORDS

freshwater ecosystem

Many different environmental conditions are found in freshwater ecosystems, and each set of conditions creates a unique living environment (Figure 12–15). The water in a particular area may be hot, cold, or loaded with dissolved materials. In each instance, different kinds of plants and animals have become adapted to the current water conditions. In some cases, organisms have become so specialized that they may be unable to survive in any other known location.

INTERNET KEY WORDS

marine food chain

Among the organisms that live in freshwater biomes is **plankton**, which consists of microscopic plants called **phytoplankton** and microscopic animals called **zooplankton**. They are found floating on the water

Figure 12–15 A freshwater ecosystem may have many different kinds of plants and animals living within its boundaries.

interest Profile

SOUTHERN CAVEFISH (*TYPHLICHTHYS SUBTERRANEUS*)

The southern cavefish has become adapted to living in the darkness of a cave environment. It has no eyes and its skin is pink-white. It has two rows of sensory feelers on the caudal fin (the back fin) to help it navigate. It lives in caves east of the Mississippi River from Indiana on the north to Georgia and Alabama on the south.

surface. They serve as food producers in the food chain and are eaten by fish and other aquatic animals (Figure 12–16).

water quality, wildlife
thermal stratification

The organisms that are capable of living in a freshwater habitat are limited by conditions such as water temperature, light intensity, concentration of dissolved materials, and the flow rate of a stream (Figure 12–17). First, water temperature determines which species of organisms can survive in a particular water habitat. Only those organisms that are capable of living within the temperature range that occurs there will be able to survive in a particular water habitat. This is because their biological functions that sustain life are disrupted by temperatures to which they have not adapted. Some organisms have little tolerance for changes in water temperature and die when the temperature becomes too hot or too cold for them. Some organisms can live in water that is quite hot, while others survive best in cold water. Few species, however, can survive in water temperatures that fluctuate rapidly more than a few

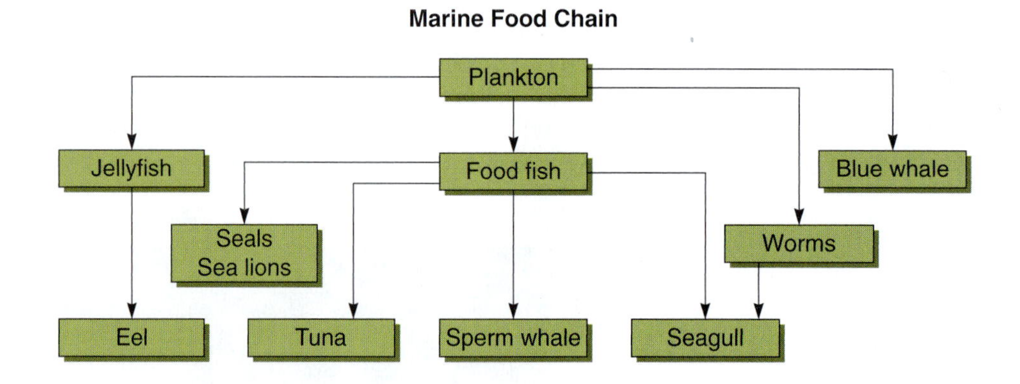

FIGURE 12–16 Plankton are the food source either directly or indirectly for nearly all marine animals.

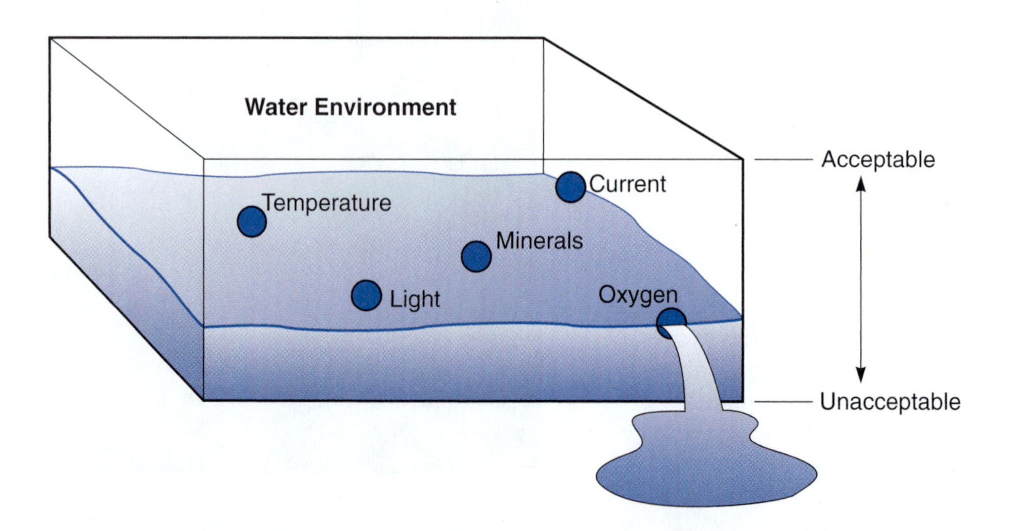

FIGURE 12–17 The most limiting factor in the water environment determines how acceptable the environment will be. Like holes in a water tank, the lowest hole determines the greatest capacity of the tank.

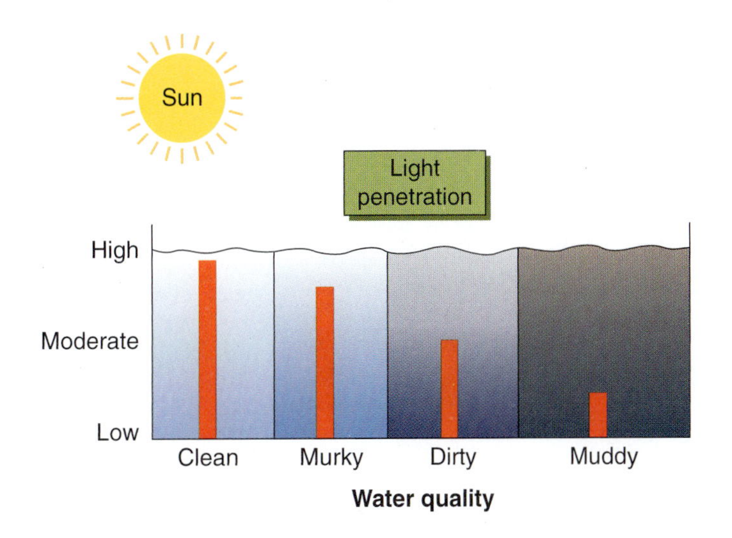

FIGURE 12–18 Light penetration in an aquatic environment

degrees. Some organisms can survive over a broad water temperature range as long as the temperature changes gradually.

Second, the amount of light that can penetrate the water affects freshwater organisms. This determines the amount of photosynthesis that can occur and the kinds of plant life that can exist in the water. When water is **turbid** or cloudy with suspended particles of silt, then photosynthesis is limited in water plants. This is because suspended particles block the sunlight and prevent light from reaching the water plants (Figure 12–18).

Third, the concentration of dissolved minerals in the water also limits the kinds of organisms that can live in an aquatic environment. When high levels of nitrates and phosphates are present, dense blooms of blue-green algae and other plants may occur in the surface water. At the same time, fish and other aquatic animals may die if the concentration of these materials in the water becomes too high. They also may die if the concentration of dissolved oxygen in the water is too low. Fish and other aquatic animals that need a rich supply of oxygen may die or be replaced by other fish that can survive in an environment with low levels of oxygen.

FIGURE 12–19 Water current in a stream may be the limiting environmental factor for some organisms. (Courtesy of Getty Images.)

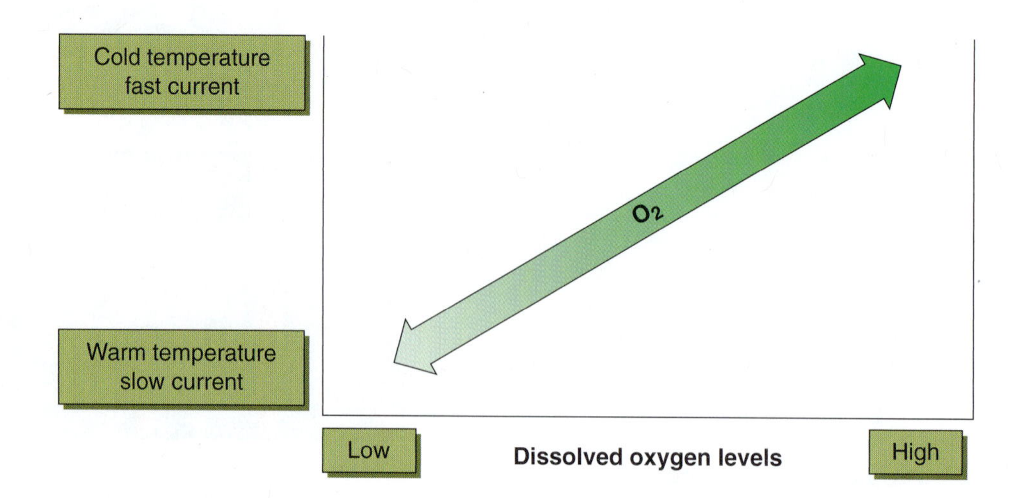

FIGURE 12–20 Dissolved O_2 in an aquatic environment

Fourth, the current or flow of the water in a river or stream may be the limiting environmental factor for some organisms (Figure 12–19). The intensity of the current contributes to the amount of force that an organism must be able to withstand to prevent being carried downstream. It also affects the availability of food.

Fifth, organisms in a freshwater habitat are affected by the amount of oxygen dissolved in the water. Rapidly flowing cold water also tends to carry more dissolved oxygen than warm water that is moving slowly through a pond or swamp (Figure 12–20). Flowing water tends to dissolve oxygen readily, and cold water temperatures help to stabilize oxygen in the water.

constructed wetlands

These five conditions interact to create each freshwater ecosystem. If any one condition becomes unfavorable, an entire population of organisms may be eliminated or replaced by other species better suited to the new conditions.

Two different kinds of habitats are found in freshwater biomes. A **lotic habitat** exists where water flows freely in streams and rivers. The dominant unidirectional flow of water tends to support little plant growth. It is typical of a detritus-based food web that food for fish and other aquatic animals is carried in by water from distant sources. In contrast, a **lentic habitat** is one in which water stands for long periods of time—as in swamps, ponds, and lakes. Lentic environments contain areas of differing light and lack a strong current. The water temperature in these habitats also differs. Deep water tends to be cold, and the water near the surface is usually warmer. This is called **thermal stratification**. Some common plants and animals that live in lentic areas are cattails, lilies, herons, and frogs (Figure 12–21).

FIGURE 12–21 Herons thrive in a lentic habitat. (Courtesy of the U.S. Fish and Wildlife Service.)

Wetlands—land areas that are flooded during all or part of the year—are currently attracting public interest. A wetland may consist of dry land for much of the time and still be classified as a wetland if it is subject to flooding during some season of the year.

The total area of wetlands in North America has declined with the advance of civilization. We have drained many acres of marshes and swamps to prepare the land for raising crops. The U.S. Department of Agriculture has reported that as much as 90 percent of the original prairie wetlands located in Iowa, Minnesota, eastern North Dakota, and southeastern South Dakota have been converted to farms. Similar trends are evident in other areas. A strong movement initiated by advocates for

FIGURE 12–22 There has been a concerted effort to restore and maintain wetlands in North America. (Courtesy of the U.S. Fish and Wildlife Service. Photo by Bob Ballou.)

waterfowl and aided by federal legislation is attempting to reverse this trend in the United States.

Organized groups of sportspeople such as Ducks Unlimited have taken a leading role in the effort to restore wetlands. They have purchased large tracts of land with donated funds and returned them to marshlands, hoping that through such efforts they can create favorable conditions for and increase the populations of migrating waterfowl such as ducks and geese (Figure 12–22). Other organizations that are involved in wetlands restoration include state and federal agencies, conservation groups such as the Nature Conservancy, and ecological researchers.

Some attempts to increase the amount of wetlands have created controversy between landowners and the government agencies that are responsible for determining and enforcing wetland policy. Laws and regulations that encourage restoration of wetlands include Executive Order 11990, the Protection of Wetlands Act of 1977, the federal Water Bank Program, state water bank programs, and the Watershed Protection and Flood Prevention Act.

The general sentiment today is that wetlands are difficult to restore and it is best not to disturb them in the first place. Some landowners believe that the law reaches too far by designating some lands as wetlands that historically did not flood. They define wetlands as some part of the land that floods only seasonally as a result of land modifications implemented on the property. Such changes include diverting water into new channels and irrigation practices that result in the ponding of water.

Regulations to modify the ways in which wetlands can be used are currently being developed by federal and state government agencies. All efforts to promote or reduce government control of wetlands will affect the freshwater ecosystems of North America.

INTERNET KEY WORDS

marine biome
marine life, temperature, salinity, light

Marine Biome

The world's largest biome is the **marine biome**. It consists of the oceans, bays, and estuaries, and makes up approximately 71 percent of the

FIGURE 12–23 The world's largest biome is the marine biome. (Courtesy of Getty Images.)

Earth's surface area (Figure 12–23). A distinguishing characteristic of the marine biome is salinity—that is, the salt concentration of the water. Oceans have a salt concentration ranging from 3.0 percent to 3.7 percent. Salinity is an important factor that affects an organism's ability to survive in a marine environment. It varies quite a bit in surface waters, especially near land where rivers begin, but it tends to be quite stable in deep ocean waters. The marine biome is also quite diverse. It covers the hot southern zones of the North American coast to the cold arctic regions of the continent's northern reaches.

The ability of organisms to live in the marine biome depends on several conditions that are necessary to sustain life in the ocean (Figure 12–24). Water temperature, for example, varies considerably in the marine biome. Deep ocean waters tend to be colder than surface waters, and arctic waters are colder than ocean waters located near the equator. Ocean currents such as the Gulf Stream, which is warm, and the Humboldt Current, which is cold, distribute ocean temperatures differently.

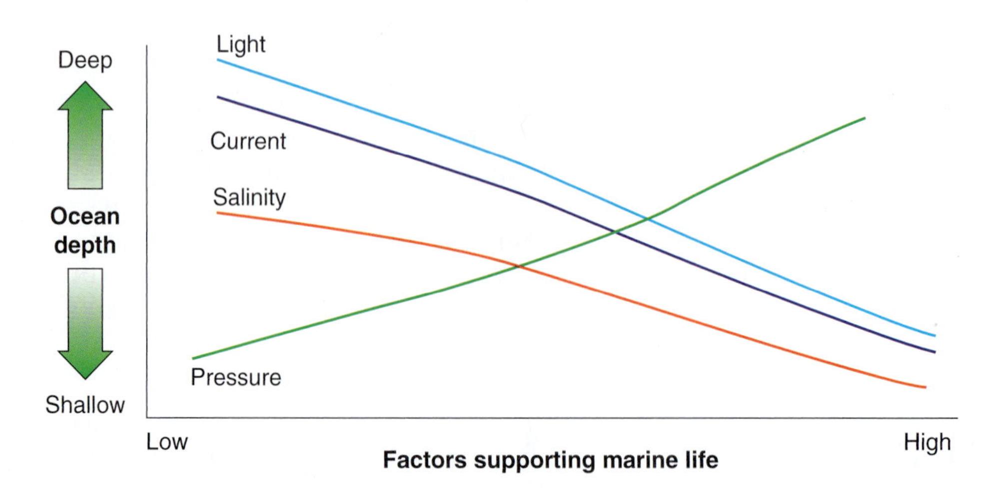

FIGURE 12–24 The characteristics of water environments in different ocean locations can vary greatly as illustrated by this graph. Each of the marine species is found in an ocean environment where the effects of light, current, salinity, and pressure fall within their range of tolerance.

The amount of light that penetrates the water's surface influences the amount of plant life found there. The presence of food-producing plants attracts fish and other aquatic organisms to waters where food is plentiful. A supply of nutrients for plants to use in photosynthesis also is important in producing food for aquatic animals. Certain areas of the ocean have strong ocean currents that circulate deep ocean waters to the surface. These currents are known as *upwelling currents*. These areas are important to commercial fishing industries because deep waters tend to be high in nutrients, and fish concentrate in these areas to feed. Ocean currents are important for moving nutrients from deep ocean waters to surface waters where photosynthesis can take place. Plankton are found in abundance where sunlight and nutrients are available, and these organisms play an important role as food producers in the aquatic food chain.

Water pressure increases with water depth because of the action of gravity on water molecules. As more water molecules are stacked above an organism, they exert greater pressure on its body. A good example of this is the pressure one feels on the eardrums during a dive in deep water. The deeper the dive, the greater the pressure. Some organisms can withstand the pressure of deep ocean water, but most aquatic organisms tend to seek and remain in surface waters where food is more plentiful and where water pressures are not extreme.

INTERNET KEY WORDS

neritic zone
intertidal zone
oceanic zone

Scientists have identified three important zones in ocean environments: the intertidal, neritic, and oceanic zones. The **intertidal zone** is located near the shore (Figure 12–25). When the tide is low, this zone is an exposed beach that is above the water. During high tide, it is covered with water. The width of this zone is greatest when the slope of the ocean floor is gradual. It is narrow when the slope of the ocean floor is steep.

Animals and plants that live in the intertidal zone must be adapted to living both in and out of water. They must be able to cope with frequent changes in the temperature of their environment. They must

Figure 12–25 An intertidal zone (Courtesy of Wendy Troeger.)

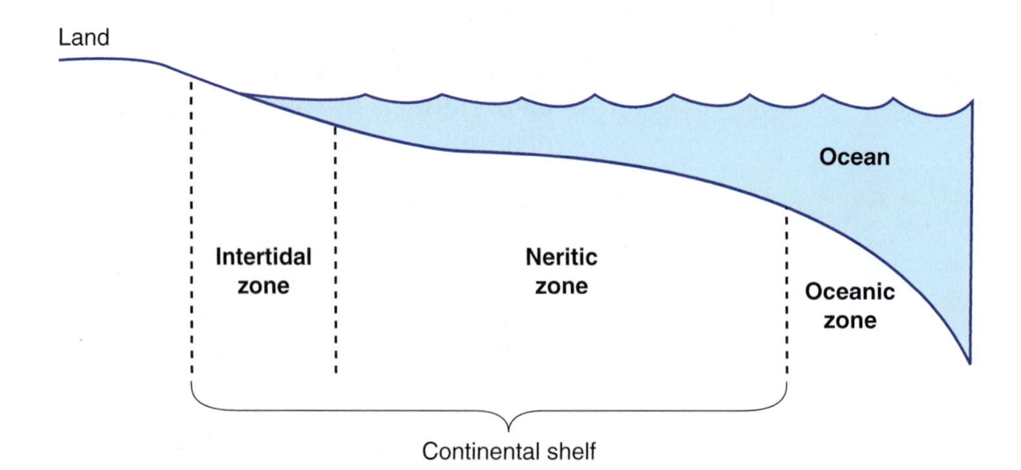

FIGURE 12–26 The continental shelf varies in width along the coastlines of the world. The neritic zone contains the most favorable marine living environments.

also be capable of retaining moisture when the tide is out as well as tolerating salinity when the tide is in. The intertidal zone is subject to strong water currents that can wash plants and animals out to sea. Organisms living in this zone must be capable of withstanding these currents by attaching themselves to rocks or burrowing into the sand. The types of animals that can survive in tidal zones depend in large part on the type of material composing the ocean floor.

The **continental shelf** is land that is submerged under the surface of the ocean (Figure 12–26). It slopes gradually away from the shore toward deeper water. The area beyond the intertidal zone that extends to the outer edge of the continental shelf is called the **neritic zone**. The most favorable living environment in the ocean is found here. Plankton and other food-producing organisms are abundant in this zone because nutrients and light are abundant. For most aquatic organisms, the neritic zone is the most productive area in the ocean.

The **oceanic zone** begins at the outer edge of the continental shelf and includes the deep ocean waters. The surface waters in this zone are productive. The presence of light and nutrients supports plankton and other food-producing organisms to a depth of some 200 yards. Below that depth, the waters become dark and the temperatures become cold. Although some production does occur to depths of 2,000 yards, the living environment below that is generally cold and less productive. Deep-ocean volcanic vents, explored in the last 10 years, provide warmth and energy for strange communities of worms, mollusks, and shrimplike crustaceans in some parts of the ocean floor (for example, off the coast of Washington).

Freshwater and saltwater mix where rivers and streams enter the ocean. Such areas are often marshy, shallow, and abundant in food-producing plants. An area such as this is called an **estuary**. The salinity of the water varies, depending on the strength of the current as well as seasonal rainfall, tides, and evaporation in shallows. These areas are prime habitats for oysters and several types of young fish. An estuary has characteristics of both freshwater and marine biomes (Figure 12–27).

Terrestrial Biomes

A **terrestrial biome** is a large ecosystem that consists of plants, animals, and other living organisms that live on land. Several distinct biomes are

FIGURE 12–27 This river opening is close to sea level. At high tide, the sea flows in and the estuary becomes salty, almost like seawater. At low tide, the water flows out and the water becomes fresh, almost like freshwater. (Courtesy of Bill Camp.)

terrestrial biome
desert biome

found on the North American continent (Figure 12–28). Among these are the desert, tundra, grasslands, temperate forests, and coniferous forests. Each biome is distinctly different from the others, and each provides a set of environmental conditions that support different kinds of living organisms.

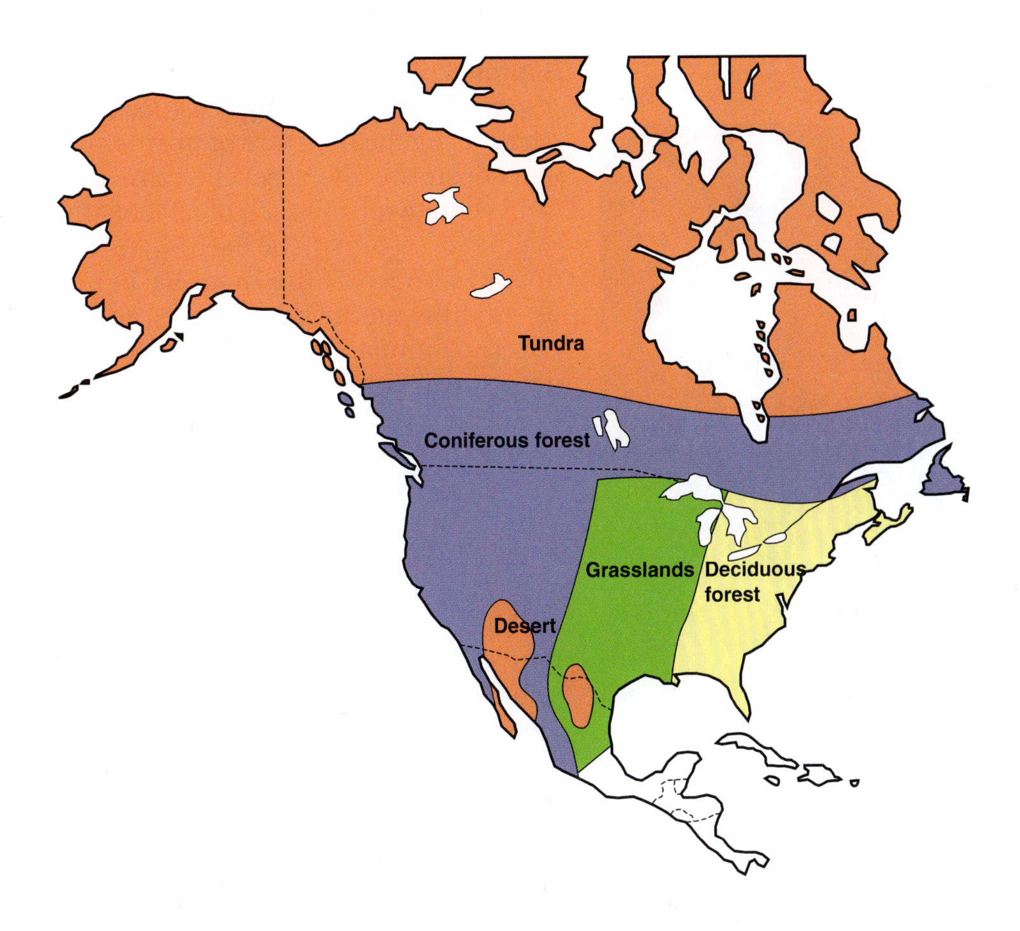

FIGURE 12–28 Several distinctive terrestrial biomes are found on the North American continent.

FIGURE 12–29 A desert biome is an extremely dry environment. (Courtesy of Getty Images.)

Desert Biome

The **desert biome** is an extremely dry environment (Figure 12–29). Precipitation is usually less than 10 inches per year, which is only enough to enable specially adapted plant and animal species to live. Daytime temperatures are usually hot, and the nights are often quite cold. The vegetation consists of exceptionally hardy plants such as cactuses and shrubs.

Some plants survive in the desert because they have developed deep taproots that can draw moisture from far down in the soil. Others have especially short growing seasons: They germinate and grow to the seed stage using the moisture of the melting winter snow or often from a single rainstorm. The desert literally "blossoms like a rose" before the hot, arid climate robs the soil of its moisture. Desert conditions are found on nearly 35 percent of the Earth's land area. In North America, desert lands make up much of the Great Basin region located in Utah and Nevada, the southwestern United States, and northern Mexico.

FIGURE 12–30 A roadrunner is capable of surviving in an exceptionally dry environment. (Courtesy of the U.S. Fish and Wildlife Service.)

The animals that inhabit the desert biome are highly specialized to survive with a limited supply of water (Figure 12–30). Some get all the water they need from the plants they eat. Others lap up the dew drops that condense on plant leaves during the cool nights. Still others get their necessary moisture from the body fluids of the animals on which they prey. In some cases, desert animals and birds travel long distances to drink at the few water holes found in these arid regions. Among the animals that are found in the desert are small rodents such as rats and mice, birds, lizards, snakes, insects, and birds of prey. In some instances, larger animals such as jackrabbits (hares), coyotes, and pronghorn antelope are found within traveling distance of water.

Tundra Biome

The **tundra biome** is located in the frozen northern regions of the continent. It extends from timberline on its southern limit to the

interest profile

THE GREAT BASIN: A DESERT PROFILE

Jim Bridger was a famed mountain man during early explorations of the American west. He was a trapper and trader who spent much of his time alone or among the Native American tribes that lived in the region. Mormon leader Brigham Young once inquired about the Great Basin as a possible location for a settlement and Jim Bridger replied that it was unfit for human habitation. He further stated that crops could not be grown there, and he was reported to have offered $1,000 for the first bushel of corn grown in the valley of the Great Salt Lake. Before Brigham Young and the Mormon pioneers brought irrigation to this desert region, the land did indeed fit Jim Bridger's description as a desert unsuitable for human settlement. Today, irrigation has made it possible to convert large desert tracts into productive farmland.

areas of permanent snow and ice cover in the north (Figure 12–31). Temperatures are well below freezing for most of the year, and the soil remains permanently frozen—that is, as *permafrost*—underneath the surface. Although less than 10 inches of precipitation falls annually in these areas, evaporation rates are low and the water cannot penetrate the frozen soil, so much of the area remains wet and spongy during the summer growing season.

Although the growing season is short, the sun shines almost 24 hours per day at these northern latitudes during summer, and the tundra plants must mature and produce seed before the winter season returns. Water ponds in the low spots, and the abundant aquatic and semi-aquatic insects support migrating waterfowl and other birds in the summer. Caribou and reindeer migrate into the tundra regions during the summer to graze on the grasses, mosses, lichens, sedges, and shrubs that grow there. As the growing season ends, these migratory birds and animals move south.

FIGURE 12–31 Alaskan range and tundra is located in a region that is extremely cold and dry.

The musk ox is a permanent resident in the tundra. Most of the precipitation in these areas comes in the form of rain during the fall months. Because snowfall is quite limited, the musk ox is able to graze year-round on the short vegetation. This animal has adapted well to the difficult climatic conditions of the tundra environment. It is protected from the cold by a long, thick coat of hair and a layer of body fat. Its hooves are useful not only as a defense against predators but also to remove snow over the musk ox's food supply.

Rodents such as lemmings are extremely abundant during some years but nearly disappear at other times. They provide food for predators such as the lynx, artic fox, ermine, birds of prey, and bears that move in and out of the tundra region. This natural cycle of the lemmings from abundance to scarcity controls the numbers of predators.

Grassland Biome

INTERNET KEY WORDS

biome, tundra
biome, grassland
biome, prairie

The **grassland biome** in North America is often referred to as the *prairie* and is characterized by a lack of trees. It is located in the middle of the continent and includes the prairies of Canada and the midwestern region of the United States and the grasslands that extend south into Mexico. Historically, the grassland areas formed the largest biome in North America (Figures 12–32 and 12–33). The term *grasslands* is misleading because many plants besides grasses are found there, including sedges, forbs, and several other types of plants. Much of the grassland in North American has now been converted into farmland where corn, wheat, and soybeans are the main crops.

The native species that covered the grassland region before the land was cultivated were well adapted to the region's climate and frequency of natural fires. Both bunch grasses, which grow in tufts, and sod grasses with matted roots have deep root systems that make good use of soil moisture. Periodic burning is characteristic of this environment. The deep roots also protect these plants against fire damage. Although fire

FIGURE 12–32 A tallgrass prairie (Courtesy of Shutterstock.)

Figure 12–33 Pronghorns are well adapted to this shortgrass prairie. (Courtesy of Shutterstock.)

may completely consume the foliage of these plants, new shoots arise from the roots when favorable growing conditions return.

The climate of the grassland biome is continental, meaning it has moderate to hot summers and cold, freezing weather in the winter. Grasslands are located in areas where severe droughts occur from time to time. These drought conditions contribute to the lack of trees in temperate grasslands. Lightning fires also are extremely damaging to trees that are not specifically adapted to burning; the fires thus keep them from becoming established.

It was in the prairie regions that the great herds of bison once ranged. Today, the bison have been replaced in these areas by herds of domestic cattle and sheep. Only in Yellowstone National Park are bison found in abundance. Other wild animals that are commonly found in the grasslands include prairie dogs, mice, snakes, rabbits, pronghorn antelope, coyotes (prairie wolves), and several kinds of birds and insects.

Temperate Forest Biome

The **temperate forest biome** is identified by broadleaf trees such as oak, maple, cherry, ash, hickory, and beech that shed their leaves in the fall (Figure 12–34). The temperate forest biome in North America begins south of the coniferous forests of Canada and Maine and extends southward along the east coast and westward until they are gradually replaced by grasslands. In its natural state, this entire area was a **deciduous forest**. As civilization moved west, many woodland tracts were cleared and replaced by farms.

Figure 12–34 The white oak tree is native to the temperate forest biome. (Courtesy of Shutterstock.)

Precipitation in this habitat generally exceeds 30 inches per year. Four distinct seasons are observed in these regions, with a bright-colored display of maple and ash leaves after the first frosts in the fall. This biome tends to be less homogeneous in its plant population than some other biomes because its climate is less homogeneous.

Several levels or layers of vegetation called **strata** are found in a deciduous forest. The tall trees form the **canopy** or ceiling of leaves at the highest levels. Smaller trees fill in the area beneath the canopy. These shorter trees make up the **understory** of the forest. Short bushes make up the **shrub layer** in the zone beneath the understory. The shortest plants, including ferns and grasses, and other flowering plants are collectively called the **herb layer**. The **forest floor** is composed of a layer of decaying plant materials that act as a mulch in preserving the soil moisture.

INTERNET KEY WORDS

forest biome, temperate
forest biome, deciduous

Mammals and birds of many kinds are native to this environment. Squirrels and many species of birds prefer to live in the forest's canopy, while other species prefer the understory. White-tailed deer, opossums, skunks, foxes, birds of prey, snakes, squirrels, and mice are all common residents of broadleaf forests.

The passenger pigeon, which was a native species in this environment, became extinct primarily because of hunting pressure and the loss of critical nesting habitats. Wild turkey and black bear are species that have suffered as deciduous forest habitat has been lost to timbering and land development. Today, the wild turkey has been introduced into many areas of the country where its habitat is available, and the species is making a dramatic comeback in numbers.

Coniferous Forest Biome

INTERNET KEY WORDS

forest biome, coniferous

The **coniferous forest biome** is an evergreen forest of pine, spruce, fir, and hemlock. It forms a broad northern belt across the continent, extending from the grasslands and temperate forests on the south to the tundra regions on the north, and from the Northeast's coastal region to the Pacific Northwest. The vegetation in this biome consists mostly of trees known as **conifers** that produce seed in cones (Figure 12–35). The foliage of conifer trees is dense; as a result, the light intensity on the forest floor is inadequate to support the growth of most plants. A heavy carpet of dead needles covers the forest floor, and few shrubs, grasses, or other plants are found there.

FIGURE 12–35 The coniferous forest biome consists of trees that produce seeds in cones. (Photo courtesy of the Boise National Forest.)

FIGURE 12–36 Forests that are harvested under good management practices will continue to prosper along with the wildlife that depend on them.

Precipitation in the region is mostly in the form of snow, which generally ranges from 15 inches to 40 inches per year. The winters tend to be long and cold, and summer temperatures are moderate with cool nights. The needle-shaped leaves of conifers are well adapted not only to cold temperatures but also to conserving moisture during dry times. The shape of the trees and the flexibility of the branches enable heavy snow loads to fall to the ground without breaking the limbs.

The coniferous forest biome is home to many birds, insects, and mammals. Large mammals such as elk, moose, mule deer, and caribou often graze in the meadows and wetlands that are scattered throughout the coniferous forests. Predatory species that are found in these regions include black bears, grizzly bears, wolverines, lynx, timber wolves, foxes, mink, hawks, and owls. Squirrels, porcupines, hares, mice, and a variety of birds also live in coniferous forests.

Coniferous forests provide much of the lumber used for construction (Figure 12–36). These forests are an important renewable resource, and forests that are harvested under good management practices can continue to provide a healthy environment for the living organisms that depend on them.

Looking Back

Population ecology is concerned with the total number of individuals of a species and their relationships with the factors within the environment. It is a study of organisms at the population level rather than as individuals. Measurements that are often used to describe a population include density, distribution, birthrate, age structure, and mortality rate.

A common factor occurs with virtually all of organisms that are listed as endangered species: the influence of humans. Many endangered species have experienced habitat degradation or destruction. The Endangered Species Act of 1973 is proving to be an effective tool in protecting species whose numbers have seriously declined.

An ecosystem is seldom isolated from neighboring ecosystems because its boundaries usually enable organisms and materials to flow in and out of the environment. Ecosystems that exist in areas with similar vegetation and climate are grouped together into biomes. A freshwater biome exists when water that makes up the living environment is not salty. Ocean water makes up the marine environment. The terrestrial biomes of North America include all land-based environments. Water habitats are sometimes called *wetlands.* Federal regulations require restoration and preservation of wetlands.

Self-Analysis

Essay Questions

1. What is the definition of population ecology and why is it important?
2. In what ways is data regarding age structure helpful in designing a management plan for a declining population?
3. What are some key factors that affect the growth or decline of a population?
4. What events are best known to contribute to a species becoming endangered or extinct?
5. What is the purpose of the Endangered Species Act?
6. How does human alteration of a habitat affect the ability of affected organisms to adapt?
7. What role does climate play in the formation of biomes?
8. What are some ways in which freshwater and marine biomes are similar and different?
9. What is plankton and why is it important in the aquatic food chain?
10. Why is wetland habitat restoration important?

Multiple-Choice Questions

1. Which of the following is not a characteristic associated with the growth or decline of an established population?
   a. latitude
   b. immigration
   c. birthrate
   d. mortality rate
2. An organism that reproduces only once in its lifetime is said to be
   a. semelparous.
   b. iterparous.
   c. promiscuous.
   d. monogamous.
3. A group of ecosystems within a region that have similar types of vegetation and similar climatic conditions is
   a. a stratum.
   b. an estuary.
   c. a habitat.
   d. a biome.
4. Which of the following factors has no effect on the climate of a region?
   a. latitude
   b. altitude
   c. distance from the equator
   d. vegetation in the region
5. The world's largest biome is the ______________ biome.
   a. marine
   b. freshwater
   c. temperate forest
   d. coniferous forest

6. Microscopic plants and animals found floating on the water's surface where they provide food to aquatic animals are called
   a. bacteria.
   b. plankton.
   c. scum.
   d. molds.
7. Water that is clouded with suspended particles of silt is described as
   a. turbid.
   b. stratified.
   c. clean.
   d. salty.
8. A water habitat in which water tends to stand for long periods of time is called a ________ habitat.
   a. lotic
   b. turbid
   c. murky
   d. lentic
9. Which of the following water habitats is not part of a marine biome?
   a. lake
   b. bay
   c. estuary
   d. ocean
10. The zone in an ocean environment that is located underwater at high tide and above the water level at low tide is called
    a. an oceanic zone.
    b. an intertidal zone.
    c. a neritic zone.
    d. an ozone.
11. Which of the following terrestrial biomes is located in the frozen northern regions of the continent?
    a. desert biome
    b. grassland biome
    c. coniferous forest biome
    d. tundra biome

Learning Activities

1. Write a research report that discusses the major factors that have contributed to a species of plant or animal being listed as threatened or endangered. Identify the steps that are being taken to reverse the population decline.
2. Create a chart that lists the distinguishing characteristics of each North American biome. It should also list the major plant and animal populations and describe the climates of each biome.
3. Take a field trip to an area near your school and perform the following tasks:
   - Describe the major characteristics of the environment.
   - List the kinds of plant life that are observed.
   - List the animal species that are known to inhabit the area.
   - Identify the insect species that are observed.
   - Develop a chart that shows how each organism might fit into the food web.
   - Discuss how human activity in the area might benefit or harm the organisms and the environment in which they live.

Note: If a field trip is not possible, then viewing a good video showing the essential elements of the environment is a reasonable alternative.

TERMS TO KNOW

industrial solid waste
municipal solid waste (MSW)
hazardous waste
radioactive waste
nonhazardous waste
percolate
leachate
landfill
natural attenuation landfill
attenuation
containment landfill
liner
recycled animal wastes
offal
recycled nutrients
recycling

CHAPTER 13

waste management

Objectives

After completing this chapter, you should be able to

- define and explain solid waste, municipal solid waste, and industrial solid waste
- explain why the disposal of municipal solid waste is such an important and growing concern in our society
- define hazardous waste and describe a way to appropriately dispose of it
- explain how natural attenuation works and when it is an appropriate means of solid waste disposal
- explain how a containment landfill works
- suggest ways in which waste products from slaughterhouses might be recycled
- discuss the importance of recycling waste materials

Waste materials have become a major problem in many of the world's industrialized nations. Industrial process by-products, solid wastes from population centers, petroleum leaks and spills, and pesticide residues from farms, gardens, and yards are only a few of the waste materials that pollute the environment (Figure 13–1). These materials do not just go away when we are finished with them.

Many so-called environmental problems are debatable. Although climatologists generally agree that global temperatures are at their highest point in several thousand years, scientists disagree on whether greenhouse gases generated by human activity are actually causing significant global warming. Even if global warming is resulting from human activity, there is no universal agreement as to whether it will be as severe and cause as much devastation as many people contend. Many other scientific and political disagreements plague the environmental movement.

One undeniable problem that faces our society is that of solid waste disposal. In fact, even with the progress we have made in water pollution and air pollution in the United States in the past 30 years, solid waste management may well be our most pressing remaining environmental problem.

SOLID WASTES

What is solid waste? Solid waste consists of nonliquid, nonsoluble materials ranging from municipal garbage to industrial wastes that contain complex and sometimes hazardous substances. Solid wastes also include sewage sludge, agricultural refuse, demolition wastes, and mining residues. Technically, solid waste also refers to liquids and gases in containers.

agricultural chemicals, wildlife
soil erosion, habitat, wildlife
industrial solid waste disposal

We have basically two major sources of solid waste: municipal waste and industrial waste. When you "take out the trash" and place

FIGURE 13–1 Surface water pollution from oil and chemical spills is a major problem and causes serious damage to the environment. (Courtesy of Getty Images.)

FIGURE 13–2 Most municipal solid waste starts with homeowners, schools, and local businesses. (Courtesy of Bill Camp.)

it in a "trash can" you are participating in the solid waste disposal process (Figure 13–2). Whether you live in a rural area and dispose of your own solid waste or live in an area served by a trash collection and disposal service, this is a traditional part of everyday life. What you are disposing of is what we refer to as *municipal waste,* even if you live in a rural area.

The first category of solid waste we will discuss is **industrial solid waste**, which consists primarily of spoilage from mining, logging, and other industrial processes that is not disposed of in landfills. Mining is the largest producer of industrial solid waste, much of it in the form of rocks, soil, and other materials scooped from the Earth and piled onto other locations. Mining waste may be as innocuous as topsoil scooped off a building site and spread onto a nearby field or as extreme as toxic waste from a copper mining operation or radioactive waste from a uranium mine. Mineral mining, coal mining, gravel mining, and other forms of natural resource exploitation produce huge quantities of spoilage, but accurate data are simply not available to document how much is even in industrialized countries like the United States, let alone worldwide. According to the World Watch Institute, 571,000 hectares (approximately 1.4 million acres) of land were mined in one recent year worldwide. Of that, approximately two-thirds was nonfuel mining (minerals, metals, gravel, etc.), and the remaining one-third was coal mining. Much coal is mined by tunneling underground, but the vast majority is strip mined. Strip mining is the process of excavating the coal seam by removing the soil and rocks above it and then digging out the coal, which is now on the surface.

Logging waste appears to be highly destructive, but in reality it consists of leaves, branches, and bark, which are all biodegradable. In the long run, solid waste from logging operations returns to the ecosystem. Indeed, one component of logging waste that has found a commercial value is ground or shredded bark. At one time bark was a solid waste, but now it is a by-product that many home and business owners prize as bark mulch (Figure 13–3).

Although smaller in total quantity than industrial solid waste, **municipal solid waste (MSW)** actually presents a much more difficult

Figure 13–3 This mulch is a by-product of logging and is useful in landscaping around homes and businesses. (Courtesy of Shutterstock.)

problem in this country. In the United States, the total amount of municipal solid waste generated has more than doubled between 1960 and 2000, and per-capita waste generation has increased by half during the same period. In 1960, each American produced an average of approximately 2.7 pounds of municipal solid waste per day. That rate increased steadily until 1990 or so when it reached 4.3 pounds per person per day. In the decade of the 1990s, the rate of waste generation in this country stabilized at approximately 4.4 pounds per person per day. With increasing population, though, the total amount of municipal solid waste generated in the United States has continued to grow, increasing some 250 percent from 88 million tons in 1960 to slightly more than 221 million tons in 2000. As we will see later, the increased generation of "garbage" over that time span has produced a problem in disposal much greater than 250 percent!

municipal solid waste disposal

Much of this MSW is generated by households, but a major source of municipal solid waste is industry. This is an entirely different category from the industrial solid waste that we discussed earlier. Think of the garbage generated by a large factory: paper, metal scraps, wood scraps, cans, plastic, and more. Imagine the massive amounts of solid wastes that might be generated in such a setting. Imagine "taking out the trash" for a factory like this one. For another example, consider a food service business such as McDonald's Corporation. McDonald's has shown responsible environmental citizenship over the years by using recyclable products and minimizing waste. Regardless of how careful managers are, though, such a commercial operation cannot avoid generating huge quantities of solid waste. In fact, according to the Environmental Protection Agency, a typical McDonald's restaurant produces approximately 230 pounds of solid waste per day, with corrugated paper shipping boxes and food waste each contributing slightly more than one-third of the total. The remaining third consists of plastic

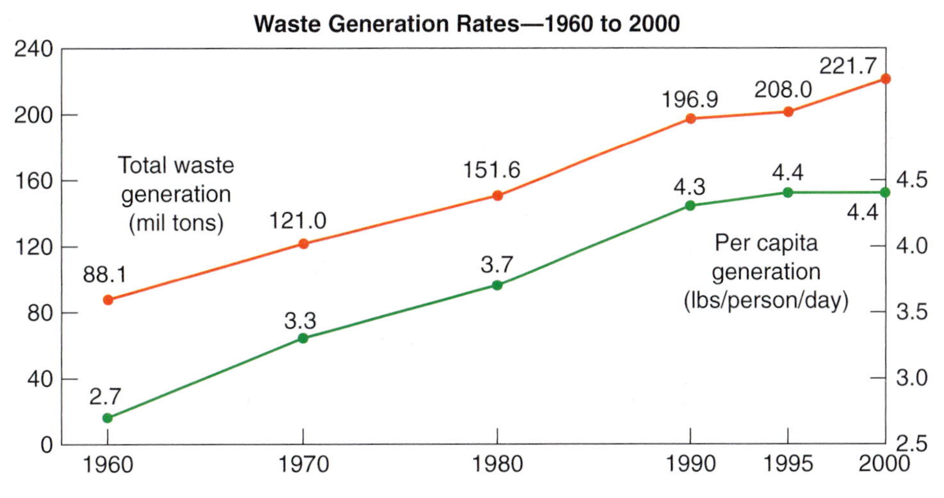

FIGURE 13–4 Waste generation rates, 1960–2000 (Source: Environmental Protection Agency, *Municipal Solid Waste Handbook.* Washington, DC: Author, 1996.)

wrappings, paper wrappers, napkins, polystyrene cups, and customer waste (Figure 13–4).

The total waste stream includes both business, industrial, and household waste (Figure 13–5). The biggest contributor to the MSW problem is still paper (39.2 percent) followed by yard waste (14.3 percent). Those two components make up more than half of the MSW generated in the United States each year, with all other materials making up slightly more than 46 percent of the total.

How do we dispose of all that municipal solid waste? The picture is getting a little brighter in this regard. In 1960, we recycled 6.4 percent of our MSW. By 2000, that recovery rate had gone up to 30 percent. That may not

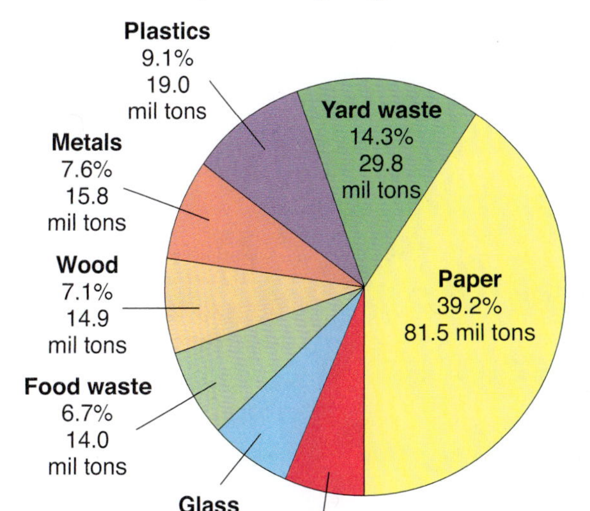

FIGURE 13–5 The total waste generation before recycling in 1995 was 208 million tons. (Source: Environmental Protection Agency, *Municipal Solid Waste Handbook.* Washington, DC: Author, 1996.)

Waste Management Practices, 1960–2000 (as a percentage of generation)

	1960	1970	1980	1990	1995	2000
Generation	100%	100%	100%	100%	100%	100%
Recovery for recycling/composting	6.4%	6.6%	9.6%	17.2%	27.0%	30.0%
Discards after recovery	93.6%	93.4%	90.4%	82.8%	73.0%	70.0%
Combustion	30.6%	20.7%	9.0%	16.2%	16.1%	16.2%
Discards to landfill	63.0%	72.6%	81.4%	66.7%	56.9%	53.7%

SOURCE: Characterization of MSW in the US: 1996 Update, US EPA, Washington, DC

FIGURE 13–6 Waste-management practices, 1960–2000 (as percentage of generation) (Source: Environmental Protection Agency, *Municipal Solid Waste Handbook.* Washington, DC: Author, 1996.)

sound like much of an improvement, but if we think of the total amount of MSW generated, that means we recycled 66.3 million tons in 2000 compared to 5.6 million tons in 1960. Again, when we discuss the changed requirements for landfills between 1960 and 2000, the importance of that change will become clear (Figure 13–6). Where do we stand internationally? The United States is the biggest culprit in terms of trash generation. The United States produced not only the largest total amount of MSW among the major industrialized nations but also the largest amount on a per-capita basis (Figure 13–7). The good news is that for the same year (1995), the United States also led the world in recycling (Figure 13–8).

Types of Solid Waste

There are three general types of solid waste: hazardous, nonhazardous, and radioactive. The state of Pennsylvania's definition of **hazardous waste** is typical: "Hazardous wastes are wastes that, in sufficient quantities and concentrations, pose a threat to human life, human health, or the environment when improperly stored, transported, treated, or disposed." In regulating hazardous waste, Pennsylvania uses a federal list of more than 600 specific wastes. Other wastes are

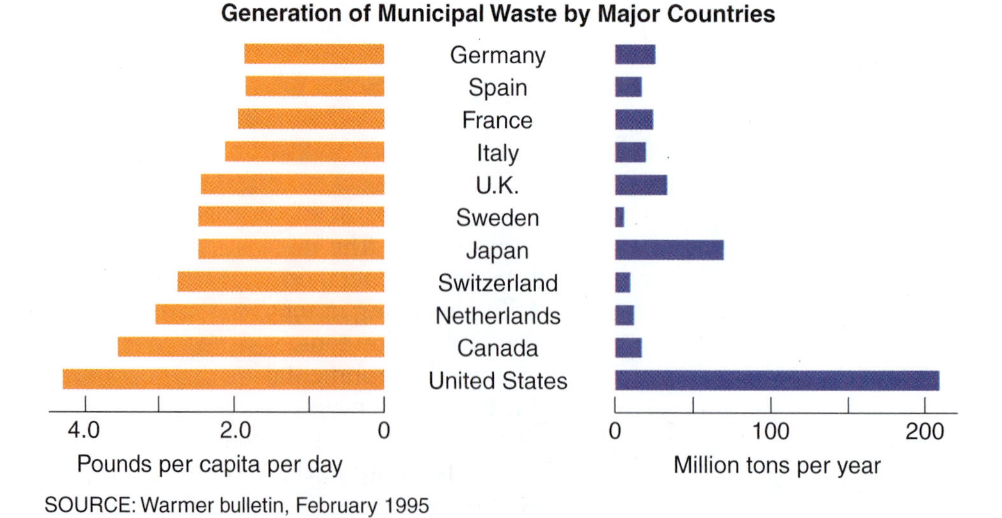

FIGURE 13–7 Generation of municipal waste by major countries (Source: Environmental Protection Agency, *Municipal Solid Waste Handbook.* Washington, DC: Author, 1996.)

FIGURE 13–8 Recycling of municipal waste by major countries. (Source: Environmental Protection Agency, *Municipal Solid Waste Handbook.* Washington, DC: Author, 1996.)

designated "hazardous" if they contain any of the following characteristics. They are:

1. *ignitable*—that is, they are combustible under certain conditions;
2. *corrosive*—that is, they are highly acidic, highly basic, or capable of corroding metal;
3. *reactive*—that is, they are unstable under normal conditions and capable of creating explosions or toxic fumes, gases, and vapors when mixed with water; and
4. *toxic*—that is, harmful or fatal when ingested or absorbed.

hazardous waste disposal
radioactive waste disposal

Mixtures of hazardous and nonhazardous waste are also labeled as hazardous. The hazardous waste designation does not include low-level radioactive waste, which is covered under separate state and federal rules.

According to the University of Maryland, **radioactive waste** is defined as:

> . . . any waste that contains, or is contaminated with any radioactive material. This includes liquids, solids, animal carcasses and excreta, used scintillation vials/cocktails, etc. A radioactive waste must not be disposed of as regular waste. Non-Radioactive waste should not be disposed of as radioactive waste. Mixed Waste is radioactive waste that also has the characteristics of a hazardous waste as defined by the State of Maryland or the EPA.
>
> Several chemicals which are specifically regulated by the State and the EPA as hazardous waste, and many more which possess the characteristics of a hazardous waste because they are corrosive, reactive, toxic, or otherwise potentially harmful to the environment. There are currently no permitted disposal options for most mixed waste, therefore care must be taken to avoid the generation of these wastes. Plans for the proper management of these materials should be reviewed with the Radiation Safety Office and the Hazardous Waste Division in the early stages of an experiment.
>
> Mixed Wastes include contaminated lead pigs or other lead shielding and radioactively contaminated organic liquid waste that contains a regulated chemical. One common example is liquid scintillation media containing toluene or xylene.

> Dry Solid Waste—Gloves, paper, plastic, glass, metal, or other solids that contains [sic] radioactive material or is contaminated with radioactive material.
>
> Liquid Waste—Organic and aqueous liquids containing radioactive material. Liquid waste must be segregated from solid waste. Organic and aqueous liquids must be stored separately in appropriate containers.
>
> Liquid Scintillation Vial Waste—Capped scintillation vials containing either organic or biodegradable scintillation fluid contaminated with radioactive material.
>
> Radioactive Sharps—Radioactive contamination of syrin-ges, needles, surgical instruments, or other articles which have the potential to cut or puncture human skin.

In general, **nonhazardous waste** is municipal waste and industrial waste that does not meet the definitions of hazardous or radioactive waste.

DISPOSAL OF SOLID WASTE

landfills, groundwater

For many centuries, human societies have managed their solid waste by burying it. In fact, archeologists, scientists who study human history, freely admit that archeology is largely the science of studying "trash" because most ancient artifacts are found either in old burial sites or in old trash disposal sites.

Until the 1950s, people generally believed that once trash was buried, it was safely disposed of. The assumption was that buried waste would eventually decompose under the soil into harmless materials and that any harmful chemicals or bacteria would be attenuated (purified and made harmless) by the soil (soil, bacteria, and fungi). As we will see later in this chapter, the soil does have several characteristics that led scientists to that belief. In addition, scientific instruments of that time were not capable of detecting the most minute concentrations of contaminants in water. Thus, the damage being done by waste decomposition and drainage to the groundwater was not easily detected.

As long as that belief prevailed, solid waste disposal was a simple process. We simply dug a hole, placed the trash in it, and covered the hole with soil. For an individual home, the waste might be burned and the residue buried in a small ditch or hole. To dispose of the solid waste from a large city simply meant hauling it to a rural area, digging a huge hole with bulldozers, dumping the trash, pushing a layer of soil over it, and packing the whole mess down by driving the bulldozers over the top. For a city, the waste might be buried in a huge open-air pit and eventually covered with soil. As rain percolated through the open trash or through the soil covering, people assumed that the waste decayed and that the soil underneath took out any harmful chemicals. The term "dump" or "trash dump" is still commonly used to describe solid waste disposal sites, although "sanitary landfill" is probably a more appropriate description (Figure 13–9).

As late as the mid-1960s, that was the way almost all solid waste disposal was done in the United States. In many parts of the world and in some parts of the United States, solid waste disposal is still

FIGURE 13–9 Waste materials must be properly disposed of to prevent serious environmental damage.

INTERNET KEY WORDS

leachate
leachate effects

accomplished using that method. The process was quick, cheap, and easy to manage. Many smaller towns and rural areas used even simpler and cheaper methods such as dumping the trash down the side of a hill or in a small natural valley and allowing it to decompose in the open.

Starting in the mid-1950s, however, several important studies revealed that significant amounts of water **percolate** through landfills. That water picks up and transports minute particles of contaminants and biological agents such as bacteria and viruses and dissolves water-soluble chemicals. The water enters the landfill, mostly from rain and melting snow; once it moves through the buried materials, it is known as **leachate**. From there, the leachate percolates downward until it either flows out of the landfill as seepage or enters the groundwater. Once contaminants are in the groundwater, they remain there indefinitely or until the water is removed for human use. Leachate from buried solid waste is one important reason that well water must be tested periodically for purity (Figure 13–10).

FIGURE 13–10 Rainfall percolates through the soil cover, collects in the buried waste, dissolves soluble materials, and seeps downward toward the groundwater as leachate. (Courtesy of Sarah Michelle Williams.)

What Is a Landfill?

In many parts of the world today, MSW is disposed of in or on any landform: ditches, open pits, hillsides, or even open areas. It may or may not be burned, and it may or may not be covered with a soil layer. A **landfill** is an open area into which garbage is placed and where it is to be covered by a layer of some other material, typically soil. Until the 1950s in the United States, landfills were typically in open fields or along a slope. Earth-moving equipment such as bulldozers might scoop out a long trench or a large flat area. Waste would be deposited in the opening. Often the waste would be burned to decrease the volume of materials. The soil that had been removed would then be pushed back over the partially burned garbage to bury it. Often heavy equipment would drive over the top of the covering to pack down the material as much as possible. With research that demonstrated the damage to the environment from leachates as well as other environmental concerns, however, much more restrictive regulatory requirements and much more sophisticated landfill designs were developed.

Landfill Design

There are two basic types of landfills today in industrialized nations such as the United States: natural attenuation and containment. We will now examine both types in some detail.

The **natural attenuation landfill** is designed to hold the waste material in a covered area and allow the natural percolation of precipitation to pass through the waste and flow through the underlying soil and rocks. The expectation in a natural attenuation landfill is that the leachate will be attenuated by microorganisms and soil particles. Although research has clearly shown that leachate can damage the groundwater and cause other environmental problems, certain kinds of MSW can be safely disposed of in natural attenuation landfills. Municipal solid waste can be of three kinds: hazardous, radioactive, and nonhazardous. Such items as paper products and yard waste are generally nonhazardous and can be safely disposed of in a natural attenuation landfill. For managing such waste, the cost is substantially lower than for more dangerous materials.

Regardless of the type of landfill, the basic shapes will be similar. Landfills may be of three basic shapes: at-grade, canyon fill, or trench fill. An *at-grade landfill* is used when waste is placed at or nearly at the normal soil level. The waste is then covered with a layer of material, typically soil. A *canyon fill landfill* involves pushing the waste over the side of a hill or slope and then pushing a covering layer over it. In both the canyon fill and at-grade landfills, a berm of soil is first constructed to contain the waste, the waste is then placed in the holding area, and, finally, a final cover is pushed over the waste. The most common kind is a *trench fill landfill* in which a trench is formed by removing the soil, the waste is placed in the trench, and the soil that had been removed is used to form the final cover (Figure 13–11).

Natural Attenuation Landfills

natural attenuation landfill

How does **attenuation** work? Six natural mechanisms occur in the soil as leachate percolates through it to attenuate the contaminants: adsorption, biological removal, ion exchange, dilution, filtration, and chemical precipitation.

Adsorption is the process by which the molecules of a chemical adhere to the surface of some other material. Clay is particularly important in this process. Because clay is made up of extremely small particles, a given volume or weight of it has a huge surface area compared to the same volume or weight of sand or gravel. The increased surface area, as well as other chemical and physical properties of clay, means that many of the contaminants in leachate simply adhere to clay particles so tightly that they are permanently removed from the percolating water.

Three Shapes of Landfill

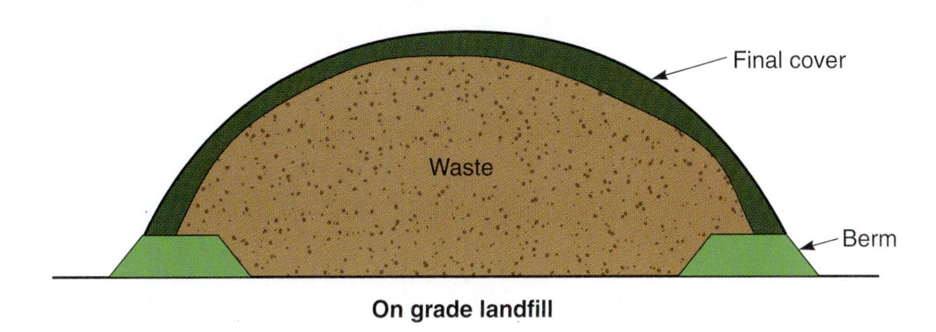

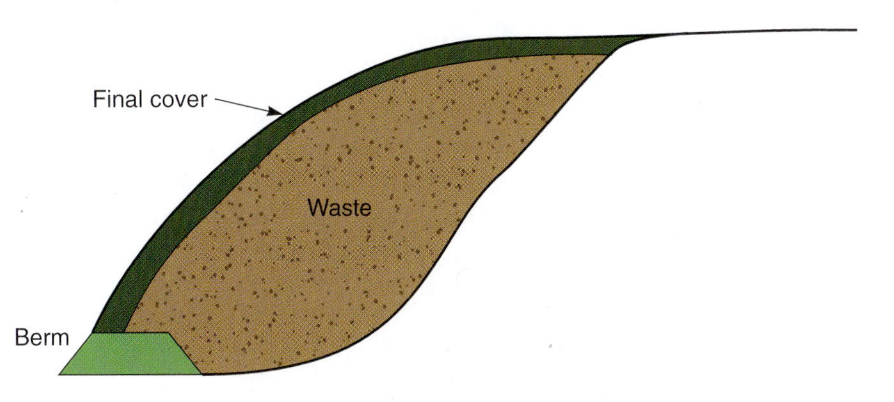

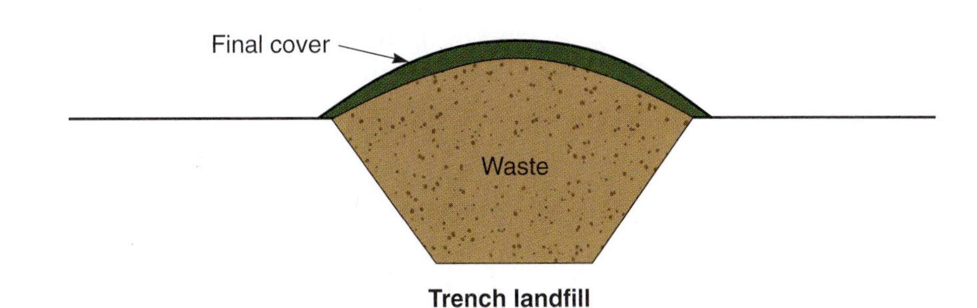

FIGURE 13–11 Three shapes of landfill (Courtesy of Bill Camp.)

Biological removal takes place when bacteria, fungi, or other soil microorganisms break down or absorb the leachate constituents. Soil microorganisms use some of the materials we would consider contaminants as food. Other chemicals are converted into materials the microorganisms need to survive. Still other materials are simply oxidized by the microorganisms into harmless products. In particular, carbonaceous wastes are broken down by microorganisms and released as carbon dioxide to become carbonic acid, which then breaks down still more of the harmful constituents of the leachate.

From basic chemistry, we know that ions are charged molecules or atoms. Because clay particles are extremely small, they are chemically active and extremely important in *ion exchange.* This process neutralizes part of the constituents in leachate by changing the molecular structure of the ions involved in the exchange. *Dilution* simply means that the concentration of the leachate is decreased by mixing it with large quantities of water. Too heavy a concentration of even the most harmless of chemicals may be dangerous. We may enjoy a little table salt (NaCl) in our food, but too much makes the food inedible, and too much salt in your diet can even be dangerous to your health.

Filtration involves the physical removal of solid constituents from the leachate by trapping them in pores in the soil. This process can remove only small quantities of materials because the pores eventually become clogged. To visualize this process, place a kitchen sieve or paper filter over a container. Pour slightly muddy water through the filter and notice that the soil particles are captured. Continue the process for awhile, and the filter will become clogged so that the water will stop passing through it and will start to pour over the top. For that reason, filtration is of only limited importance in the attenuation process.

Chemical precipitation is the process of a phase change in the leachate. To understand what a phase change is, let us consider a familiar example. A phase change occurs in the atmosphere when a gas, such as water vapor, changes to a liquid (such as rain) or a solid (such as snow). In the leachate, a phase change occurs when a dissolved mineral crystallizes. The liquid contaminant in the leachate then becomes a solid material that is removed from the leachate by adsorption or, to a lesser extent, filtration.

Containment Landfills

containment landfill

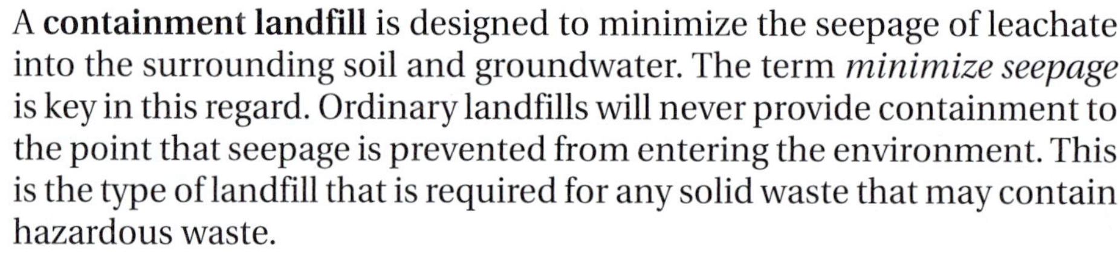

A **containment landfill** is designed to minimize the seepage of leachate into the surrounding soil and groundwater. The term *minimize seepage* is key in this regard. Ordinary landfills will never provide containment to the point that seepage is prevented from entering the environment. This is the type of landfill that is required for any solid waste that may contain hazardous waste.

Total containment is necessary for certain extremely hazardous and radioactive waste. The discussion in this section does not apply to total containment waste management. These extremely hazardous materials are disposed of in deep shafts, salt beds, and other extreme locations.

A natural attenuation landfill can be made into a containment landfill by the addition of some sort of **liner** that restricts water from percolating in a natural way. The liner can be natural, such as a heavy compacted clay or bentonite soil. It can also be constructed by using a synthetic material such as butyl rubber, chlorinated polyethylene, chlorosulfonated polyethylene, ethylene-propylene rubber, low-density polyethylene, high-density polyethylene, polyvinyl chloride (PVC), or a combination of several materials. In theory, any liner will eventually allow some leachate to seep through, but the amount of seepage can be cut to near zero in several ways.

The first way to minimize seepage is by using multiple layers. A layer of extremely fine compacted clay or bentonite at the bottom will cut the seepage down dramatically. If a PVC or other synthetic layer is applied over the clay, the synthetic material will stop most of the seepage. The clay will have to restrict only the seepage that gets through the synthetic layer. Naturally, the synthetic material will have to be covered by a layer of clay, bentonite, or even sand to prevent damage from the

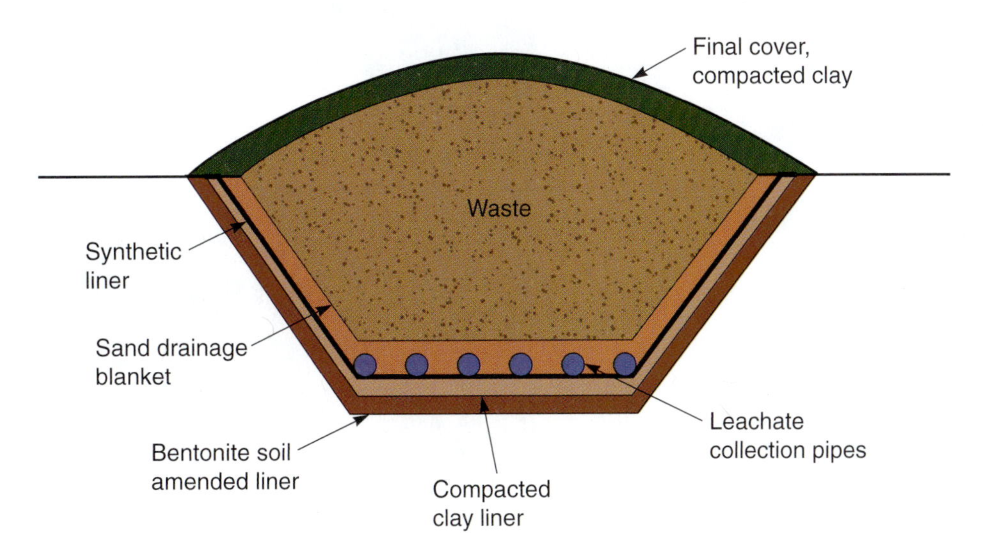

FIGURE 13–12 A multiply lined landfill with a single leachate collection system (Courtesy of Bill Camp.)

large equipment used to move the waste and to apply the final cover. In general, the more layers, the less seepage will occur over time.

In addition to multiple layering, leachate collection pipes can be used to capture the seepage collecting above the liners and drain it away. That leachate is typically treated in a wastewater treatment plant (similar to a sewage treatment plant) until it is harmless and can be disposed of in some environmentally safe way. In some cases, not only multiple liners but also multiple leachate collection systems are used (Figure 13–12).

Why should we care how landfills are constructed? It should be obvious by now that the least expensive way to dispose of municipal solid waste is simply to dump it on the ground and leave it there. Although that solution was acceptable for thousands of years, it is no longer a solution that a modern society can live with. The next cheapest way is to push it over a canyon wall and cover it with a layer of soil or bury it in a trench. Although that may be acceptable for limited kinds of solid waste, the mixed waste that comes from households, factories, and other businesses (in other words, almost all the municipal solid waste produced in this country) can no longer be safely disposed of in natural attenuated landfills. Containment landfills are the only acceptable method of disposing of the majority of municipal solid waste today and in the future. Containment landfills are expensive to build, and the leachate must be captured in the drainage pipes and treated for 30 to 50 years before the landfill is safe to abandon.

FIGURE 13–13 Organic waste can be composted to create a useful product for use as fertilizer or mulching material. (Courtesy of Shutterstock.)

Recycled Animal Wastes

Recycled animal wastes include such products as manure (Figure 13–13) and animal body parts from meat-processing plants, also known as **offal**. Such materials have often been of little value in the past, but new uses have been found for many waste products from animals.

FIGURE 13–14 Methane gas is recovered from animal manure and used to generate heat and electrical power.

INTERNET KEY WORDS

recycled animal wastes

During digestion, animals tend to use fewer nutrients than they have available to them in their feed. **Recycled nutrients** are feeds that use these wasted nutrients in livestock diets. For many years, farmers have recycled wasted nutrients by raising a few hogs in the pens of fattening cattle. The hogs gathered and ate kernels of corn that had gone through the digestive tracts of the cattle without being digested. Small farm flocks of chickens were common sights on many farms until recent years. They were allowed to run freely around the farm to gather their own feed from any source they could find.

Recycled animal wastes have recently become much more important as sources of income. Animal wastes can be recycled as livestock feeds. Methane gas can also be recovered from animal manure and used to generate heat or electricity (Figure 13–14). Clean fuel-grade gas is injected into natural gas lines as a supply source for public utilities. It is also used for commercial processing or manufacturing.

Several methods have been used to recycle animal waste on farms. Operators of large cattle feedlots have constructed large concrete slabs on which feed bunks are placed. Manure is "harvested" each day from these areas and mixed with fresh chopped forage. This mixture is then placed in a silo for curing and storage. By the time the feed has fermented, the unpleasant odor is gone, and the mixture of animal manure and fresh forage has been transformed into pleasant-smelling and palatable cattle feed.

A growing source of recycled nutrients also is obtained from poultry processors. Feathers provide a source of protein and other nutrients for ruminant animals. Once the feathers have been processed and pelleted, they become palatable to cattle and provide a reasonable source of protein and other nutrients. Poultry manure, which is high in nitrogen, can also be used by the rumen bacteria of cattle as a source of protein.

The agricultural industry has used the offal from processed fish, poultry, and livestock as part of livestock rations for many years. The industry also depends on waste animal protein from slaughterhouses for components of fish rations. Another industry that is heavily dependent on recycled animal protein is the fur-bearing animal industry. Domesticated fur-bearing animals, such as mink, require meat in their diets. It is usually obtained from waste animal protein because this source is much less expensive than other meat products. Finally, recycled nutrients from animal sources are important to the livestock industry as sources of high quality dietary protein.

Why Recycling?

INTERNET KEY WORDS

recycle, benefits

Given the obvious costs of solid waste disposal, the argument for **recycling** should be clear. We do not really recycle paper to save trees. As you can read in the chapters that deal with forestry, there is no shortage of trees for paper. As you will see from the chapter that deals with energy sources and use, there is no real shortage of energy sources. Given the undeniable fact recycling does not really benefit our natural resources in any substantial way, recycling is important because of the cost of solid waste disposal.

Looking Back

Waste materials have become a major problem in many of the industrialized nations of the world. They do not just go away when we are finished with them. Solid waste consists of nonliquid, nonsoluble materials ranging from municipal garbage to industrial wastes that contain complex and sometimes hazardous substances. The generation of solid wastes in the United States continues to increase as a result of a growing population. Many of these waste materials are placed in landfills.

There is no shortage of land on which landfills could be constructed. There is, however, a severe shortage of communities in which people are willing to allow a landfill to be constructed. Would you want a landfill in your backyard? Most people would not. Even when a politically acceptable location for a landfill can be found, the cost of constructing a multiple lined containment landfill is tremendous. Even after it is constructed, when the landfill is closed, the costs of capturing the leachate and treating it can go on for as much as half a century. Recycling waste materials is an alternative to landfills and is becoming an important way to reduce the amount of solid waste.

Self-Analysis

Essay Questions

1. What is solid waste?
2. What are the two major sources of solid waste?
3. What are the differences between municipal solid waste and industrial solid waste?
4. Which types of wastes are most abundant in the municipal solid waste stream?

5. Suggest some ways in which significant amounts of municipal solid wastes might be diverted from landfills.
6. Define "hazardous waste" and list some of its characteristics.
7. Define "radioactive waste" and explain why it should not be disposed of in a landfill.
8. What type of waste includes hospital waste such as used needles and disposable surgical blades?
9. Why do natural attenuation landfills work for some kinds of waste but not others?
10. How is a natural attenuation landfill converted into a containment landfill?
11. Why do so many local governments try to avoid constructing their own landfills and instead ship their municipal solid waste elsewhere for disposal?

Multiple-Choice Questions

1. Which of the following is a source of municipal solid waste?
   a. household garbage
   b. mine waste
   c. logging waste
   d. waste from industrial processes
2. The largest producer of industrial waste is
   a. households.
   b. mining.
   c. timber industry.
   d. manufacturing industry.
3. In the United States, the amount of waste generated per person each day is approximately
   a. 1.8 pounds.
   b. 3.1 pounds.
   c. 4.4 pounds.
   d. 7.2 pounds.
4. The biggest component of municipal waste is
   a. yard waste.
   b. plastics.
   c. metals.
   d. paper.
5. Which nation in the world generates more trash than any other?
   a. England
   b. France
   c. Brazil
   d. United States
6. Which of the following is not a characteristic that defines hazardous waste?
   a. biodegradable
   b. corrosive
   c. reactive
   d. toxic
7. The most important characteristic of a containment landfill is
   a. maximum degradation of waste materials.
   b. minimum seepage.
   c. proximity to population centers.
   d. soil type.
8. The most promising practice for reducing the amount of solid waste deposited in landfills is
   a. reducing the consumption of packaged goods.
   b. recycling waste materials.
   c. eliminating the production of radioactive wastes.
   d. eliminating paper and plastic packaging materials.

Learning Activities

1. As a class, find and visit a location in your community where people are dumping trash improperly.
2. Organize a cleanup campaign for a community park or other recreation facility.
3. Collect all the trash discarded at your home for a week. Lay it out on a large plastic sheet and do an inventory to determine the kinds of waste materials included such as paper, food waste, containers, and so on. Weigh the various types of waste and figure out the percentages of the total. Be sure to use rubber gloves when handling the trash and be careful not to get cut.
4. As a class project, organize an adopt-a-park program to keep some community area clean on an ongoing basis.
5. Invite a knowledgeable official who represents the local landfill or recycling center to speak to your class or arrange a class field trip to visit the landfill or recycling center.

TERMS TO KNOW

- fossil fuel
- lignite
- bituminous coal
- anthracite coal
- strip mine
- cut
- shaft mine
- slope mine
- drift mine
- room and pillar
- longwall
- gravimeter
- magnetometer
- seismograph
- derrick
- jack-up rig
- drillship
- semisubmersible rig
- fixed platform
- tar sand
- oil shale

CHAPTER 14

Fossil Fuels

Objectives

After completing this chapter, you should be able to

- define fossil fuel and explain how such fuels are formed
- name the three most used kinds of fossil fuel
- explain the various ways coal is mined from the Earth
- discuss oil exploration and drilling techniques
- explain how natural gas is obtained and distributed
- discuss oil shale, tar sands, and the petroleum potential
- explain why conservation of fossil fuels is important

FIGURE 14–1 The oceans absorb carbon dioxide from the atmosphere when the carbon concentrations are high and release carbon dioxide to the atmosphere when concentrations are low.

When plant and animal tissues decay, carbon dioxide is either released back to the atmosphere as a gas or, if the tissues are buried, compressed and converted over a long period of time to form **fossil fuels** such as natural gas, crude oil, coal, tar sands, and oil shale. Humans mine or extract fossil fuels from the Earth's surface to burn as fuels and for other purposes. When these materials are burned, the combustion process releases carbon dioxide to the atmosphere.

The oceans absorb large amounts of carbon dioxide from the atmosphere when the carbon content of the atmosphere is high, and they release carbon dioxide to the atmosphere when atmospheric carbon dioxide decreases (Figure 14–1). Until recent years, the carbon content of the atmosphere has remained nearly the same due to this action by the oceans. Over the last 100 years, however, the large amounts of fossil fuels that have been burned have increased the levels of carbon dioxide in the atmosphere. As a result, we are burning these fuels faster than the oceans can absorb the extra atmospheric carbon.

Why did the United States develop into a leading nation of the world while many nations of a similar size remained relatively undeveloped? Among the reasons why is the vast amount of natural resources this country contains. Of those resources, fossil fuels have played a dynamic role in how our country operates. Without those fuels to provide energy, critical industries such as agriculture, transportation, manufacturing and processing would have to shut down (Figure 14–2). These resources, however, are nonrenewable, so it is critical that they be managed properly and wisely if our country is to remain a leading power and influence in the world.

coaL

Coal is a black or brown rock that was formed primarily from swamp plants that died between 1 million and 400 million years ago. The extreme pressure in the Earth concentrated the carbon matter until it eventually formed the coal. Coal is ranked and classified according to the carbon it contains. The same basic process also produces diamonds (with somewhat higher heats and at specific locations).

Figure 14–2 Modern trucks rely on fossil fuels to transport raw products to markets and consumer-ready products to population centers. (Courtesy of Shutterstock.)

INTERNET KEY WORDS

coal formation

The coal with the lowest carbon content is also the first step in the coal formation process: **lignite**. Lignite contains about 33 percent carbon and is formed from peat deposits. As the rock and soil above the lignite increases, the high pressures cause the coal to become harder. This next step in coal formation is called *sub-bituminous coal.* With additional pressures, the sub-bituminous coal is transformed into **bituminous coal**. The hardest and oldest coal is **anthracite coal**, which contains as much as 90 percent carbon.

Most coal was formed approximately 300 million years ago during the Carboniferous period. Tall ferns growing in swamp areas were the predominant plant life of this period. To understand how extensive this plant life was, consider that more than 7 feet of plant matter are needed to make a 1-foot vein of bituminous coal.

Most of the coal found today occurs in veins or seams sized from as small as 1 inch to as large as several hundred feet thick. The known coal reserves in the world are estimated to total more than 2 trillion tons. Approximately 479 billion tons are in the United States, and most of this coal lies in the eastern half of the country. Nearly all the anthracite deposits are found between the Appalachian Mountains and the edge of the Great Plains. The principal coal-producing states include West Virginia, Pennsylvania, Kentucky, Illinois, Ohio, and Virginia.

Coal has a great variety of uses. According to a recent estimate of the U.S. Department of Energy, 68 percent of the coal is used in the production of electricity, 13 percent in steel production, 9 percent each in general industries and exports, and 1 percent for home heating. Coal that has a high heating value is used to drive steam turbines to generate electricity. In recent years, concern has developed over the use of coal because of air pollution. As coal burns, sulfur dioxide is released. This poisonous gas irritates the body's respiratory tract. Many scientists believe it also contributes to the growing problem of acid rain. Considerable research has been done to design filtering systems to trap this gas. Another concern is the powdery ash material called *fly ash* that is also emitted from smokestacks. The adoption of screens has limited this problem.

Coal is mined from two types of mines: surface (or strip) mines and underground mines. Most **strip mines** (Figure 14–3) are within 50 feet of

FIGURE 14–3 Strip coal mine (Courtesy of Shutterstock.)

the Earth's surface. Strip mines produce approximately 60 percent of the coal in the United States. Underground mines may extend hundreds of feet under the Earth's surface and supply approximately 40 percent of our coal.

Strip mining coal requires huge machinery to remove the overburden, or top earth, from the coal vein. This soil is placed in large piles, which can cause vast environmental problems. In past years, the strip mines were left as they were on the last day coal was removed. Huge pits and tall piles of soil were left behind to erode and wash away.

INTERNET KEY WORDS

coal, strip mining
coal, underground mining

The federal government now requires a strip mining company to replace or reclaim the land before moving to a new location. The company can have no soil slopes steeper than 33 percent. Today, when an area is selected for a strip mine, large earthmovers peal a strip of land away until the coal is exposed. This ditch is called a **cut**. The coal is removed from the seam by large trucks. Once the coal is removed, the earthmovers make another cut, dropping soil on the first cut. When the coal supply has been exhausted, only small ridges need to be leveled in the reclamation process.

In underground mining procedures, the work of getting coal from the Earth is extremely hazardous. There are always chances of cave-ins,

Interest Profile

UTAH CRANDALL CANYON MINE COLLAPSE

On August 6, 2007, six miners were trapped inside a Utah coal mine when a large section of the mountain being mined collapsed into the mine shaft. For two weeks, the nation watched as holes were drilled into the mine from above in an attempt to find and rescue the miners. Down in the mine shaft, workers were frantically moving and removing debris in an attempt to get into the isolated section of the mine where the men were thought to be located. Cameras were lowered 1,500 feet down the drill holes to look for any evidence of life, and air samples were taken to determine whether oxygen levels were adequate. Then a second collapse killed three rescue workers, leading to the suspension of rescue efforts. The six trapped miners may never be found, as once again, the attention of the nation focused for a few weeks on the dangers of underground mining.

explosions, and exposure to poisonous gases. The underground mines require more human labor than do strip mines. Recent advancements have produced better machines to help underground miners.

There are three main types of underground mines: shaft mines, slope mines, and drift mines. In **shaft mines**, access passageways are vertical to the coal seam. These passageways are equipped with elevators to bring machinery and personnel to and from the coal. In a **slope mine**, the access tunnel is on a slope from the surface to the coal seam. These mines are not as deep as those with shafts. A **drift mine** is used when a passageway is bored into a hill or mountain. These tunnels may be parallel to the surface.

Once the access tunnels are constructed, mine experts must decide how the coal is to be removed. There are two common methods: the room and pillar system and the longwall system. The **room and pillar** system uses the idea of leaving pillars of coal to help support the ceiling. Seams of coal are cut that are 40 feet to 80 feet wide. Once the parallels are cut, the blocks of coal are left to support the ceiling. The underground mine may then be expanded into rooms. The first rooms serve as main entries and coal transport rooms.

The **longwall** system mines coal from one face that is approximately 300 feet to 700 feet long. As the coal is cut from the seam, it falls into a conveyor belt and is transported to the surface. This method is more common in Europe than in the United States. Mine safety laws require longwall mines to have both subentries and main entries.

Because mining is extremely dangerous, mine safety is of the utmost importance. Most mining accidents come from roof and wall failures, accumulation of gases, concentration of coal dust, and accidents with machinery. In the early history of mining, approximately three in every 1,000 miners were killed annually. Today, that rate has dropped to approximately one in every 2,000. The federal Bureau of Mines has the responsibility for setting standards. In 1969, the U.S. Congress enacted the federal Coal Mine Health and Safety Act to strengthen the standards for ventilation, coal dust concentration, and mining structure. The act also gave financial benefits to miners who were disabled by black lung disease. The standards and safety measures are enforced through the Mine Safety and Health Administration, the U.S. Department of the Interior, and the Environmental Protection Agency.

mining cave-ins
mine safety

The newest research in the coal industry is involved in finding new uses of coal and developing more economical methods of converting coal into a liquid or gaseous fuel. Coal gasification, for example, converts coal into fuel. The fuel can then be used as raw product of chemicals and fertilizer. Large-scale production is still in the experimental stage, but there seems to be promise in new uses for high-sulfur grade coal.

Oil or Petroleum

oil, petroleum, formation
oil, petroleum, uses
gravimeter, magnetometer, seismograph

Petroleum has long been called "black gold," especially because the price of its products has skyrocketed in recent decades. Like coal, it is a vital backbone of our industries. It keeps the United States and much of the industrialized world moving. Today's research has found a large variety of uses for oil.

Oil is formed in ways that are essentially the same as for coal. As plants and animals living in the water died, they settled to the bottom of the ocean, pond, or swamp. Although coal requires several million years to form, oil usually requires only 1 million years or so. As pressures were exerted on these organic materials, oil was formed and squeezed into the rock openings or into specialized rocks called reservoir rocks. These reservoir rocks are porous, allowing them to fill with oil. Many times shifting of great blocks of earth trapped oil deposits in the rock layers. It is the job of petroleum engineers now to find those oil deposits.

The most widely used machines today for finding oil are the gravimeter, the magnetometer, and the seismograph. The **gravimeter** uses the principle that the gravitational pulls of oil-filled rocks differ from those of rocks that contain no oil. The **magnetometer**, on the other hand, measures differences in the magnetic pull of the Earth to find oil-bearing rocks, enabling the geophysicist to locate rock layers that might contain oil. The **seismograph** uses sound waves to identify various layers and formations under the Earth's surface. This method is useful on offshore drilling rigs.

Once oil is thought to be present, the oil company must lease the land before drilling. The landowner is usually paid royalties if oil is found on the property. After obtaining the lease, the drilling site is prepared. Roads may be built, land leveled and cleared, and, possibly, site reinforcements made. A power plant must be readied as well as a water supply system. After much preparation, the drilling rig is ready for construction. A hoisting apparatus called a **derrick** is constructed first. The drills, tanks, and pumps then follow.

Actual drilling can be completed in one of three ways: cable-tool, rotary, or directional. The early oil crews used the first method, cable-tool drilling. The drill contains a tool such as a chisel that is lifted and dropped to loosen the soil and rock. Fresh water is then poured into the hole to loosen the soil. This process is extremely slow. Rotary drilling uses a bit similar to a carpenter's wood bit. As the drill is lowered, breaking loose rock fragments, water is pumped down the pipe to carry the sediments to the surface. Several drilling teams work 8- to 10-hour shifts to keep the rig working 24 hours a day (Figure 14–4). Directional drilling involves drilling the shaft at an angle. Special drills called *turbodrills* and *electrodrills* rotate and bend, directing the cutting bit. This method is used when the well cannot be drilled directly over the deposit.

For several decades, offshore drilling rigs have been constructed to access oil under the ocean surface. They are more expensive and more dangerous than land rigs. These offshore operations include **jack-up rigs**, **drillships**, **semisubmersible rigs**, and **fixed platforms**. Most exploratory rigs are one of these first three. Once oil is discovered, the well is completed with a fixed platform. When the oil is brought to the surface, it is piped to supertankers and taken to a refinery.

FIGURE 14–4 Oil drilling rig. (Courtesy of Shutterstock.)

At the oil refinery, the crude oil is distilled into various products. These may be fuels, lubricants, or petrochemicals. Fuels include aviation gasoline, diesel fuel, gasoline, jet fuel, kerosene, home heating oil, and distillate oils. The lubricants include greases, road oils, technical oils, and medicine oil. The petrochemicals include alcohol, ammonia, ink, paint, plastic, synthetic rubber, and food additives, to name a few.

Oil companies and geologists have been estimating how much oil is left around the world to be recovered. As technology has changed over the years, it has become possible to recover crude oil that could not have been recovered earlier. Also, as prices of crude oil rise, it becomes economically feasible to recover oil that was previously left in the ground at lower prices. Finally, as oil exploration techniques have improved, new sources of oil have been found that were unknown previously.

oil, EUR
natural gas production

The key estimate of the oil supply is called *estimated ultimately recoverable (EUR)* oil. Over the last 50 years, the world's EUR oil reserve has actually increased. Estimates in the 1950s averaged 1.2 trillion barrels of EUR oil. Today, the estimates generally range between 2.0 and 2.4 trillion barrels of EUR oil. The countries in the former Soviet Union are estimated by the U.S. Geological Survey (USGS) to have 0.3 TGG and China is estimated to have 0.23 TGG of recoverable oil. The USGS estimates that much of the oil has not even been found yet.

Of the 2.0 to 2.4 trillion barrels, only 1 trillion barrels are actually proven to be available. Of that, two-thirds are in the Persian Gulf, with approximately one-half of that as proven reserve in Saudi Arabia. Most of the rest is more or less equally divided among Kuwait, Iraq, and Iran. Using these estimates of EUR oil, we can expect world production of crude oil to peak between sometime between now and 2019. Nobody believes that we will "run out" of oil in the foreseeable future. When crude oil production capacity falls, as it must some day, prices will rise. When oil prices rise, it will become economically feasible to extract oil from coal, **tar sands**, heavy oil, and **oil shale**. Huge reserves of oil are stored in those resources. The only limit on oil extraction from those sources is the immense cost of building the extraction facilities. If oil prices get high enough, then oil can be removed from coal, tar sands, heavy oil, and oil shale for centuries.

How long will the petroleum last? We know it cannot last forever. It is time to start a conservation program to reduce our overuse of oil. The petroleum industry is working on methods to recover more oil from the wells we now have as well as to develop more precise location methods. The consumer can save oil as well. This can be done by driving automobiles less and lowering thermostats in the winter and raising them in the summer. Lower speed limits, greater use of mass transportation systems, and the use of fuel-efficient automobiles are other ways to conserve oil. Each person must play a part in a conservation program.

Natural Gas

Natural gas is one of the most perfect and most highly demanded fuels in the United States. We use it to heat our homes as well as to cook our food. We use it in the production of plastics, detergents, and drugs.

The gas industry can be divided into three areas: production, transportation, and distribution. In the exploration of gas, most gas and oil companies work together. The gas is usually located above an oil deposit. One of the deepest wells ever drilled was in Oklahoma in 1972 to a depth of 30,000 feet at a cost of approximately $5 million.

Gas taken from a well must be cleaned in an extraction unit. This unit removes impurities such as water, sulfur, and dust. At the processing plants, butane, propane, and gasoline are removed from the gas. The finished gas is then fed into a transportation pipeline where it is compressed to approximately 1,000 pounds per square inch. It travels through the pipeline at 15 miles per hour or so. When it reaches a city, distribution lines carry it to the consumers. The consumers receive the gas from individual service lines connected from the main lines. The United States has some 1 million miles of gas lines in operation.

One of the main problems with natural gas is supplying the gas when the consumer needs it. The peak usage time for natural gas is during winter months. The amount used on a cold winter day can be more than five times the summer usage. The oil and gas companies are currently storing gas in huge underground reservoirs that are actually old and unused oil wells. When the demand for gas is low, the excess is pumped into the well and sealed for later use when demand for gas increases. In the early days, this excess gas would be burned off and wasted. Some countries in the Middle East continue this practice.

Gas can also be stored in a liquid state by lowering its temperature to –260° F (–168° C) and changing it to a liquid. This fuel is termed *liquefied natural gas (LNG)*. LNG takes up approximately 600 times less space, so it is a better form for shipping overseas. Simply by raising its temperature, the LNG changes back into a gaseous form.

Natural gas is now the most rapidly growing source of fossil fuel energy. Improving exploration technology has rapidly uncovered much larger and previously unknown natural gas reserves. In spite of that, the known reserves of natural gas are considerably less than known reserves of crude oil. Russia has the largest natural gas reserve at 48.2 trillion cubic yards (1993 estimates). That represents approximately one-third of all the world's known reserves (141 trillion cubic yards) and more than 10 times the U.S. reserves (4.6 trillion cubic yards).

natural gas reservoir
oil shale, tar sands

The largest amounts of U.S. natural gas are found in Texas and Louisiana. The increased use of natural gas has caused concern over our future supplies. With the passage of the Natural Gas Act of 1938, the Federal Power Commission regulated the industry to prevent it from exploiting consumers. This act established a cost rate base system of controlling oil companies' operating and capital costs. Recently, the deregulation of natural gas has led to rapid increases in costs to consumers and gas producer profits have climbed in proportion.

Oil Shale and Tar Sands

When oil, natural gas, and coal were considered inexhaustible, oil shale and tar sands were of little interest to Americans, especially because they were not economically feasible to mine and develop. In recent years, however, interest has increased and more research is being applied to this area.

The U.S. Geological Survey has published survey estimates of the large oil shale deposits in Colorado, Utah, and Wyoming: The reserve contains 80 billion barrels of oil. The deposit was estimated to be 25 feet thick, yielding 25 gallons of oil per ton of shale. On further examination and extension of the surveyed region, additional deposits estimated at 3 trillion barrels of oil were found. In Canada, tar sand was discovered with a reserve of 300 billion barrels of oil.

If there is so much oil in these reserves, then why all the public concern about an energy shortage? First, there are problems in getting the oil from the oil shale and tar sands. The main problems deal with the vast cost differences in recovering oil from shale and sand rather than from natural crude. The shale and sand are solids and require huge mining equipment. Furthermore, most shale and tar sands contain only 13 percent to 16 percent oil. Second, the technology at our disposal is extremely limited when it comes to extracting the oil economically. To release the oil, the shale must be heated to 800° F to 1,000° F (427° to 538° C).

For a plant to be commercially important, a capacity of 50,000 barrels per day is required. This means the plant would have to process approximately 75,000 tons of shale per day—and 60,000 tons of that would be waste. On a yearly basis, the plant would have to dispose of approximately 15 million tons of waste product. Because water is also required in the process, careful consideration must be given to this resource. Rock disposal and water pollution are only two of the critical environmental factors that need to be considered.

Fossil Fuel Conservation

Conservation efforts directed toward fossil fuels—mainly oil, natural gas, and coal—have been weak at best. For many years, the United States literally had oil and gas to burn. Natural gas from oil wells was routinely burned off. No serious conservation efforts were aimed to reduce our use of these fuels. The world, and the United States in particular, enjoyed abundant, relatively cheap fuel.

fossil fuel conservation
fuel conservation, cars, trucks

In 1960, four Persian Gulf states and Venezuela formed the Organization of Petroleum Exporting Countries (OPEC). OPEC was created to halt the decrease in crude oil prices that the seven largest oil companies had attempted to impose on these oil-producing countries. Although OPEC succeeded, it was unable to significantly raise the prices these countries received for their crude oil. This changed after the 1973 Arab–Israeli war when the Arab oil-producing countries cut back production and stopped oil shipments to the United States, sharply limiting supplies and dramatically raising prices. The price for a gallon of gas more than doubled. Supplies were limited, resulting in long lines at gas stations. Significant shortages occurred again in the mid-2000s, and prices doubled once again (Figure 14–5).

The oil embargo, along with real and artificial domestic oil shortages, have caused several things to happen. Americans had been accustomed to spending 30 cents to 35 cents for a gallon of gas. When prices shot up and supplies became limited, people were shocked. The big, heavy American cars that used gas at rates of 8 to 10 miles per gallon suddenly became much more expensive to drive. Fuel economy became more important.

Until this time, few small cars were sold in the United States. U.S. automobile companies were unprepared for the crisis and had few small and gas-efficient models to offer the public. This allowed the foreign automobile makers, primarily the Japanese, to acquire a share of the U.S. market. The oil embargo also demonstrated to leaders in Washington just how dependent we are on foreign oil imports, and some efforts at conservation were initiated. One result of these efforts is the vastly improved fuel economy of most American automobiles. Another is the increased use of ethanol as a fuel.

Figure 14–5 Gas prices escalated after the first oil embargo in the early 1970s. (Courtesy of Kevin Deal.)

Although transportation represents only some 25 percent of the total energy used in the United States, it accounts for 60 percent of U.S. oil consumption. Congress passed a law in 1975 that required the doubling of new car fuel efficiency by 1985. Gasoline shortages in 1974 and 1979, coupled with much higher prices, forced the increase in fuel efficiency of the average U.S. car by 40 percent by 1990. Improvements have continued since then, but these have been largely offset by the growing numbers of vehicles on the road. We have also seen an increase in car pooling and the use of public transportation. Although both methods are fairly efficient, the sprawling design of most U.S. cities makes their use difficult for many citizens.

We are beginning to see alternatives to gasoline in the form of ethanol, electricity, propane, and natural gas. City governments in particular are using vehicles powered by propane and natural gas because there is no significant decrease in power, they generate less pollution, and they are cheaper than gasoline. Until these fuels are available to the average driver, however, their use will be limited.

We still rely on large amounts of foreign oil, but the situation has improved somewhat in the past 35 years. Although the amount of oil we use continues to increase, it is increasing at a slower rate because of such factors as better fuel economy in automobiles, improved insulation in our homes and offices, more efficient air conditioning and heating units, and the use of alternative energy sources.

Looking Back

For many years, fossil fuels—materials formed over time from compressed vegetation such as coal, oil, natural gas, tar sands, and oil shale—were believed to be our only source of energy. The United States has reserves of each, but we must remember that they are nonrenewable resources. Someday they will run out. When fossil fuels are burned, the combustion process releases carbon dioxide to the atmosphere. The oceans absorb large amounts of carbon dioxide from the atmosphere when the carbon content of the atmosphere is high and release carbon dioxide to the atmosphere when atmospheric carbon dioxide content decreases.

Coal is a black or brown rock that developed from plants. The extreme pressure in the Earth caused the concentration of carbon matter that eventually formed coal. Most of the coal found today occurs in veins or seams. Oil is formed in essentially the same way as coal. As plants and animals living in the water died, they settled to the bottom of oceans, ponds, and swamps. As pressures were exerted on these materials, oil formed. Natural gas is usually located above an oil deposit.

When coal is removed from strip mines, tons of soil are also moved, thus interrupting the balance of the ecosystem. When oil is piped over frozen ground or carried in huge tanker ships, there is a chance for a spill or slick. Procedures have been instituted and are continually monitored to ensure that other segments of the environment do not suffer because of our mining operations.

Self-Analysis

Essay Questions

1. What are fossil fuels?
2. How is coal mined?
3. What are the differences between shaft mines, slope mines, and drift mines?
4. What is the most dangerous coal mine? Why?
5. Who governs coal mine safety standards?
6. What are some differences in the ways in which coal and oil formed?
7. What are the three devices used to find oil deposits? What do they tell the geophysicist?
8. What are the three oil drilling methods? Briefly explain each.
9. How do offshore drilling methods differ from land drilling methods?
10. List at least 10 uses of oil.
11. Where is natural gas found?

12. What unit of measure do we use for natural gas?
13. List the steps that prepare natural gas for home use.
14. What are the problems associated with oil shale and tar sands?
15. Explain why it is becoming more important to conserve fossil fuels.

Multiple-Choice Questions

1. Which of the fossil fuels is deposited as a liquid?
   a. coal
   b. oil
   c. natural gas
   d. shale
2. What is causing atmospheric carbon dioxide levels to increase?
   a. coal-fired generating plants
   b. burning gasoline by internal combustion engines
   c. wood-burning stoves
   d. all of the above
3. The coal containing the lowest amount of carbon is called
   a. anthracite.
   b. bituminous.
   c. lignite.
   d. nebulacite.
4. The coal reserves in the world total approximately how many tons?
   a. 2 trillion
   b. 2.9 billion
   c. 479 billion
   d. 4.1 trillion
5. A serious problem associated with the use of coal is
   a. low heat output.
   b. acid precipitation.
   c. low combustion rating.
   d. high cost.
6. Strip mines produce what percent of the coal mined in the United States?
   a. 60 b. 91 c. 22 d. 42
7. Which of the following is not a type of underground coal mine?
   a. strip
   b. shaft
   c. slope
   d. drift
8. A measuring device used to explore for offshore oil is a
   a. gravimeter.
   b. magnetometer.
   c. hygrometer.
   d. seismograph.
9. What equipment is used to clean natural gas before it is used?
   a. extraction unit
   b. seismograph
   c. gravimeter
   d. magnetometer
10. The temperature at which natural gas is converted to a liquid is
    a. –260° F. b. –180° F. c. 32° F. d. –112° F.

Learning Activities

1. Prepare a field trip to a strip mine site or an oil field, if available. If those are not available, perhaps a petroleum refinery, pumping station, or storage facility is nearby.
2. Prepare a report on the safety procedures in mining. How have mining procedures changed over the years? What steps still need to be taken?

TERMS TO KNOW

- solar energy
- solar cell
- semiconductor
- solar panel
- nuclear energy
- atom
- chain reaction
- meltdown
- nuclear waste
- radiation
- geothermal energy
- biodiesel
- methane gas
- methane digester
- hydropower
- turbine
- tidal power
- wind turbine
- biomass

CHAPTER 15

Energy and Alternative Fuels

Objectives

After completing this chapter, you should be able to

- explain how rising gasoline prices have increased the urgency to develop alternative sources of renewable power
- discuss the use of solar energy as an alternative energy source
- describe the arguments for and against the expansion of nuclear power
- discuss the potential value of geothermal energy
- speculate on public acceptance of alcohol and biodiesel as fuels for transportation
- explain the process for generating methane gas from sewage and animal wastes
- discuss the benefits and problems associated with hydropower
- speculate on the potential for the widespread adoption of tidal power
- explain the trend toward greater use of wind to generate electricity
- discuss the past and present use of wood as an alternative energy source

Energy Sources	
Nonrenewable	**Renewable**
Coal	Solar (sun)
Petroleum	Wind
Natural gas	Geothermal
	Ethanol (grains)
	Hydro-Power (water)
	Nuclear
	Methane gas
	Wood fuel
	Tidal power

FIGURE 15–1 Energy comes from both renewable and nonrenewable sources.

As the prices of fossil fuels increase, many Americans are searching for other ways to supply their energy needs. When gasoline was 25 cents per gallon, few people were concerned with examining other fuel sources. In this chapter, we review the types of fuels that can be used as an alternative to the fossil fuels (Figure 15–1). The main types of alternative energy sources include solar power, nuclear power, alcohol, geothermal energy, methane, wind, hydropower, tidal power, and wood. Considerable research has been done to develop each source into a reasonable alternative to fossil fuels, and individuals can adapt one or all of them to fill part of their energy needs.

SOLAR ENERGY

Solar energy is energy obtained directly from sunlight. It is more abundant, less exhaustible, and more pollution-free than any other energy source. Every day, the sun floods the Earth with energy that is 100,000 times greater than the entire world's electric power capacity.

As a potential energy source, the sun has interested humans for a long time, but it was not until the extreme rise in cost of conventional energy sources that a closer, more detailed look at solar energy came about. Now, increasing numbers of houses are being built that employ solar energy as their primary energy source.

solar energy
solar panel, energy

The agricultural industry has used energy from the sun in south-facing livestock structures for many years. Heat was radiated into the buildings. Modern energy-efficient buildings derive heat from the sun in the same way today. They are constructed to allow the sun to shine into the structure. Masonry products or water are often used to absorb the heat from the sun for use during the night. During hours of darkness, the heat radiates back into the room from the storage materials.

A greenhouse is a structure designed to capture energy from the sun in the form of heat (Figure 15–2). This extends the growing season, allowing plants to grow during unfavorable seasons of the year. A greenhouse allows plants to capture light energy using the process of photosynthesis. Although we do not often think of photosynthesis and solar energy as being closely related, all food energy is captured from the sun and stored in plant tissues.

A **solar cell** uses the nonmetallic element silicon to generate electricity directly from light. Silicon occurs in nature in sand and other

FIGURE 15–2 A greenhouse provides a protected environment for plants. It traps heat from the sun and makes it possible to raise plants in artificial environments when outside conditions are not favorable for plant growth. (Courtesy of Shutterstock.)

compounds. It is a **semiconductor**, meaning that it is a poor conductor until it is acted on by heat, light, or electricity. Silicon and some other materials give up electrons when light strikes them. It is this characteristic that makes a solar cell function. When radiant heat in the form of sunlight strikes the surface of a solar cell, an electrical current is produced. It is reasonable to expect that a time will come when solar cells will become major suppliers of electrical energy.

A **solar panel** is a heat-exchange device that has been developed to trap heat from sunlight (Figure 15–3). Such panels are often used to heat

FIGURE 15–3 Solar panels are used to convert the sun's energy to electricity or to capture the sun's heat. (Courtesy of Shutterstock.)

FIGURE 15–4 A solar collector (Courtesy of Getty Images).

water or other liquids from which the heat can be extracted for other uses. Some solar panels are used to generate electricity, storing it in batteries for later use. Electricity from solar sources is often used in remote locations to provide power to pump water for livestock and to provide power along freeways for traffic-control devices and lights, among many other uses.

Solar energy systems can be divided into two major types: active and passive. The active systems are those that capture, store, and distribute the energy from the sun. Passive systems provide avenues for the sun to enter but rely on natural airflow to distribute energy in the form of heat.

An active solar system includes a collector (Figure 15–4), a storage mechanism, and a distribution device. The active solar systems overcome the age-old problem of the entire solar concept: The energy is most needed when the sun is not shining—for instance, at night. This system collects the sun's energy in the form of heat and stores it for future use. The storage area can be water, such as an indoor swimming pool, or stones and bricks located in the basement. Once the heat is in the storage area, it can be distributed as the need arises. The heat is extracted by means of pumps or fans, depending on the storage method used.

A passive solar system (Figure 15–5) is less expensive to construct than an active system because it has only a collection device. This device consists of a south-facing solar panel that collects the sun's rays. This method works well in greenhouses and homes equipped with another backup heat source. Construction techniques must be carefully followed to ensure that the heat is held once it has been captured. Passive solar systems are used to heat water, dry materials such as grain, and to distill water as well as in cooking.

The newest research into solar energy concerns converting the sun's rays (radiant energy) into electricity (electrical energy). The second law of thermodynamics tells us that when we change the form of energy, heat is lost. Research indicates that approximately 99 percent of the radiant energy from the sun is lost in the form of heat leaving the Earth's surface. The main goal of using solar energy to produce electrical energy is to more efficiently use the escaped heat of the sun. This method can

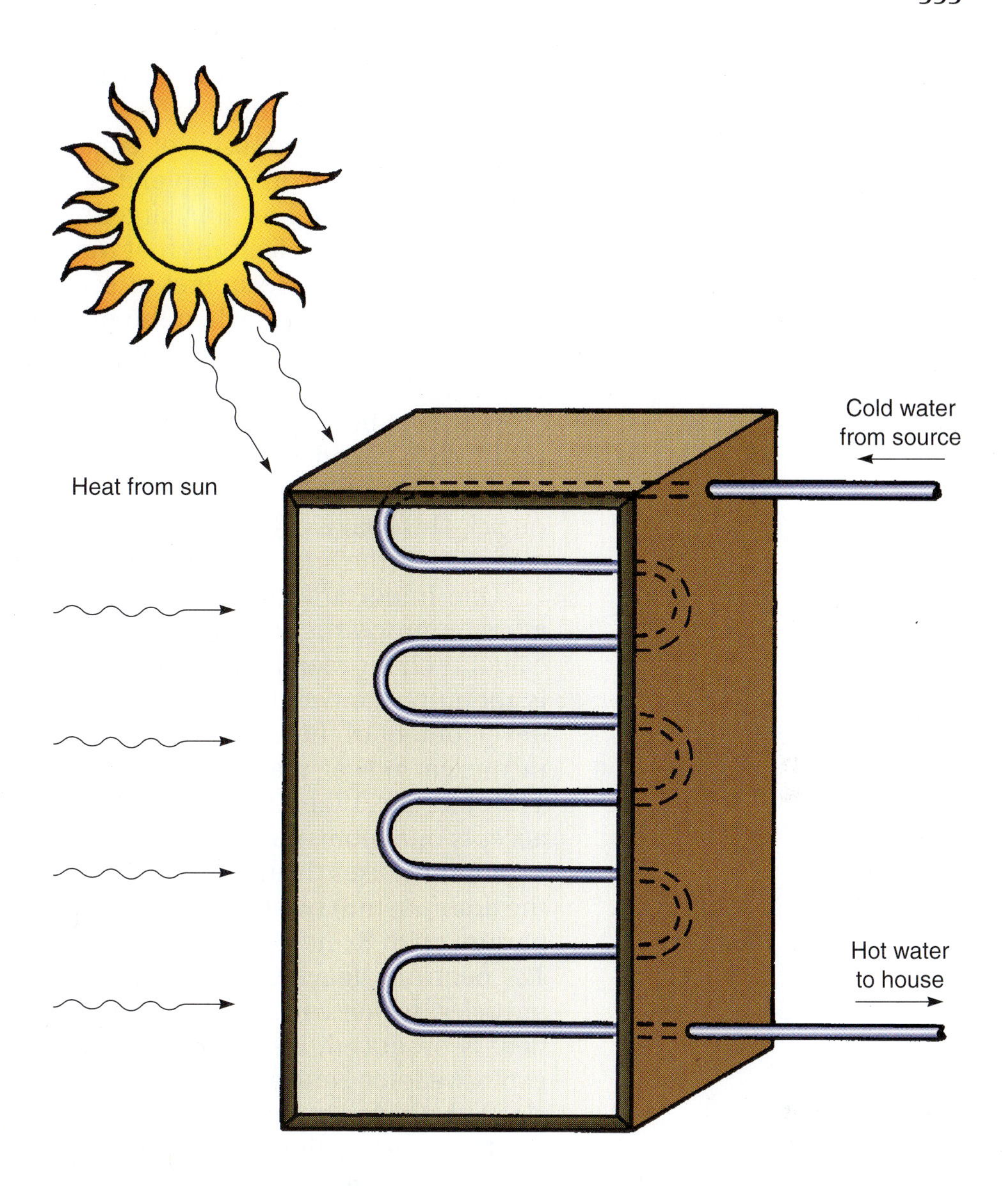

FIGURE 15–5 Passive solar system (Courtesy of Bill Camp.)

be considered a rather indirect way of converting solar energy to electrical power. Research is also being conducted on the direct conversion of sunlight to electricity by the use of solar cells. Solar cells need further development before they are readily available to the public.

NUCLEAR POWER

solar energy, active, passive
nuclear power
nuclear power, benefits
nuclear power, risks

Nuclear energy is heat that is released from an atom during a nuclear reaction. It is frequently used to generate electricity. The electricity that is delivered from a nuclear reactor is not different from any other electricity and is used to perform many tasks in agriculture, industry, cities, and private homes. It provides power to do everything from processing consumer products to pumping irrigation water.

Nuclear power will probably continue to provide electrical energy for many decades to come. Despite opposition, its use has continued in the United States as a safe, clean, and reliable source of energy, but it does require vigilance to ensure that safety controls are always in place. Nuclear plants have been the focus of considerable news coverage in recent years, mostly because of fears of a nuclear disaster. Here we examine what nuclear power is, where the power comes from, and what some of its problems are.

All matter is composed of small, submicroscopic particles called **atoms**. The atom is made up of a central nucleus of positively charged protons and neutral (or uncharged) neutrons. Outside the nucleus are negatively charged electrons that travel at incredibly fast speeds.

Henri Becquerel

In 1896, Henri Becquerel, a scientist working with photographic plates, discovered that when the plates were subjected to radium they were exposed even though they had been covered. It was later discovered that the exposure of the plates was the result of radiation emitted from the radium atom (thus the term *radioactive*). Radium emits positively charged alpha rays, negatively charged beta rays, and neutral X rays. When rays interact with other compounds, they split atomic nuclei. This split is called *fission*. During the fission process, heat is given off, which is the important component in a nuclear power plant.

One important characteristic of radioactive material is that once a fission process is started, it can continue on its own. This process is called a **chain reaction**. Uranium is a good example. Uranium oxide is the important material found in the Earth's crust that allowed the development of the atomic bomb. It is also the one material most used in nuclear power plants. The nucleus of the uranium atom contains 146 neutrons. When the uranium nucleus is flooded with neutrons, it accepts one more, making a total of 147 neutrons. This highly unstable condition causes the uranium to disintegrate. The disintegration breaks the uranium into two different elements: krypton, with 47 neutrons, and barium, with 82 neutrons. Between krypton and barium, we have used 129 neutrons, leaving 18 unattached neutrons to flood other uranium particles, which causes additional reactions. Each time a split occurs, heat is produced. Fissioning one pound of uranium quickly yields an explosive force equivalent to 10,000 tons of the explosive TNT. If released slowly, however, it can produce 12 million kilowatt hours of power.

The heart of the nuclear power plant (Figure 15–6) is the reactor. The reactor uses a mixture of two isotopes—uranium 235 (U-235) and

FIGURE 15–6 Nuclear power plant (Courtesy of Shutterstock.)

uranium 238 (U-238)—in the form of uranium pellets as the fuel source. This makes up the reactor core. As discussed earlier, the chain reaction is triggered by free neutrons in the fuel mixture. To control or stop the reaction, the reactor uses cadmium rods to absorb neutrons. The rods can be inserted or withdrawn around the core. As the reaction progresses, heat is produced. Water in tubes surrounding the core turns to steam, which is sent through turbines to turn electrical generators. An auxiliary water system is available to maintain the core at approximately 1,000° F (538° C). The main concern with the increased use of nuclear power is the fear of an explosion or an uncontrolled heat buildup causing a **meltdown**. High public pressure has resulted in strict controls designed to prevent a nuclear disaster.

Nuclear power plants generate great amounts of hot water, and this water must be cooled before reentering the reactor. If the cooling is completed in a nearby stream or river, there can be environmental damage through thermal pollution. Another main concern with nuclear power plants is the disposal of radioactive wastes from the reactor. The uranium in the reactor core will last as long as two years before it must be replaced.

nuclear waste, disposal

The most contentious and difficult issue raised in the debate over nuclear safety is what to do with the spent fuel. This **nuclear waste** material still emits **radiation**, or high-energy particles that emanate from the nuclei of atoms of certain elements such as plutonium, uranium, and radium, among others. These particles are known to be dangerous to living organisms because exposure to even low levels of radiation is known to contribute to cancer and birth defects. High levels of exposure cause severe burns and sometimes death. Unlike most other waste materials, nuclear waste is slow to degenerate into harmless debris. Nuclear waste generated in our lifetimes will still be unsafe to our grandchildren.

Nuclear wastes are first sent to a processing plant that recovers any unused uranium. Once the recovery process is complete, the wastes are packaged in stainless steel containers and buried. The wastes are still somewhat active, however, and continue to generate heat. The containers last only 100 years or so, after which time they must be repaired or replaced. A better technique may be to place the containers in aboveground concrete bunkers. Wherever they are placed, extreme care must be used to ensure that the containers are not broken open.

If we are to continue using nuclear power, then permanent storage methods and sites must be developed for nuclear wastes. Sites have been constructed or are currently under construction in New Mexico and Nevada as permanent storage facilities for radioactive materials, although there has been strong political opposition to their use. Despite the issues surrounding nuclear power, it is a source of power that will still be available to us long after the fossil fuels are gone. Some countries such as France derive substantial amounts of their domestic energy from nuclear sources.

Geothermal Energy

Geothermal energy is energy from heat derived from the Earth's interior. It involves tapping the underground reservoirs in volcanically active areas (Figure 15–7). When the molten material—magma—that

FIGURE 15–7 Geothermal energy occurs when water is heated by the hot core of the Earth's molten center. Using a variety of technologies to capture the water's heat, energy can be recovered. (Courtesy of Shutterstock.)

forms the Earth's core is near the surface, underground water is heated to high temperatures. Geysers and hot springs are the result of this condition. Sometimes the water is hot enough to produce large amounts of steam that can be used to generate electricity. Iceland, which lies on the fissure between two tectonic plates, derives considerable energy from geothermal sources. The steam that is obtained is piped through the ground to turbines that turn electric generators. The largest geothermal plant in the United States, which uses steam from geysers, is located in northern California. Operated by Pacific Gas and Electric Company, this geothermal plant supplies electricity to the city of San Francisco.

INTERNET KEY WORDS

geothermal energy

The two main disadvantages of geothermal energy are that (1) the energy is not uniformly located around the country and (2) the minerals in the steam are extremely hard on machinery. The development of machinery to withstand the abrasives is needed to make this energy source economically feasible. In areas where geothermal energy is available, it is cheap and clean. It creates no pollution, and setting up a plant requires only a small investment.

ALCOHOL

INTERNET KEY WORDS

alcohol fuels

Alcohol fuel has the potential to solve at least two major problems associated with petroleum fuels: nonrewability and pollution. Petroleum fuels will be used up someday, and they are a source of atmospheric pollution when they are burned. Evidence of air pollution is most easily seen in populated areas, but pollution also occurs in rural areas.

Gasoline and diesel fuels come from crude oil, which is a nonrenewable resource. Once it is used up, there will no more to replace it. Although vast supplies of oil are still available for our use, fuel costs

FIGURE 15–8 Alcohol fuel is produced by fermenting plant materials that contain carbohydrates and sugars. This ethanol plant uses potato waste from a nearby potato-processing plant as its raw material source. The alcohol is mixed with gasoline to make gasohol. (Courtesy of Shutterstock.)

are much higher today than they were just a few years ago, so we must begin to develop new and renewable sources of fuel. With the increased petroleum prices in the 1970s and again in the early 2000s, a renewed interest in alcohol as a fuel has emerged.

Alcohol fuels come from renewable resources (Figure 15–8). Every new generation of plants produces another supply of the raw materials needed to produce alcohol fuels, and many plant by-products can be used for this purpose. Alcohol is produced by growing yeasts in carbohydrate-based solutions such as those provided by grain and sugar crops. The yeasts take in sugar, proteins, vitamins, and minerals and give off carbon dioxide and ethanol. The grain solution can be anything from moldy corn to the waste from cheese production. Supplies will never run dry like an oil well and need no fancy refineries. Alcohol can be used in a blend with gasoline called *gasohol.*

In addition to the renewable alcohol fuels, **biodiesel** is a diesel substitute that is derived from plant oils and animal fats to extend diesel fuel, kerosene, and heating fuel oil. It is usually mixed with diesel in the proportion of 80 percent diesel and 20 percent biodiesel.

Alcohol production produces some useful by-products. The carbon dioxide can be used to carbonate beverages, dry grain, as a fertilizer ingredient, and in fire extinguishers, refrigeration systems, and the manufacturing of dry ice. The residue from the alcohol can be used as feed for livestock. Research has shown an increase of 15 percent to 30 percent in feeding efficiency when the alcohol residue is used.

Air pollution is a serious problem in our world. Although much of that pollution comes from cars and factories, some of it comes from the agricultural industry's use of fossil fuels, which are used to operate the many large engines used in farming operations. They are also used to generate electricity for farms and cities. When alcohol fuels are used, air pollution is minimal because alcohol burns cleanly (Figure 15–9).

FIGURE 15–9 Alcohol fuels are likely to gain favor for agricultural and commercial uses as the cost of petroleum products increases and more emphasis is placed on reducing pollution to the environment. (Courtesy of Shutterstock.)

Methane

Methane gas is a component of natural gas and is formed when vegetable matter decomposes. It is produced in natural settings such as marshes and swampy areas where plant materials build up and begin to decay. The damp, warm conditions that are found in these areas during summer are favorable to the production of methane, and large amounts are released into the atmosphere.

INTERNET KEY WORDS

methane energy

Methane is used most efficiently as a fuel for heating. This odorless gas has a heating rate of 600 to 700 British thermal units (BTUs) per cubic foot of gas (1 BTU is approximately the amount of heat needed to raise 1 pound of water 1 degree F). Under natural conditions, decomposing wastes produce both methane and a compound called *hydrogen sulfide,* which has a distinctly unpleasant odor like rotten eggs and is sometimes referred to as *sewer gas.* Through devices called *scrubbers,* this odor-bearing compound can be removed from the methane.

As cities continue to grow, the problem of waste management also increases. One method of putting the waste to good use is by using anaerobic bacteria—that is, bacteria that grow and prosper in the absence of oxygen—to decompose the wastes. Methane, which also sometimes results from the production of *biogas,* is also a by-product of anaerobic decomposition. The decomposing wastes can be from humans, animals, or plant materials.

Methane also can be produced artificially in a device called a **methane digester** (Figure 15–10), which is an airtight container that holds wastes for decomposition. Anaerobic bacteria break down the wastes, giving off biogas as a by-product. Biogas contains approximately 60 percent methane, 35 percent carbon dioxide, 3 percent nitrogen, 0.1 percent oxygen, and a trace of hydrogen sulfide. For optimal gas production, the digester should be gently agitated and maintained at a temperature of 100° F (38° C). The digester will produce approximately 2 cubic feet to 5 cubic feet of gas per pound of waste. Although more than 50 percent of

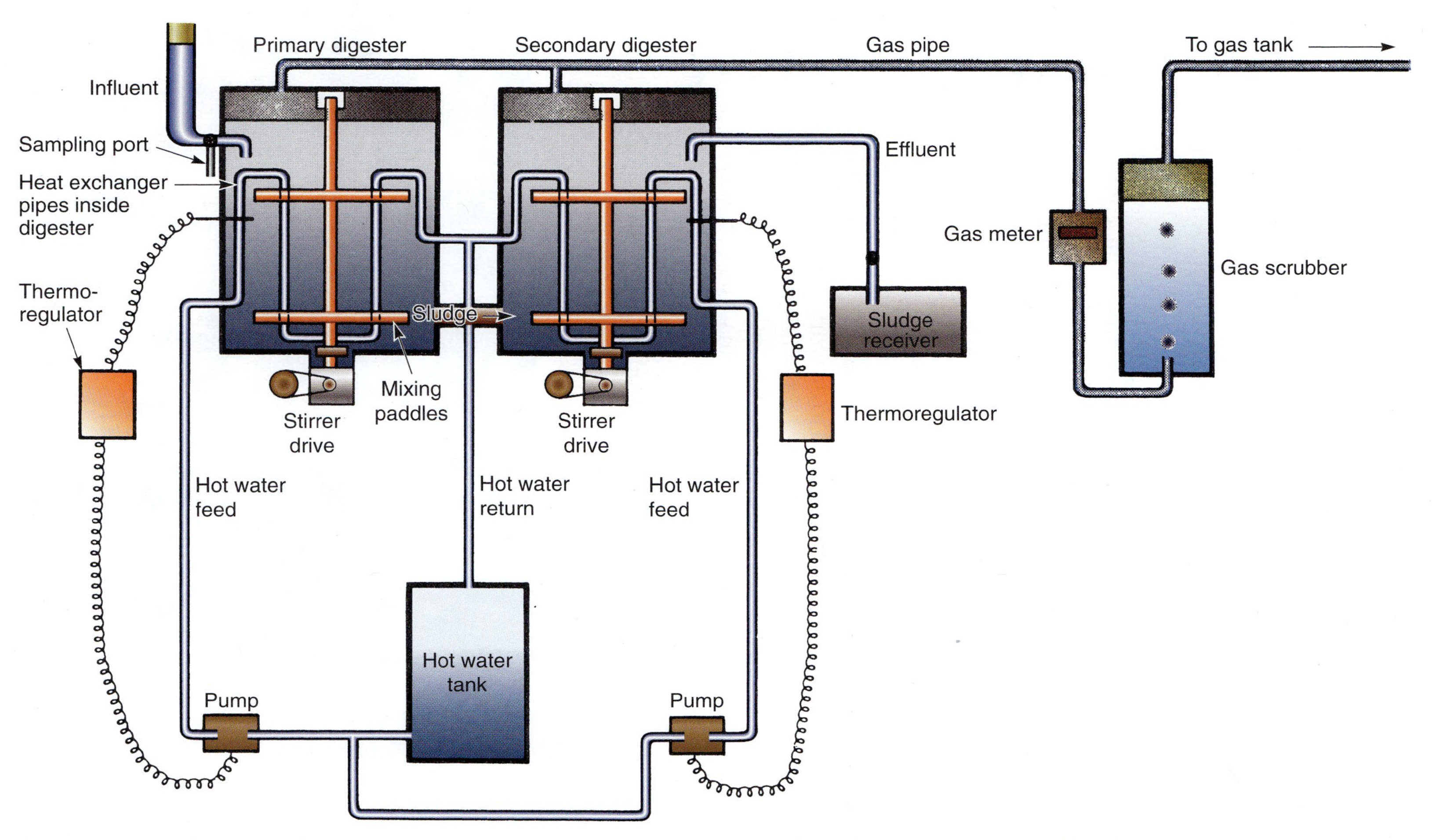

Figure 15–10 Diagram of an experimental methane digester that uses livestock waste (Courtesy of Bill Camp.)

the gas is extracted from the waste in the first two weeks of production, gas can be obtained for as long as six weeks.

After the wastes have been digested, solids remain. This sludge is an important by-product of methane production because of its use as a fertilizer that contains nitrogen, phosphorus, potassium, boron, calcium, copper, iron, magnesium, sulfur, and zinc, all of which are essential for plant growth. One problem with the digester is the gas storage technique. Because of its characteristics, methane cannot be highly compressed like other gases. Currently, storage structures must necessarily be large and bulky.

The heat from burning methane can be used in many ways. Home methane production can be done at extremely low cost, and the wastes can be used on home gardens, thus lessening fertilizer costs. Some farmers are constructing methane digesters to be used as a heat source for the production of alcohol from grain. The methane digester also gives farmers a way of getting something extra from livestock wastes. So methane is one more method of lessening the heavy use of fossil fuels in this country.

Hydropower

hydropower, benefits
hydropower, risks

Hydropower is water power. In its simplest form, the power from moving water is used to do work, whether it is grinding wheat or producing electricity. Water power has been used for many centuries. To gain more control over water, whether for flood control or electrical generation, dams are constructed in water pathways (Figure 15–11). When it is released through openings in the dam, falling water strikes the curved vanes of a **turbine**, causing it to turn an electrical generator and thus produce electricity. Hydroelectric power is a relatively inexpensive form of renewable energy.

FIGURE 15–11 Hydropower dam (Courtesy of Shutterstock.)

Hydroelectric power is a relatively inexpensive and clean form of energy that is renewable. It is usually reliable, but when drought conditions occur, electrical power output is severely limited by reduced stream flows. The use of hydroelectric power is also limited in areas that are unsuited for the construction of dams or where the creation of large reservoirs and lakes would be inappropriate.

Despite the dependability of this form of power, however, some environmental groups oppose continued development of hydropower as an alternative energy source. When a dam is constructed, the dammed waters are slowed and backed up, often destroying wilderness areas. Legislators have already stopped future hydropower projects on the Colorado, Columbia, and Snake rivers. People will eventually have to decide between environmental beauty and electricity.

Hydroelectric power also has been blamed in recent years for many of the problems associated with declining populations of migrating fish. Steelhead trout and several species of salmon live in rivers and streams for several months after they are hatched. In their second year, they move down the streams to the ocean where they live until they are mature adults. Eventually they migrate back up the rivers to the streams where their lives began and where they will spawn and die.

The construction of dams has disrupted the flow of the major rivers through which these fish migrate. Fish ladders have been constructed at some dams to accommodate the inland migration of mature fish, but the juvenile fish populations are frequently subjected to heavy losses on the outward migration. Many of these fish are killed when the water current carries them through the blades of the turbines that provide power to electrical generators. Others find their way to the ocean in the quiet backwaters of the dams.

River currents keep the young fish moving in the right direction as they make their outward migration to the ocean. Fast river currents also move them quickly past such predators as squawfish and sea lions. When the river currents are slowed by dam construction, the migrating fish lose the advantage of speed in their trip to the ocean. Some salmon populations have become so small that the fish have been placed on the endangered and threatened species list.

Efforts have been made to maintain populations of these migratory fish by capturing the adults, collecting their eggs, and raising their offspring in fish hatcheries. This is done when they return to their native streams to spawn, but these efforts have been minimally successful. Too many of the young fish have died as they migrated down the rivers. Large amounts of stored water have been released in recent years in efforts to flush the young fish through the backwaters of lakes and reservoirs, but success has likewise been minimal. This practice uses large amounts of water that is also needed for irrigation, generation of electricity, and maintenance of commercial shipping lanes on the inland rivers.

Tidal Power

tidal power, how to
wind energy, pictures

Another alternative energy source now being researched is **tidal power**. The tides work like clockwork and carry huge amounts of force behind them. Finding a way to harness this energy source has been intriguing to researchers.

FIGURE 15–12 Massive wind turbines are in use in many areas of North America where there is a history of consistent wind currents.

The method most often used is the building of small basins that collect the water during high tide. When the tide waters recede, the water is released through openings containing turbines that drive electric generators. The potential of tidal power is estimated at 2.9 million megawatts of power.

The principal drawback of tidal power is that massive and expensive dams and levees must be constructed. Once the construction is complete, however, the operation is relatively inexpensive and power is always available.

wind

Wind has been used as an energy source for many years. The early settlers relied on the wind to bring water up from deep wells. Many parts of the country continue to use wind power to provide water for livestock where electricity is not available. The incredible power of the wind can be seen by looking at a tornado in motion.

In recent years, scientists have been experimenting with a new and more efficient windmill. The new structure, which It is called a **wind turbine**, appears to be more like a tall "windtower" equipped with propeller blades. The main use of the new towers is to turn generators that produce electricity (Figures 15–12 and 15–13). The rotating generator may produce either alternating current (AC) or direct current (DC), depending on the consumer's needs. Some states have enacted legislation that requires public electrical utilities to purchase power generated by privately owned wind farms, water power, and other generating technologies.

As with most energy sources, this alternative has its limitations. The wind does not blow all the time, and the speeds at which it blows vary. This creates a need for some sort of storage mechanism. The most widely used storage device is the DC battery. Another problem is in the construction of the high towers. With aircraft traffic continuing to increase, builders are restricted on where towers can be placed. Building codes and zoning regulations also must be followed.

FIGURE 15–13 Regardless of the direction the wind blows, the Darrieus windmill will turn, thus driving a generator to produce electricity. (Photo courtesy of Shutterstock.)

wood

Wood is one of the oldest energy sources known to humans. Until the twentieth century, wood was usually the only supply of fuel available for factories and home heat. As with other sources discussed, wood has become more attractive with the rise of traditional energy prices (Figure 15–14). In recent years, an industry has grown up based on a fuel known as **biomass**. This fuel consists of wood that has been chipped and dried for the purpose of generating heat or electrical power.

As a heating source, wood has several advantages. It is widely available, so transportation costs can be low, and it comes from a renewable natural resource. Considerable research has gone into developing faster-growing trees. With effective forest-management practices, the supply of wood should be abundant. Wood also gives us an energy edge in case of a disaster. Water pipes can be prevented from freezing, cooking can

FIGURE 15–14 Wood continues to be a widely used fuel for heating homes.

continue, and lives can be saved if this alternative energy source is available. Most people who enjoy heating with wood usually like the independence from other expensive fuels. Wood energy does have disadvantages, of course. It is less convenient to burn than other energy sources and is bulkier and less efficient than oil and gas. The use of wood stoves in some areas of the country is also restricted during periods when air quality is low.

Looking Back

Because energy costs are high, people are always looking for alternate sources of energy, whether for their home, office, or automobile. Each alternate source examined in this chapter can be used. Each individual, however, must look carefully at each source to make sure the right selection is made. Among the alternatives to fossil fuels are solar energy, nuclear power, geothermal energy, methane gas, hydropower, tidal power, wind power, and wood. Not all alternative sources are for everyone. If each person uses only one, however, the dependence on high-priced nonrenewable fuels will be lessened.

Self-Analysis

Essay Questions

1. What is the potential of solar energy in the United States and the world?
2. Explain the difference between an active and passive solar energy system.
3. How can solar energy be stored?
4. What are the strengths and weaknesses of dependence on nuclear energy?
5. What is fission?
6. Why is the chain reaction important in a nuclear power plant?

7. How is a nuclear chain reaction started? How is a chain reaction controlled?
8. What is the fuel that is used in a nuclear power plant?
9. Where is geothermal energy obtained? What are the problems related to geothermal energy?
10. What is the process for producing fuel-grade alcohol?
11. How is alcohol used as a fuel? How are alcohol by-products used?
12. What is the source of methane and how is it produced?
13. How are the by-products of methane production used?
14. What is hydropower? What are the environmental concerns about hydropower?
15. What is the potential of tidal power?
16. How has modern technology changed wind power?
17. What are the advantages and disadvantages of using wood and biomass as alternative energy sources?

Multiple-Choice Questions

1. The source of solar energy is
   a. solar system.
   b. heat bursts from meteors.
   c. sunlight.
   d. uranium.
2. What material is used in a solar cell to generate electricity?
   a. silicon
   b. carbon
   c. fiber glass
   d. uranium
3. A device that is used to generate electricity in remote locations is
   a. a remote sensing device.
   b. a solar panel.
   c. a GPS.
   d. a radial gyberometer.
4. Nuclear energy is produced from what fuel source?
   a. water
   b. nucleic acid
   c. ribonucleic acid
   d. uranium
5. An energy source obtained from the heat of the Earth's molten core is
   a. nuclear power.
   b. geothermal power.
   c. solar energy.
   d. methane.
6. A combustible gas that is obtained from decaying vegetation, sewage, and animal waste is
   a. ethanol.
   b. gasohol.
   c. helium.
   d. methane.
7. An alcohol fuel called ethanol is
   a. derived from the fermentation of grains and other plant materials.
   b. nonrenewable.
   c. considered to be a serious atmospheric pollutant.
   d. produced in nature from decaying plant materials in swampy environments.
8. The waste materials obtained from the production of methane are used for
   a. animal feeds.
   b. formulating asphalt.
   c. lining municipal waste sites.
   d. crop fertilizer.

9. Some opponents of hydroelectric power have called for the breaching of hydroelectric dams to save
   a. water.
   b. migrating fish.
   c. energy.
   d. prime agricultural land.
10. Which energy source contributes most to atmospheric pollution?
   a. wood
   b. ethanol
   c. methane
   d. solar power

Learning Activities

1. Choose one alternative energy source. Prepare a list of the changes you would have to make in your home to implement the new energy source.
2. Invite a representative from your local electric company or gas company to speak to your class about energy sources for the future.
3. Design and build a passive solar energy collector. Plans should be available in your school or public library.

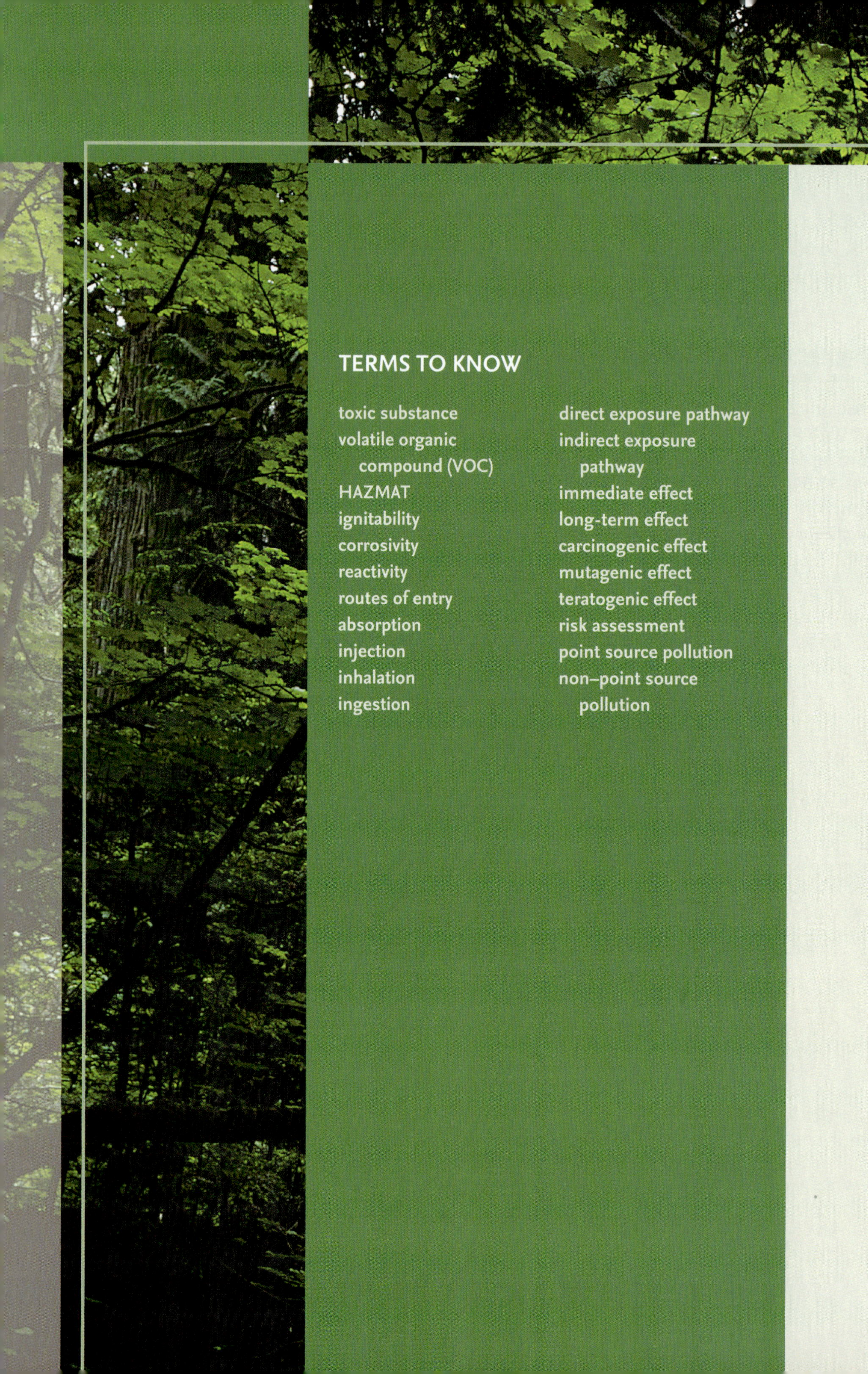

TERMS TO KNOW

- toxic substance
- volatile organic compound (VOC)
- HAZMAT
- ignitability
- corrosivity
- reactivity
- routes of entry
- absorption
- injection
- inhalation
- ingestion
- direct exposure pathway
- indirect exposure pathway
- immediate effect
- long-term effect
- carcinogenic effect
- mutagenic effect
- teratogenic effect
- risk assessment
- point source pollution
- non–point source pollution

CHAPTER 16

Toxic and Hazardous Substances

Objectives

After completing this chapter, you should be able to

- define toxic and hazardous materials
- list and describe the characteristics of hazardous materials
- explain the need for safety regulations and procedures for handling toxic and hazardous substances
- relate the lessons learned from the Love Canal incident to the need to deal with toxic materials in appropriate ways
- discuss routes of entry as they relate to hazardous materials exposure
- contrast the differences between direct and indirect exposure pathways
- clarify the different ways in which long-term exposure to toxic materials may affect an individual's genetic code
- identify some critical questions that are helpful in conducting a risk assessment of exposure to hazardous materials
- explain why the Superfund legislation is so significant to the process of cleaning and remediating sites that have been contaminated with hazardous and toxic materials
- distinguish between point source pollution and nonpoint source pollution

FIGURE 16–1 Waste materials must be properly disposed of to prevent serious environmental damage. (Courtesy of Shutterstock.)

INTERNET KEY WORDS

volatile organic compound

The human population generates waste materials that are not readily disposable, some of it persisting in the environment almost indefinitely and some of it harmful or toxic to living organisms. In fact, waste materials have become a major problem in many of the world's industrialized nations (Figure 16–1). By-products of industrial processes, solid wastes from our population centers, petroleum leaks and spills, and pesticide residues from farms, gardens, and yards are only a few of the waste materials and pollutants that affect the environment (Figure 16–2). To prevent serious environmental problems, we must properly dispose of all waste materials.

FIGURE 16–2 Hazardous and toxic materials are sometimes present in smoke and fumes and are subject to laws and regulations that govern toxic substances. (Courtesy of Digital Vision.)

Toxic and Hazardous Materials

Household, agricultural, and industrial wastes have been a serious problem for many years. They include a variety of harmful chemicals, poisonous metallic compounds, acids, and other caustic materials. In recent years, we have learned that highly toxic liquids tend to ooze out of landfills and pollute local soils and water supplies.

Government safety regulations define a **toxic substance** as material that contains one or more of 39 specific contaminants such as heavy metals or organic carbon-based compounds known as **volatile organic compounds (VOCs)**, or substances that contain hydrogen and carbon and may contain oxygen, nitrogen and other elements. Some common materials that meet the definition for a VOC, such as carbon monoxide and carbon dioxide, are excluded from the VOC list. Most of the materials on the list are gaseous fuels or industrial solvents.

Many substances meet the definition of a hazardous material or **HAZMAT**. In some cases, the hazardous material may be a byproduct of a legitimate process, event, or activity; as such, it is called *toxic waste*. A hazardous material is any gas, liquid, or solid substance that is capable of harming humans or other animals, property, or the environment (Figure 16–3). Descriptions of hazardous materials include such terms as *explosive*, *corrosive*, *toxic*, *flammable*, and *radioactive*. Such materials may also be *oxidizers*, *allergens*, or *asphyxiants*. In every instance, such material is harmful and dangerous. For this reason, safety regulations are in place regarding the use, storage, disposal, and especially the

Hazardous Material

Any gas, liquid, or solid substance that is capable of harming people, animals, property, or the environment.

FIGURE 16–3 Hazardous material

transport of such materials. When such materials are spilled, specially trained HAZMAT response teams are called in to implement proven safety precautions.

Some hazardous materials will burst into flame at low temperatures. **Ignitability** is the tendency for a substance to have a flash point less than 140° F. In comparison with boiling water (212° F), such materials begin to burn at a much lower temperatures. Examples of such products include paint thinners, lighter fluids, and similar substances. Federal regulations define **corrosivity** as substances with a pH of greater than 12.5 or less than 2.0, making these materials extremely strong acids or extremely strong bases. Only a small amount of corrosive material can be severely damaging to living organisms.

The tendency of a material to be unstable at normal temperatures and pressures or to explode or react violently when mixed with air, water, or other chemicals is known as **reactivity**. Such reactions are often accompanied by the release of dangerous fumes or gases such as ethers, cyanides, and other acutely toxic gases.

Love Canal

Interest Profile

LOVE CANAL

Love Canal was originally intended to be a human-made waterway connecting two branches of the Niagara River where it was separated by Niagara Falls. It was never completed. Instead, the area became a landfill for disposal of chemical waste from the petrochemical industry. Later, it was bought and used by Hooker Chemical and Plastics Corporation, a subsidiary of Occidental Petroleum. This company buried approximately 22,000 tons of toxic waste in the area and sealed the site with what was thought to be impenetrable clay.

The local school district wished to buy property for a new school, and eventually the property was deeded to the school district. A school was built on the site despite written warnings from Hooker about safety issues. Low-income and single-family residences were built on adjacent property. The city of Niagara breached the clay seals on many occasions to construct sewer lines, and this allowed chemical contaminants to seep out of the area. This was followed by odors in the area and health problems among the residents, including a cancer rate that was extremely high.

Scientific studies did not identify a direct link between the health of the residents and the chemicals from the landfill, but the residents' blood tests showed chromosome damage that could result in reproductive problems, genetic damage, and increased cancer risk. In August 1978, President Jimmy Carter declared a federal emergency at the site. Meanwhile, both homes and the school had been abandoned and demolished.

Officially known as the Comprehensive Environmental Response, Compensation, and Liability Act, the legislatively-created Superfund was expressly designed to help clean up toxic wastes throughout the country. For the Love Canal incident, the federal government relocated more than 800 families from the area. Occidental Petroleum was sued under the Superfund law, and in 1995 the company agreed to pay $129 million in restitution. The Love Canal remains one of the largest environmental disasters in U.S. history.

FIGURE 16–4 Spills of petroleum or chemicals in water environments severely damage populations of fish, wild animals, and plants. (Courtesy of Shutterstock.)

INTERNET KEY WORDS

hazardous materials, body, effects

Harm from Hazardous Materials

Hazardous or toxic materials are able to enter the bodies of humans and other animals in several ways, called **routes of entry**. One route is **absorption** through the skin or the eye by contact with the contaminant (Figure 16–4). Symptoms include irritation, burns, ulcerated skin, sores, and even blindness. All of these symptoms are caused by the reaction of the contaminant with living tissue as the material moves through the skin and into the capillaries that deposit it in the bloodstream. It is important to wear gloves when working with these materials and to avoid touching the eyes.

Another route of entry is **injection**, which usually occurs when the skin is cut or punctured and the contaminant is carried into the cut. Entry into the bloodstream is immediate in such an instance. **Inhalation** is also a route of entry for some hazardous materials. Small droplets may become suspended in the air or attached to dust particles that are breathed into the lungs. The lungs are interlaced with capillaries that aid in the exchange of oxygen and carbon dioxide. They work equally well in allowing contaminants to be absorbed into the blood stream. The body often reacts by creating mucous for the purpose of capturing such particles, but high concentrations of inhaled contaminants can quickly overwhelm the body's defense system.

A fourth route of entry is **ingestion**. This form of contamination occurs when something contaminated is eaten. Failure to properly wash hands and food products before eating is one way in which a person might ingest a contaminant. Another possibility might come from keeping the food in a place where contaminated dust or droplets suspended in the air can get directly on the food or its container. In such an instance, the contaminant is easily transferred to the hands and then the mouth during eating. Young and curious children sometimes pick up contaminated objects and put then into their mouths. This can be a

FIGURE 16–5 Hazardous materials are sometimes discovered in accessible places near populations that are in harm's way. (Courtesy of Getty Images.)

serious problem when children are growing up near an area contaminated with lead or other toxic substances.

Consider what might happen if you live near an area where toxic waste has been discarded (Figure 16–5). If the waste material gives off fumes, then you might inhale or absorb the contaminant through your skin or lungs. This is known as a **direct exposure pathway**. Suppose that dust particles become contaminated by the waste material and that they get on a crop that is part of the food chain for humans or other animals. A person could ingest the food itself or meat from an animal that ingested the contaminant. This is an **indirect exposure pathway**. Over time, the amount of some contaminants builds up in the organs of the body and achieves a level of toxicity that is harmful. Heavy metals such as lead and mercury are among the contaminants that react with the body in this manner.

The effects that toxic substances have on the body can vary greatly. Exposure to some toxic materials results in reactions during or shortly after exposure. The reactions include one or more of the following: vomiting, irritation of eyes and sinuses, or any number of other body responses. These are known as **immediate effects**. In contrast, **long-term effects** usually require multiple exposures that may occur over a period of years. Long-term effects are usually associated with a buildup of the toxic contaminant in body tissues and are often triggered by larger and larger accumulations in the body.

Some long-term effects of exposure to toxic materials result in changes in an individual's genetic code. These genetic changes occur in three different ways. The first is a **carcinogenic effect** in which an affected individual becomes more susceptible to cancer. The second category is a **mutagenic effect**. In this instance, the DNA of an individual undergoes a permanent change that can be passed along to successive generations. The third category is the **teratogenic effect**, which is an increase in the risk of physical defects in a developing embryo.

Assessing Risk

When a source of contamination is identified, the risk is assessed by evaluating several characteristics of the contaminated material as well as pathways that might lead to the exposure of plants, humans, and other animals. Among the relevant questions are:

- What is the concentration of the contaminant, and how much of the substance is there?
- How mobile is the substance? Is it a solid, liquid, or gas?
- What are the contaminant's physical properties? What effects does the contaminant have on living organisms?
- What is the location of the contaminant relative to the population?
- Is the contaminant contained or exposed? If contained, is the container adequate?
- What are the physical and chemical properties of the contaminant? Is it easily dissolved in water or vaporized into the atmosphere?
- What are the exposure pathways that link the contaminant to the population?
- What is the duration of exposure?

The answers to these questions will help in evaluating the degree of danger to the environment and the organisms that live within it. An evaluation of this kind is a form of **risk assessment**. Once the degree of risk is known for a specific contaminant, it is much easier to develop a plan of action to minimize the danger it poses to the environment.

Toxic and Hazardous Substances Legislation

Although necessary, treating and remediating toxic substances is complicated and expensive. Humans as well as all other forms of life are at risk when hazardous materials are allowed to accumulate in open dump sites. In too many instances, our society has tended to ignore such problems, believing they will go away over time. They do not. We are accumulating toxic wastes at a much faster rate than they are being degraded. Most hazardous materials require long periods of time to be degraded, which is why they require processing at a toxic waste site.

toxic waste, body, effects

Federal and state legislatures have passed laws that provide for the regulation of hazardous materials. Funds, such as the Superfund, have been appropriated for the purpose of cleaning up dangerous sites where hazardous materials have been illegally disposed.

In addition to federal laws, states have approved legislation to control toxic and hazardous material transport and disposal within their borders. In many instances, these state standards are stricter and more comprehensive than federal standards, allowing the states to address troubling local problems. States are not allowed to reduce federal standards that apply to toxic and hazardous substances.

The national Superfund priorities list identifies contaminated sites in every U.S. state and territory. It is revised as new needs arise and

Interest Profile

LAWS AFFECTING TOXIC AND HAZARDOUS SUBSTANCES

- Atomic Energy Act (1954)
- National Environmental Policy Act (1969)
- Clean Air Act (1970)
- Toxic Substances Control Act (1976)
- Resource Conservation and Recovery Act (1976)
- Clean Water Act (1977)
- Comprehensive Environmental Response, Compensation, and Liability Act (also known as the Superfund Act) (1980)
- Superfund Amendments and Reauthorization Act (1986)

as cleanup efforts are completed. An examination of the priority list of Superfund sites reveals the staggering magnitude of problems with seriously contaminated sites. A current listing of sites within states and territories is available on the Internet at <http://www.epa.gov/superfund/sites/npl/npl.htm>.

One of the most difficult challenges in dealing with hazardous materials is finding the source of contaminants in the environment. Sometimes, a contaminant gets into a water supply, polluting surface water or groundwater by leaching through the soil. This problem is extremely difficult to correct. Pollutants must be tracked to their source in order to solve the problem. **Point source pollution** is pollution that can be traced to a specific source. In some instances, it may be discovered that an underground fuel or chemical tank has leaked some of its contents into the soil. In other instances, a chemical spill may be to blame. In each instance, it is important for the pollutant to be cleaned up and isolated from the water supply.

A far bigger threat to water quality is **non–point source pollution**. This kind of pollution occurs each time water flows over or through soil, absorbing or carrying soil particles and pollutants and leaving them suspended or dissolved in surface and groundwater. Other mediums that carry pollutants include air and soil. When the source of a pollutant cannot be identified, it becomes nearly impossible to improve the quality of contaminated air, soil, or water. Surveys of U.S. waterways have concluded that 40 percent of rivers, streams, lakes, and estuaries are no longer safe for outdoor activities such as swimming and fishing. Most of these problems are due to non–point source pollution. The most common contaminants of this type are nutrients such as nitrates and phosphates.

The use of agricultural chemicals is often cited as a practice that has extensively damaged the environment. Some chemicals, such as DDT, are known to have side effects that were not known when they were approved for use. Damage to the environment through misuse of agricultural chemicals sometimes occurs, just as it occurs when industrial, lawn, or garden chemicals are used improperly. When such chemicals are misused, it is right for citizens to be concerned. Each of us should speak up when we observe misuse of hazardous materials.

Agricultural chemicals have been tested carefully and can be safely applied when used according to the manufacturer's directions (Figure 16–6). Agricultural chemicals that are left over from a job are

Figure 16–6 Carefully follow the manufacturer's directions when applying chemicals to ensure that they are used safely. (Courtesy of Shutterstock.)

considered to be toxic waste because they are poisonous to living organisms. Laws prescribe how they should be properly handled and disposed. These materials should be safely stored until they can be delivered to a toxic waste treatment center or properly degraded. Chemical abuses must be corrected for the safety of living plants, animals, and humans.

INTERNET KEY WORDS

disposal of chemicals

Chemicals that are applied for their intended purposes according to the recommendations of the manufacturer can usually be safely used (Figure 16–7). Agricultural chemicals that are in use today are subject to extensive scientific testing in an effort to ensure that they pose no threat to the environment.

The incorrect disposal of unused chemicals and chemical containers poses as much a threat to the environment as approved uses. People who use agricultural, garden, and industrial chemicals must assume responsibility for applying them properly and appropriately disposing of

Figure 16–7 A pest-control specialist ensures that chemicals will be used safely and wisely when they are needed.

Figure 16–8 Laws and regulations for disposing of hazardous waste materials should be strictly obeyed.

any waste associated with chemical use (Figure 16–8). Most of the laws and regulations that apply to the disposal of hazardous materials apply equally to agricultural chemicals and they should be strictly obeyed.

The improper use of agricultural chemicals poses greater dangers to farm families than to anyone else, so most farmers carefully avoid misusing toxic materials. They live on the land with their families and drink the water from farm wells. Their own children would be the first to suffer from chemical abuses because they work and play in the fields. Many farm families still eat meats, fruits, and vegetables that are homegrown, and the improper use of chemicals would surely affect these foods.

Looking Back

Some of the materials that humans create persist in the environment almost indefinitely. To prevent serious environmental problems, we must dispose of all waste materials properly (Figure 16–9). Industrial

Figure 16–9 Biotechnology has modified bacteria so that they break down crude oil. These bacteria have become important in reducing the harmful environmental effects when oil spills occur. (Courtesy of the U.S. Department of Agriculture.)

wastes have been a serious problem for many years. They include a variety of harmful chemicals, poisonous metallic compounds, acids, bases, and other caustic materials that are left over from various manufacturing processes.

Many substances meet the definition of a *hazardous material*, also known as HAZMAT. A hazardous material is any gas, liquid, or solid substance that is capable of harming humans and other animals, property, or the environment. Descriptions of hazardous materials include such terms as *explosive, corrosive, toxic, flammable*, and *radioactive.*

Hazardous or toxic materials are able to enter the bodies of humans and animals in several ways called *routes of entry.* Exposure to some toxic materials results in reactions during or shortly after exposure. These are known as *immediate effects. Long-term effects* usually require multiple exposures that may occur over a period of years.

When a source of contamination is identified, the risk is assessed by evaluating several characteristics of the contaminated material as well as pathways that might lead to exposure of plants and humans and other animals. Legislation has been passed by federal and state legislative bodies that provides for the regulation of hazardous materials. Pollutants must be tracked to their sources to solve the problem. *Point source pollution* is identified as pollution that can be traced to a specific source. *Non–point source pollution* cannot be tracked to a specific source.

Self-Analysis

Essay Questions

1. What is the definition of a toxic substance?
2. How is a toxic substance defined by government safety regulations?
3. What is a volatile organic compound (VOC)?
4. What kind of material is considered to be toxic waste?
5. Why is it necessary to notify a HAZMAT team to clean up a spill of hazardous material?
6. With respect to hazardous and toxic materials, what is the importance of routes of entry?
7. List examples of a direct exposure pathway and an indirect exposure pathway.
8. What kind of information is needed to assess the risk of exposure to toxic or hazardous substances?
9. What are some examples of federal legislation that have affected the ways that U.S. citizens deal with issues related to hazardous and toxic materials?
10. Why must empty chemical containers be disposed of in the same way as hazardous waste?

Multiple-Choice Questions

1. Which of the following is not another name for hazardous material?
   a. VOC
   b. HAZMAT
   c. PSP
   d. NPSP
2. The tendency of a particular hazardous substance to burst into flame at a temperature of less than 140° F is related to
   a. corrosivity.
   b. toxicity.
   c. ignitability.
   d. volatility.

3. The pH of a toxic material is related to which of the following terms?
   a. corrosivity
   b. toxicity
   c. ignitability
   d. reactivity
4. The tendency of a material to be unstable at normal temperatures and pressures or to explode or exhibit violent tendencies when mixed with air, water, or other chemicals is called
   a. corrosivity.
   b. toxicity.
   c. ignitability.
   d. reactivity.
5. The legal and sensible way to deal with a spill of a suspected hazardous material is to
   a. immediately call for a HAZMAT team to assess and clean up the material as necessary.
   b. treat the spill as you would any other spill.
   c. collect samples of the material and send them to a laboratory for analysis.
   d. call the local sanitation department to clean up the spill.
6. Which of the following is not considered to be a route of entry for hazardous material?
   a. injection
   b. digestion
   c. ingestion
   d. inhalation
7. Which of the following fits the definition of a direct exposure pathway?
   a. inhalation of toxic gas
   b. eating a plant that has been contaminated
   c. eating the meat of an animal that has eaten contaminated food
   d. drinking contaminated water
8. Which of the following effects of exposure is not known to cause changes in the genetic code of an individual who has experienced long-term exposure to toxic material?
   a. skin irritation
   b. carcinogenic
   c. mutagenic
   d. teratogenic
9. Which effect of exposure to toxic material is known to increase the risk of physical defects in a developing embryo?
   a. teratogenic
   b. carcinogenic
   c. mutagenic
   d. cryogenic
10. Pollution that can be traced back to a specific source is known as
    a. mutagenic.
    b. non–point source pollution.
    c. carcinogenic.
    d. point source pollution.

Learning Activities

1. Identify a Superfund priority in the area where you live, and learn what caused the problem. Also seek information about the steps that are required to clean it up. Information is available on the internet for each state and territory in the United States; see <http://www.epa.gov/superfund/sites/npl/npl.htm> for more information.
2. Visit with a local government official to see what laws and regulations control the disposal of hazardous and toxic wastes in your local area. You may want to invite the official to visit your class as a guest speaker. Be sure to prepare appropriate questions before you visit so that you can guide the discussion and keep it relevant to toxic waste disposal.

CAREER OPTIONS

- air quality control
- biologist: aquatic
- biologist: fish and wildlife
- biologist: marine
- botanist
- dendrology, silviculture, and forestry
- ecologist
- educator: forestry
- entomologist
- environmental analyst
- environmental engineer
- environmental quality technician
- environmental scientist
- forester
- game bird farm manager
- herpetologist
- ichthyologist
- inspector: environmental safety
- mitigation specialist
- oceanologist
- ornithologist
- predatory animal control officer
- science teacher
- soil conservationist
- soil specialist
- taxonomist
- technology specialist
- water treatment specialist and wastewater treatment specialist
- wildlife conservation officer
- wildlife technician

CHAPTER 17

careers in Environmental science

Objectives

After completing this chapter, you should be able to

- list some careers available in environmental science
- describe the educational requirements for environmental science careers
- identify the characteristics of some environmental science careers
- explain the importance of careers related to the environmental sciences
- identify one or more career interests that correlate with personal strengths
- develop an educational plan leading to a career of choice

Preparing for a career is one of the most important things you will ever do. Many high schools and community colleges offer career counseling services, and it is a good idea to spend time with counselors to narrow your choices and identify your interests and strengths. There are many unhappy people in the world who chose a career for the wrong reasons. Going to work every day to a job you don't find challenging and fulfilling is extremely hard. Earning a good living wage is important, but having a passion for your work is important, too. You should be able to find a way to enjoy or even thrive in your work while being compensated at a level that allows you to support your family financially.

Most people who have careers in environmental science made this career decision early in their educational experience. This is important because developing an educational plan that accumulates experience in mathematics, science, and communication is critical. You need a solid beginning in high school followed by rigorous classes in college as you prepare for careers in environmental sciences. This chapter will help you explore some of those careers.

selecting a career

The final choice of a career should be made only after much thought and research. In the conservation and environmental fields, one factor to consider is whether you wish to work indoors or outdoors (Figure 17–1). Almost everyone enjoys being outdoors in the spring and early summer, but many conservation positions require being outdoors virtually all year. Pleasant spring temperatures often give way to summer temperatures of 100° F or more. The mild weather of fall often gives way to bitter winter cold. Being outdoors day in and day out means coping with rain, snow, cold, and heat. If you would rather be dry and comfortable all the time, then you might be better off choosing a conservation or environmental career that allows you to spend the bulk of your time indoors.

FIGURE 17–1 Conservation and environmental research are fields in which work is performed both indoors and outdoors. (Courtesy of the U.S. Department of Agriculture.)

Another important factor to consider when selecting a career in the conservation and environmental field is whether you are a "people" person or prefer to work with plants and wildlife. Many of us are more skilled at working with plants and animals than we are at dealing with people. Other people enjoy working with others and possess good communication skills. In fact, in today's world, most employment opportunities require good communication skills and the ability to get along with a variety of other people. Employers expect employees to work hard and work smart. Employees must be well mannered and well groomed, regardless of the place of employment or job description. A professional work ethic means arriving to work on time, dressed and groomed in an appropriate manner, and being prepared to work long hours if necessary.

Safety can also be an issue in selecting a career, particularly when working with wildlife. In many cases, the animals being worked with are sedated, restrained, or both. Although most wild animals would not normally injure a person, they might do just that when they feel threatened or cornered. This is the type of situation a wildlife biologist or technician may encounter. Whenever you handle wildlife, whether to attach a radio transmitter, weigh the animal, or take blood samples, the risk of injury is present for the person and the animal. Paying special attention to safety training and precautions is critically important. Most wild animals are not accustomed to being near people, much less being touched and handled by them. They become frightened and are more likely to defend themselves in such situations. Working with animals of any kind, wild or domestic, requires great patience. These are important factors to consider when making a career choice.

Another factor to consider when making a career choice is the level of education required for a particular career. Many of the careers in the environmental sciences require education beyond high school. Some careers require a four-year college degree or graduate degree, and others require only two years of college or technical school. With the great variety of different occupations within the environmental science field, selecting one that suits you should not be difficult (Figure 17–2).

FIGURE 17–2 Technician Jeff Nichols collects a water sample from the Walnut Creek watershed in Ames, Iowa. Samples are collected weekly from this area and surrounding watersheds to study the effects that farming practices have on water quality. (Courtesy of the U.S. Department of Agriculture's Natural Resources Conservation Service. Photo by Keith Weller.)

career options

Careers in the environmental sciences are extremely varied, including research scientist, field biologist, engineer, and animal caretaker. Many clerical and trade positions are available with the U.S. Fish and Wildlife Service and most state wildlife agencies. Professional positions such as wildlife biologist, fishery biologist, engineer, and research scientist require at least a four-year college degree. Most enforcement careers, such as conservation officer and customs inspector, require at least some training in law enforcement.

There are also many jobs available in the computer field. With the increased use of technology in environmental science, increasing numbers of jobs are becoming available for computer programmers, analysts, and specialists. Jobs are available for personnel specialists and public affairs officers in many state, federal, and private organizations. Most state environmental agencies, as well as the U.S. Fish

FIGURE 17–3 Air quality specialists use specialized equipment to gather and process data. (Courtesy of the U.S. Department of Agriculture's Agricultural Research Service.)

INTERNET KEY WORDS

air quality control, roles, responsibilities
aquatic biologist, roles, responsibilities

FIGURE 17–4 Aquatic biologists study aquatic organisms such as fish, plankton, clams, and snails in the water environments they inhabit. (Courtesy of the U.S. Department of Agriculture.)

and Wildlife Service, hire people in a variety of technical positions, most often as aides or assistants. Educational requirements for these positions vary widely. Most federal and state agencies employ a variety of craftspeople such as carpenters, electricians, and plumbers to keep their facilities in proper working order. A great variety of jobs are available, with various degrees of education required. We will now take a more in-depth look at a few of the careers available in environmental science.

Air Quality Control

Careers in air quality are available with the weather services of local, state, and national agencies. Local and network radio and television stations all use weather reporters and meteorology forecasters. Of equal importance are those who monitor and help to improve air quality. Technicians collect and chemists analyze samples of air taken from various places in the atmosphere, buildings, and homes. Employees with environmental protection agencies and environmental advocacy groups also are important links in our efforts to maintain a healthful environment.

Air quality specialists advise and assist industry in reducing harmful emissions from motor vehicles and industrial smokestacks. Not to be overlooked are the entomologists who monitor the winds for signs of invading insects, as well as plant pathologists who watch for airborne disease organisms. Career opportunities in air quality maintenance and improvement undoubtedly will increase in the future (Figure 17–3).

Biologist: Aquatic

An education in aquatic biology prepares a person for many different occupations that study relationships among and between aquatic organisms such as fish, plankton, clams, and snails and the water environments in which they live (Figure 17–4). All of these careers require college degrees with emphasis on advanced graduate degrees.

An aquatic biologist who specializes in saltwater aquatic life is known as a *marine biologist*. A person who chooses a similar career specializing in freshwater aquatic life is a *limnologist*.

Biologist: Fish and Wildlife

A biologist who works with fish and wildlife is a person who makes a career of learning about the basic needs of animals. He or she studies the living habits of different animal species to determine the kinds of food and shelter that are needed. A fish and wildlife biologist also studies other characteristics of organisms such as reproductive habits and territorial ranges.

Fish and wildlife biologists must have a strong background in the biological and environmental sciences. A person who plans a career in this field will need a four-year degree from a good university with graduate study recommended. Often this type of biologist will need to conduct field studies to determine management alternatives for wild animals (Figure 17–5).

FIGURE 17–5 Fish and wildlife biologist (Courtesy of the U.S. Department of Agriculture.)

INTERNET KEY WORDS

biologist, fish, wildlife, roles, responsibilities
biologist, marine, roles, responsibilities
botanist, roles, responsibilities
dendrologist, silviculturist, roles, responsibilities

Biologist: Marine

A career in marine biology involves the study of animals and plants that live in saltwater environments. It includes studies of the effects that environmental conditions such as light intensity, salinity, temperature, and pollutants, as well as other factors, have on marine organisms. This career requires an advanced science degree with emphasis on the biological sciences. Field work is often an important activity in this career.

Botanist

A botanist must have a good understanding of the anatomy and physiology of plants. A strong science background is required, and a person in this career will use tools, microscopes, and scientific instruments of many kinds to study the internal and external structures of plants (Figure 17–6). He or she will study the effects of temperature, precipitation, climate, soil, elevation, and other environmental factors on plants. This career will require a college or university degree in botany or a related science.

FIGURE 17–6 A botanist is a scientist who specializes in the study of plants. (Courtesy of Shutterstock.)

Dendrology, Silviculture, and Forestry

Forestry is known as a career area for rugged individuals who prefer the outdoors and like to work in relative isolation (Figure 17–7). Many jobs in forestry, however, are in urban areas and involve considerable indoor work. The U.S. Forest Service hires large numbers of forestry technicians and managers. Although many forestry jobs do involve an extensive amount of outdoor work, most jobs provide a desirable mix of outdoor and indoor work.

Forestry includes the work of several types of specialists: *dendrologists*, who engage in the study of trees; *silviculturists*, who specialize in the care of trees; *forestry consultants*, who advise private forest land owners; and lumber industry workers, government foresters, loggers, national and state forest rangers, and firefighters. A relatively new position is that of the *urban forester*, who is responsible for the health and well-being of the millions of trees found in parks, along streets, and in other areas of our cities. An *arborist* is an urban forester whose work may include planting, transplanting, pruning, fertilizing, and tree removal.

FIGURE 17–7 The forest industry supports many different career opportunities ranging from harvesting timber to litigating timber sales. (Courtesy of FFA.)

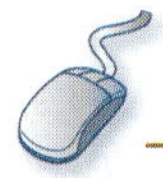

INTERNET KEY WORDS

ecologist, roles, responsibilities
educator, forestry, roles, responsibilities
entomologist, roles, responsibilities

Ecologist

A scientist who studies relationships between living organisms and their environments is known as an *ecologist*. A person in this career will need a strong science background and should plan on additional graduate work after college. Ecology is a relatively new employment field that is continually evolving. With the emphasis on protecting the environment that is expected in the years ahead, a person who is educated as an ecologist can expect a high demand for his or her skills (Figure 17–8).

Educator: Forestry

An educator who specializes in forestry is usually a person who has completed a university graduate program and successfully worked in the forest industry or a related field. Forestry educators teach at high schools, technical schools, and colleges and universities. Some educators enter teaching after successful work experience in the forest industry (Figure 17–9). A professor of forestry at a college or university is a teacher and advisor of students who plan to engage in forestry careers. In addition to teaching, a forestry professor is expected to spend some of his or her time doing research in forestry or in a closely-related field of knowledge.

FIGURE 17–8 An ecologist studies relationships between living organisms and their living environments.

FIGURE 17–9 Forestry science educators instruct college and university students as well employees in environmentally sensitive industries.

FIGURE 17–10 Environmental analysts collect and test materials to determine compliance with environmental laws and to test the feasibility of projects planned for environmentally sensitive areas. (Courtesy of the U.S. Department of Agriculture's Agricultural Research Service.)

Land grant universities employ extension foresters whose duty is to work with the public and private forests to hands-on instruction in the field. They also provide services such as identifying insect and disease problems. Strong writing and communication skills are needed along with a strong science background including chemistry, botany, biology and social sciences.

Entomologist

An entomologist is a person who specializes in the branch of biology that studies insects. A career in this field usually requires an advanced degree in a biological science with a specialty in entomology. The work of an entomologist will involve gathering research data through fieldwork and using the data to learn more about the relationships of insects to the environments in which they live. Entomologists use their knowledge of insect anatomy, feeding habits, and life cycles to discover ways of strengthening populations of useful insects while controlling or reducing populations of harmful insects.

Environmental Analyst

An environmental analyst uses science and engineering principles to find ways to remove pollutants from the environment. This career requires good research skills and an ability to gather data and analyze them properly.

Data are gathered from samples of air, soil, water, plants, animals, and other sources (Figure 17–10). A college degree is required that combines the disciplines of science, engineering, and mathematics (statistics).

INTERNET KEY WORDS

environmental analyst, roles, responsibilities
environmental engineer

Environmental Engineer

The job of an environmental engineer is to design ways to prevent, control, and remediate (clean up) environmental problems that may threaten the health of people, trees, wildlife, and other living things (Figure 17–11). These engineers design plans and systems that are capable of isolating and removing pollutants. Most of these professionals

FIGURE 17–11 Environmental engineers design and build sophisticated research equipment that is capable of measuring and recording changes in environmental conditions, often in remote locations.

have specialized in such areas as water pollution, landfill design, or air pollution. They identify pollution sources and prepare recommendations on ways to correct problems.

A knowledge and understanding of engineering, environmental safety, environmental law, and site development are also required. The ability to prepare accurate technical reports and to communicate them to the public and government agencies is important. One of the following degrees is usually required before a person can enter this career: environmental, civil, or chemical engineering, technical safety, geology, or environmental science.

Environmental Quality Technician

A person who works as an environmental quality technician is responsible for monitoring the environment for pollutants and contaminants. He or she also does the work that is required to remediate the environment as needed. Environmental quality technicians collaborate with process engineers to devise ways to monitor wastewater outputs and smokestack emissions. They gather air, water, and soil samples and do laboratory testing (Figure 17–12). They work to ensure that government standards are met for air and water quality. A college or university degree is usually required for employment in this career.

Environmental Scientist

environmental quality technician, roles, responsibilities
environmental scientist, roles, responsibilities

Environmental scientists conduct research related to environmental pollution. They gather and analyze data to determine how the environment is affected by different approaches to controlling pollution (Figure 17–13). They also use experimental results to determine pollution standards for government regulations, and they propose improved practices for managing pollution problems.

A career as an environmental scientist requires an advanced professional degree in chemistry or biology with emphasis on environmental science, wildlife, or natural resources. High school preparation should

FIGURE 17–12 Environmental quality technicians gather samples of air, water, soil, and other materials and test them in the laboratory for the presence of contaminants.

FIGURE 17–13 Environmental scientists participate in projects that help protect and enhance the healthful management of diverse habitats.

include a strong curriculum in science and mathematics with a broad experience base in agriculture and natural resources.

Forester

A forester is a person who is educated in the sciences related to the propagation, growth, management, and harvesting of trees (Figure 17–14). He or she is responsible for administering all of the activities that occur in the forest. This includes maintaining forest health, planning and managing timber sales, conducting environmental impact studies, and ensuring that forest activities are compatible with the wildlife that live in the forest environment. A degree in forestry with education in related sciences is needed along with broad experience in dealing with public issues and resource management.

INTERNET KEY WORDS

forester, roles, responsibilities
game bird farm manager, roles responsibilities

Game Bird Farm Manager

In some ways, fish and game agencies and private hunting preserves supplement the natural populations of game birds with birds that have

FIGURE 17–14 A forester is responsible for administering all of the activities that occur in a public forest.

FIGURE 17–15 Game bird farms help supplement wild game bird populations. (Courtesy of the U.S. Fish and Wildlife Service.)

been raised on game farms (Figure 17–15). Managers of these farms are required to have administrative and managerial skills to direct the farm employees.

In most cases, a college degree related to wildlife management is necessary. A strong background in the biological sciences is the foundation for all careers in fish and game management. A good understanding of avian nutrition and diseases is also essential to success in this career.

INTERNET KEY WORDS

herpetologist, roles, responsibilities
ichthyologist, roles responsibilities
inspector, environmental safety, roles, responsibilities

Herpetologist

A herpetologist is a scientist who specializes in the study of reptiles and amphibians, which are studied in their natural settings as well as in the laboratory. A person who engages in this career sometimes devotes an entire lifetime to learning about the interrelationships between reptiles or amphibians and other living and nonliving elements found in their environments. This career requires a graduate degree in the zoological or biological sciences.

Ichthyologist

Ichthyology is the branch of zoology that deals with the classification, structure, and life cycles of fishes. An ichthyologist is a scientist who studies fishes (Figure 17–16). Ichthyologists are employed for fish management in marine and freshwater fisheries, hatcheries, and commercial

FIGURE 17–16 A ichthyologist studies the classification, structure, and life cycles of fishes.

fish farms. Other ichthyologists may serve as curators of scientific collections of fish in museums and educational institutions. Employment in this field requires a graduate degree in the biological sciences. A master's degree will sometimes suffice in fish management positions, however, a doctorate is required of curators.

Inspector: Environmental Safety

Every industry witnesses more regulation by government, often initiated by consumer advocacy groups. This creates inspection or monitoring jobs to issue permits and check for compliance with regulations in such areas as water quality, chemical use, and processing-facility standards. Examples of occupations in inspection and monitoring include soil conservation officer, game conservation officer, food inspector, and water tester.

INTERNET KEY WORDS

mitigation specialist, roles responsibilities
oceanologist, roles, responsibilities

Mitigation Specialist

A mitigation specialist must be well-versed in ecology, life cycles, population trends of wildlife, and their habitat requirements. This career requires developing and implementing plans to reduce the negative impacts of human activities on the environment and every living thing in it. This is done by identifying and evaluating habitat losses and then modifying tracts of land to create critical wildlife habitat. A university degree in ecology or a related science is required.

Oceanologist

The science of oceanology is focused on the environments in the world's oceans. An oceanologist works with water quality, plants, animals, reefs, and any other features that affect ocean environments. This career field also includes the exploration of the underwater environment.

A person who qualifies to become an oceanologist will need a university degree in science with a strong emphasis on biological science. He or she will usually need a graduate degree in a specialty that is within the science of oceanography or closely related to it.

Ornithologist

The branch of zoology that deals with birds is called *ornithology*. A scientist who studies birds is an *ornithologist*. A career in this field requires a graduate degree in zoology, biology, or a related science—and a specialty in ornithology.

INTERNET KEY WORDS

ornithologist, roles, responsibilities
predatory animal control officer, roles responsibilities
teacher, science, roles, responsibilities

The work of an ornithologist will require gathering research data through fieldwork and using the data to learn more about the relationships of birds to the environments in which they live. Ornithologists also use their knowledge of bird migration and behavior to assist in the evaluation of hunting regulations and the restoration of endangered species. This career will usually require an advanced degree from a university.

Predatory Animal Control Officer

The killing instincts of predatory animals sometimes bring them into conflict with humans. Animals that develop appetites for domestic livestock are eliminated or removed to areas with limited or reduced access to livestock.

Government agencies that manage public lands hire officers whose duties include controlling predation on livestock. One method is live trapping for relocation. When this fails, or when large numbers of predators are concentrated in a region, they may be poisoned or shot. A strong background and understanding of the habits and behaviors of wild animals is required.

Science Teacher

One of the most interesting careers in the field of zoology is that of a teacher. Science education is important to every student because so much that we do in life has a scientific basis. Teachers are needed desperately in the science discipline, and working with students is interesting and rewarding. A career as a science teacher requires at least a bachelor of science degree from a four-year college or university, and a graduate degree is preferred. Science is a laboratory-based discipline—a teacher must become skilled in conducting hands-on lab exercises. In addition, a teacher also needs to be the kind of person who enjoys working closely with people.

INTERNET KEY WORDS

soil conservationist, roles, responsibilities
soil specialist, roles, responsibilities

Soil Conservationist

A soil conservationist develops plans and recommends practices for controlling soil erosion. Other duties include land-use planning activities, developing soil management plans such as crop rotations, and reforestation projects. They also establish permanent vegetation and develop other practices that are related to soil and water conservation.

A career as a soil conservationist requires a bachelor's of science degree in soil science, agronomy, forestry, or agriculture. A majority of time also is spent doing outdoor fieldwork.

FIGURE 17–17 A soil specialist uses GPS equipment to identify exact field locations and map soil types in the field. (Courtesy of the U.S. Department of Agriculture.)

Soil Specialist

A person who enters a career as a soil specialist will work in the field classifying soils according to soil types (Figure 17–17). These data are then used to develop soil maps that are used by managers as they make soil management decisions. The soil specialist is involved in solving problems related to soil damage resulting from erosion, fire, or other causes. A specialization in soils along with a university degree in soil science or a related field of study is a common approach to preparing for this career.

Taxonomist

A taxonomist is a scientist who classifies living organisms into related groups. This is a highly specialized field that requires a person to be able to observe and distinguish small but distinct differences among organisms. An advanced graduate degree is required from a reputable university. Much of the work of a taxonomist involves collecting specimens of organisms and accurately observing the features that make one organism distinctly different from other similar organisms.

Taxonomists often find careers at colleges and universities and in museums. Some are also involved in field research and collecting expeditions.

FIGURE 17–18 The technology specialist is one of the new careers that has emerged in the technology age. There are many different specialties in this field, such as computers, lasers, satellites, and electronic control systems.

Technology Specialist

INTERNET KEY WORDS

taxonomist, roles, responsibilities
technology specialist, roles, responsibilities

A person who understands the operation of computers, lasers, satellites, and electronic control systems can find a career in nearly any sector of environmental science (Figure 17–18). Modern technologies are used to organize data and monitor the environment in a variety of ways. Computers, for example, are invaluable in developing models for environmental management.

Job preparation for this field is available through technical colleges, community colleges, universities, and private training organizations. This career option requires continual education and training to stay current with new and emerging technologies. A person who enters this career must be proficient in setting up, troubleshooting, maintaining, testing, and repairing high-tech equipment.

Water Treatment Specialist and Wastewater Treatment Specialist

A water treatment specialist operates facilities that purify water so that it is safe for human consumption. Wastewater treatment specialists treat sewage and other wastewater to remove pollutants and neutralize pathogens so that the water is safe to return to the hydrologic cycle. Both water treatment and wastewater treatment specialists may work in industrial plants, small town treatment plants, or large city plants.

INTERNET KEY WORDS

water treatment specialist, roles, responsibilities
wastewater treatment specialist, roles, responsibilities
wildlife conservation officer, roles, responsibilities

Operators must be prepared to operate sophisticated systems. In both positions, the actual operation of the equipment is becoming computerized, but operators must understand the processes completely so that they can monitor the operations responsibly. Training for this career is provided by colleges as well as by agencies responsible for water quality. Workers are required to maintain government-issued certifications through participation in regular certification classes.

Wildlife Conservation Officer

A wildlife conservation officer is a law officer who is responsible for patrolling an assigned area to prevent fish and game law violations, investigating complaints about crop damage by wildlife, and gathering biological research data. Game wardens also apprehend violators of fish and game laws, issue citations, and make arrests when appropriate. They present evidence in court hearings, investigate hunting accidents, offer educational programs to the public, and work with community groups to improve fish and game habitat. Education requirements for this career include training in law enforcement and a strong curriculum in biological sciences.

wildlife technician, roles, responsibilities

Wildlife Technician

A career as a wildlife technician will involve in-depth study and an education to develop specialized skills that deal with wildlife problems. Most careers in this field require a bachelor's of science (BS) degree. Technicians are usually skilled in gathering appropriate testing materials such as blood or tissue samples and in conducting laboratory tests. They may specialize in the causes and treatments of diseases or find ways to overcome reproductive problems in certain animals. Whatever the wildlife problem might be, technicians work to find ways to identify causes and discover solutions.

The careers that have been discussed in this chapter do not represent all of the careers in environmental science. New careers are constantly evolving, and new skills find demand frequently as technology changes and science advances.

Learning Activities

1. Visit with your school counselor and request help in finding and using those measurement tools that will help you identify your career strengths and interests. Discuss the results with the counselor and obtain a list of careers that most closely match your aptitudes and skills.
2. Attend a career fair. Concentrate your time on the careers that interest you most and most closely match your aptitudes and skills. Gather literature from the vendors and thoroughly study the opportunities that are available to you.

GLOSSARY

A

abiotic: Nonliving.
abiotic disease: An unhealthy state of being caused by a nonliving factor or condition.
absorption: Process by which water and dissolved substances pass into cells.
acid: Material with a pH of less than 7.0.
acid precipitation: Also called *acid rain*; precipitation that has become polluted with sulfur or nitrogen compounds that form weak acids when they are dissolved in water and that may damage plants and animals.
adaptive behavior: Changes in the habits of animals that they have learned for the purpose of increasing their chances of survival.
aeration: The mixing of air and water by wind action or by air forced through water; generally refers to a process by which oxygen is added to water.
age structure: Describes a population's makeup in terms of maturity.
A horizon: Undisturbed soil profile located near the surface; includes mineral matter and organic matter.
air: The mixture of gases surrounding the Earth; consists of approximately 78 percent nitrogen, 21 percent oxygen, 0.9 percent argon, 0.03 percent carbon dioxide, and minute quantities of helium, krypton, neon, and xenon, as well as water vapor.
alfisol: One of 12 soil orders; mineral soil, usually formed under forest, that is common to northern and midwestern states; has leached E horizon with accumulation of bases and clays in B horizon.
alien species: A species that is not native to an environment and competes with native species for food and shelter.
alkaline: A material that is basic—that is, having a pH greater than 7.0.
alluvial deposit: Soil deposited by moving water.
alluvial fan: Geological formation of gravel, clay, sand, and silt that has been deposited by water, and is often seen near a location where a stream slows down as it enters a plain or where a tributary joins with a main stream; fan-shaped alluvial deposit formed where flowing water slows down and spreads out at the base of a slope.
anaerobic: A process or organism that does not require oxygen.
anchorage: Soil function of holding plant firmly in place.
animal equivalent unit (AEU): Amount of forage (grazing) required to feed a 1,000-pound animal (such as a steer) for a given period of time.
anion: Negative ion.
annual ring: A visible circle in the cross section of a tree trunk produced in one year by rapid, soft spring growth followed by slower, denser summer growth.
annual weed: A weed that completes its life cycle within one year.
anthracite: Oldest and hardest form of coal.
aquatic: Growing or living in or on water.
aquifer: Underground supply of water; an underground formation that holds water; it is porous enough that water can flow through it to a well so it can be a source of groundwater.
arachnid: A living creature such as a spider or mite that can be distinguished from insects by having eight legs.
artificial regeneration: Forest renewal that occurs when seeds or seedlings are planted at a harvest site.
asbestos: Heat-and friction-resistant material.
atom: Single unit of matter containing a central nucleus surrounded by electrons.
attenuation: The process of lessening the impact of an agent, such as attenuation of water contamination by any number of natural processes or human intervention; purification or neutralization of impurities.

B

bedrock: Solid, or consolidated, rock lying under the soil; may be, but is usually not, the parent material of the soil lying above it.
B horizon: Level of soil below the A horizon; generally referred to as *subsoil.*
biennial weed: A plant that takes two growing seasons or two years from seed to complete its life cycle.
biodiesel: Fuel similar to diesel that is obtained from vegetable oils.
biological control: Pest control that uses natural control agents.
biological succession: The change that occurs as one kind of living organism replaces another in an environment.
biological synthesis: Any change in the composition, shape, size, or structure of the plants or animals in an ecosystem.
biological value: The relative worth of the life forms that populate an area, taking into account economic values and effects on climate, watersheds, water temperature, soil erosion, wildlife, and so on.
biomass: Total living weight; vegetation and waste materials that contain large amounts of vegetable matter that is rich in cellulose.
biomass power: Electrical power generated from the energy obtained from burning plant materials.
biome: Group of ecosystems within a region that have similar types of vegetation and climatic conditions; a major land area characterized by dominant plant life forms, such as tundra and grasslands.
biosphere: Consists of all of the ecosystems of the Earth that are capable of supporting life.
biotechnology: A scientific field that applies principles of science to organisms that modify them to meet specific purposes; use of cells or cellular components to make beneficial products or processes.
biotic: Living organisms.
biotic disease: A disease caused by living organisms.
biotic potential: A measurement of the ability of an organism to reproduce sufficient numbers of offspring in order to maintain a stable population.
bituminous coal: The most common type of coal; softer than anthracite.
bog: Land area that is especially wet, usually with evergreens present, and covered with moss and peat; surface provides a spongy walk or sticky mud.
botany: The branch of biology that deals with plants.
buffer: Any substance in a solution that tends to resist pH change by neutralizing any added acid or alkali; a chemical that sustains pH within a narrow range by taking up or giving up hydrogen ions.

C

cambium: The growth layer in a tree root, trunk, or limb.
canopy: The highest level of vegetation in a forest, consisting of the crown branches and foliage of the tallest trees that receive direct sunlight.
capillary water: The moisture that remains after gravitational water leaves the soil; plant roots can absorb or take up this moisture.
carbon cycle: The recurring circular flow of carbon from living plant and animal tissues to nonliving materials.
carbon monoxide: Colorless, odorless, and highly poisonous gas; carbon dioxide and water combine to make plant food and release oxygen.
carcinogenic effect: Long-term effect from exposure to toxic substances resulting in an increased susceptibility to cancer.
carnivore: An animal (or plant) that eats animals.
carrying capacity: The population, number, or weight of a species that a given environment and its resources (food, shelter, and water) can support for a given time.
cation: Positive ion.
causal agent: An organism that produces a disease.
chain reaction: Continuing fission process such as that found in nuclear fission reactors.
chemical control: The use of pesticides for pest control.
chemical energy: Energy that is stored in plant tissues as sugars, oils, and starches during the process of photosynthesis.
chlorofluorocarbons: Molecular compounds consisting of chlorine, fluorine, carbon, and hydrogen.
C horizon: Soil below the B horizon; composed mostly of parent material.
clay: The class of smallest soil particles—smaller than 0.002 millimeter in diameter; the textural class highest in clay.
clean culture: Any practice that removes pests' breeding or overwintering sites.
climax community: The plants that occupy an environment when the succession of species is complete and the plant populations has become stable.
coarse-textured soil: Soil texture with traits largely set by the presence of sand; includes sands, loamy sands, and most sandy loams.

commensalism: Relationship in which one type of wildlife lives in, on, or with another without either harming or helping it.
community: Group of animal and plant populations that live together in the same environment.
competition: Relationship in which two or more individuals of a species or types of wildlife compete for the same resource (food, shelter, etc.).
competitive advantage: When one organism is more able than another to survive in an environment.
competitive exclusion principle: The loss of an organism from a specific niche or habitat because it cannot compete with another species for a single scarce resource.
compound: A chemical substance composed of more than one element.
conifer: An evergreen tree with needlelike leaves that produces seeds in cones.
coniferous forest biome: Group of evergreen forest ecosystems made up of larch, pine, spruce, and other cone-bearing trees; located mostly in northern regions of North America.
conservation: Practice of protecting natural resources against waste.
conservationist: Person who advocates the conservation of natural resources.
containment landfill: Landfill designed in such a way as to minimize the percolation of water through buried materials.
continental shelf: Land submerged under the ocean surface that slopes gradually away from the shore toward deeper water.
contour: An imaginary line across a slope that stays at the same elevation.
coppice method: A form of asexual reproduction of a forest in which all of the trees are cut and new forest growth is generated from the stumps of the harvested trees; also known as the *sprout method.*
corrosive: Defined by federal regulations as a substance with a pH of greater than 12.5 or less than 2.0, including materials that are extremely strong acids or bases.
cover crop: A crop planted to prevent erosion on a soil; such crops can be planted on soils not currently being farmed, between crop rows, or after main crop harvests.
crop: An organ located in the digestive tracts of birds and some other organisms where food is stored before it is digested.
cultivar: A particular species that has been cultivated and is distinguished by one or more characteristics. Through sexual or asexual propagation, it will keep these characteristics.
cultural control: Pest control that adapts farming practices to better control pests.
cut: Seam of coal that has been removed.

D

deciduous: Trees that shed their leaves annually; generally hardwood trees.
deciduous forest: Tracts of deciduous trees.
decomposer: Microbes (bacteria and fungi), insects, worms, and so forth that obtain nutrients from dead organic materials and are capable of breaking down complex substances to form simple elemental components, thus making them available to plants.
decomposition: The process by which complex organic compounds are broken down into simpler compounds with the release of energy and usually carbon dioxide and water as by-products.
decreaser: Vegetation in a grassland area that tends to be easily depleted by moderate grazing.
defoliate: To strip a plant of its leaves.
denitrification: A biochemical reaction in which nitrate (NO_3) is reduced to NO_2, N_2O, and nitrogen gas.
density: The number of individual organisms living within a defined area.
derrick: A hoisting apparatus used to hold an oil drilling machine.
desert biome: Terrestrial environment with less than 10 inches annual precipitation inhabited by drought-tolerant species of plants and animals.
direct exposure pathway: Inhalation or absorption of a contaminant through skin or lungs.
direct seeding: Planting tree seeds to generate new forest growth.
disease: Any departure from health; an unhealthy condition; a particular destructive process in an organ or organism with a specific cause and consistent symptoms.
disease triangle: Term applied to the relationship of the host, pathogen, and the environment in disease development.
dissolved oxygen: The amount of elemental oxygen, O_2, in solution under existing atmospheric pressure and temperature.
distribution: Characteristic of a population that describes the spacing or dispersion of individuals.
domestic: A condition in which a plant or animal is raised in captivity or under controlled conditions.
drainage: methods of draining a pond; surface water runoff.

drift: Movement of a pesticide through the air to nontarget sites.
drift mine: Coal mine in which a passageway is bored into a hill or mountain.
drillship: Offshore drilling rigs used to look for oil.

E

ecological resource: Living resources such as forests, wetlands, and wildlife.
ecological wetland: Defined by the U.S. Fish and Wildlife Service as lands transitional between terrestrial and aquatic systems where the water table is usually at or near the surface or the land is covered by shallow water. According to this definition, wetlands must have one or more of three attributes: (1) At least periodically, the land supports predominantly hydrophytes; (2) the substrate is predominantly undrained hydric soil; and (3) the substrate is nonsoil and is saturated with water or covered by shallow water at some time during each year's growing season.
ecologist: Scientist who studies relationships between living organisms and their environments.
ecology: The branch of biology that describes the complex relationships among living organisms and the environments in which they live.
economic threshold level: The level of pest damage that justifies the cost of a control measure.
economic value: The inherent monetary value of something.
economy of scale: Increases in production efficiency that result from increases in size (e.g., farms).
ecosphere: All of the Earth's ecosystems as a whole.
ecosystem: A system of interrelated organisms and their physical and chemical environment.
electrical energy: Energy in the form of electricity.
element: A uniform substance that cannot be further decomposed by ordinary means.
elemental cycle: The recurring circular flow of elements from living organisms to nonliving materials and back again.
emigration: Movement of an animal out of a population.
encroachment: Trespassing or advancing beyond normal, natural, or proper limits.
endangered species: A legal designation under the Endangered Species Act in which a species or subspecies is in immediate danger of becoming extinct because of the small numbers of survivors in the population. This classification is assigned for the purpose of providing protection to such organisms.
energy: The ability to do work or to cause changes to occur.
entomophagous: Insects that feed on other insects.
environment: All the conditions such as air, water, and soil that affect life; physical factors that influence the existence or development of a living organism.
environmental impact statement: A science-based study of the harvest area that describes the expected effects of human activities on an area's environment and wildlife.
Environmental Protection Agency (EPA): Federal agency charged with protecting and maintaining the environment for future generations.
environmentalist: A person who undertakes some political or social activity with the intention of affecting some aspect of the environment, normally in a way that he or she perceives as beneficial or positive. In one real sense, every person is an environmentalist.
environmental science: The study of interactions and relationships among living organisms and the physical and chemical features of the surroundings in which they live.
eradication: Removal of all recognizable units of an infecting agent from the environment.
erodability: The tendency of soil aggregates to break apart and make the soil susceptible to erosion.
erosion: Loss of topsoil from a region because of flowing water or strong winds; wearing away of land surface by wind and water; can occur naturally and by land-use activities; process of wearing away or removing surface layer of anything; especially in natural resources management, refers to loss of surface soil to water and wind.
estuary: Aquatic environment in which freshwater and saltwater mix in areas where rivers and streams flow into oceans.
evaporation: The dispersion of water into the atmosphere in the form of water vapor.
evergreen: A plant that does not lose its leaves with the changing of the seasons.
evolution: Process in which physical changes occur in species of organisms over long periods of time; during those periods, physical traits that help organisms survive in the environment become predominant and are expressed more frequently.
exhaustible resource: A natural resource that exists in a limited quantity and cannot be replaced when used.
exoskeleton: Hard outer shell that protects the body of an organism such as an insect or such crustaceans as crawfish, crabs, and lobsters.

extinct: No longer existing because all the members of a population have died.

F

fecundity: How fertile, prolific, or productive an animal or species is.
fertility: Ability to produce viable offspring.
fertilization: The process in which the male and female gametes join together in a single cell.
fine-textured soil: Soil with a large amount of clay; usually includes clay, sandy clay, clay loam, silty clay, and silty clay loam.
first law of energy: A law of science stating that energy cannot be created or destroyed, only converted from one form of energy to another (for example, light to heat).
fixed platform: Permanent oil rig for offshore drilling purposes.
flood hazard: the probability that a flood of a given degree of severity will occur in a known period of time.
floodplain: Land near a stream or river that is commonly flooded when water is high; soil is built from sediments deposited during flooding.
food: Material needed by an organism to sustain life.
food chain: A series of steps through which energy from the sun is transferred to living organisms; members of the food chain feed on lower-ranking members of the community; transfer of energy from one living thing to another in the form of food.
food pyramid: A series of organisms that are arranged in ranking order according to their dominance in a food web.
food web: A group of interwoven food chains.
forage: Food for animals, especially that taken by browsing or grazing.
forb: Legume, herb, or other nongrass vegetation eaten by herbivores.
forest: A thick growth of trees and underbrush covering an extensive tract of land; a large group of trees and shrubs.
forest floor: The layer of decaying plant materials on the soil surface. It acts as a mulch, preserving moisture.
forester: A professional who plans, manages, or supervises a forest.
forest land: land at least 10 percent stocked by forest trees.
forestry: Industry that grows, manages, and harvests trees for lumber, poles, posts, panels, paper, and many other commodities.
fossil fuel: A fuel that comes from deposits of natural gas, coal, and crude oil that are formed in the Earth from plant or animal remains.
free water: Water that drains out of soil after it has been wetted.
freshwater: Water that is not high in salt content.
freshwater biome: Set of similar ecological communities found in or near water that is not salty.

G

gamete: A haploid reproductive cell.
genetic engineering: Human modification of the genetic makeup of organisms.
geothermal energy: Heated groundwater used as an energy source.
germination: The process by which a seed sprouts and begins to grow.
gizzard: A muscular organ in the digestive tracts of birds, reptiles, and other organisms that uses small rocks and pebbles to grind food into small particles.
grassland: Rangeland that has grasses as its primary naturally-occurring vegetation.
grassland biome: Terrestrial environment located in the middle of the continent and lacking tree cover; sometimes called a *prairie*. Dominant plants are grass and broadleaved herbs.
gravimeter: A device that uses changes in the gravitational pull to find oil-bearing rock.
gravitational water: Water that moves through the soil under the influence of gravity.
grazing capacity: Number of animal equivalent units that can be supported by a given grazing area without unacceptable damage to the grass cover.
greenhouse effect: A buildup of heat at the Earth's surface caused by energy from sunlight becoming trapped in the atmosphere.
gregarious: Sociable; tending to associate with others of one's kind.
groundwater: Water located under the Earth's surface in underground streams and reservoirs; water stored underground in a saturated zone of rock, sand, gravel, or other material.
growth impact: A calculation of the extent of insect damage, taking into account the timber losses from reduced growth rates and tree deaths.

H

habitat: An environment in which a plant or animal lives.
hardwood: Any tree yielding a tough, compact, tight-grained wood; usually broadleaf trees.

hazardous materials: Materials treated as threats to the environment; their disposal is controlled by law.
hazardous waste: Waste that, in sufficient quantities and concentrations, poses a threat to human life, human health, or the environment when improperly stored, transported, treated, or disposed.
HAZMAT: *See* hazardous materials.
heartwood: The inactive core of a tree trunk or limb.
heat capacity: Characteristic of water that makes it resistant to temperature changes.
herbicide: A substance for killing weeds.
herbivore: An animal or other organism that eats plants.
herb layer: The bottom layer of vegetation in a forest consisting of ferns, grasses, and other low-lying plants that grow on the forest floor beneath the shrub layer.
humus: Decay-resistant residue of organic matter decomposition; darkly colored and highly colloidal.
hydric soil: Soil that is saturated, flooded, or ponded long enough during the growing season to develop anaerobic conditions in its upper part.
hydrocarbons: Organic compounds that contain hydrogen and carbon.
hydrology: Science dealing with the properties, distribution, and circulation of water.
hydrophyte: Water-loving or water-tolerant plant that survives in soil that is underwater or frequently saturated.
hydropower: Water power.
hygroscopic water: Water held so tightly by adhesion to soil particles that plants cannot use it; remains in soil after air drying but can be driven off by heating.

I

ignitability: Tendency for a substance to have a flash point less than 140° F.
immediate effect: Exposure to some toxic materials that results in reactions during or shortly after exposure.
immigration: Movement of an animal into a new habitat.
imprinting: A learning process whereby young animals learn to mimic the behavior of a parent or trusted caregiver to establish a behavior pattern, such as recognition of and attraction to its own kind.
increaser: Vegetation that tends to prosper under moderate grazing pressure.
indirect exposure pathway: Contamination of a crop that is part of the food chain for animals or humans and which results in human ingestion of a contaminated food product such as meat.
industrial solid waste: Waste materials that consist primarily of spoilage from mining, logging, and other industrial processes that are not disposed of in landfills.
industrial waste: Harmful chemicals, poisonous metal compounds, acids, and other caustic materials that are left over from manufacturing processes.
ingestion: To eat or take by mouth; a contaminant that enters the body when it is eaten on food.
inhalation: Route of entry for some hazardous materials whereby small droplets may become suspended in air or attached to dust particles that are breathed into the lungs.
injection: Method of introducing a drug or a vaccine into a muscle or a body cavity.
inner bark: Contains the cambium from which the tree grows and the bark is renewed; transports food within a tree.
inorganic compound: Any chemical compound that does not contain carbon.
insect: Large class of small arthropod animals whose adults have bodies divided into head, thorax, and abdomen; three pairs of legs are found on the thorax.
insecticide: A pesticide that is used to kill insects.
integrated pest management (IPM): Pest-control program based on multiple-control practices.
intertidal zone: Area near the ocean shore covered with water during high tide and exposed above water level during low tide.
invader: Plant species that tend to take over a grassland area under either heavy grazing pressure or undergrazing.
ion: Electrically charged atom, radical, or molecule; atoms or molecules that are electrically charged because of gaining or losing electrons.
IPM: *See* integrated pest management.
iterparous species: Group of individuals that reproduce numerous times during their lifetimes.

J

jack-up rig: Temporary offshore drilling rig.
jurisdictional wetland: Defined by the U.S. Army Corps of Engineers as an area that has frequent flooding or saturation, is covered by hydrophytes, and includes hydric soils; according to this definition, wetlands exhibit wet hydrology, hydrophytes, and hydric soils.

K

key pest: A pest that occurs on a regular basis for a given crop.
kinetic energy: Energy that is associated with motion and movement in animals.

L

land: The part of the Earth's surface that is not covered with water and includes minerals deposited as rocks, gravel, and sand as well as organic materials.
landfill: An open area into which garbage is placed to be covered by a layer of some other material, typically soil.
law of conservation of matter: Matter may change from one form to another, but it cannot be created or destroyed by natural physical or chemical processes.
leachate: When connected with landfills, leachate is any liquid—typically water—that contains contaminants, percolates from a landfill, and moves into and through the underlying or surrounding soil.
lentic: Aquatic environment characterized by still water—such as a marsh, swamp, pond, or lake.
lignite: Low-carbon soft coal; the first step in the coal formation process.
liner: In conjunction with landfills, any layer of synthetic or natural material that is relatively impermeable and intended to minimize the seepage of leachate from the landfill into the surrounding soil.
load-bearing capacity: Ability of soil to bear weight from roads and structures.
loess deposits: Wind-deposited silt or windblown soil deposits.
logging: The process of harvesting trees.
long-term effect: Multiple exposures to toxic contaminants that may occur over a long period and are associated with a buildup of the contaminant in body tissues; often triggered by larger and larger accumulations in the body.
longwall: Method of mining in which coal is removed from one face and then transported by a conveyor.
lotic: Aquatic environment characterized by actively moving water—such as a stream or river—where flowing water restricts plant growth and food for fish and other aquatic animals is transported from a distant source.

M

magnetometer: An oil-finding device that uses the Earth's magnetic pull.
marine biome: Earth's largest biome (approximately 71 percent of the planet's surface area); consists of oceans, bays, and estuaries.
marine deposit: As used in this text, a layer of parent material deposited over long periods of time over the floor of an ocean or sea; forms the basis for the subsequent development of a soil.
marsh: Wetland continuously or frequently covered by freshwater, tidal water, or standing saltwater. Marshes do not rely on rainfall for their water supply, and soft-stemmed plants are the dominant plant type.
mast: Nuts such as acorns and beechnuts that are used as food by animals.
medium-textured soil: Soil intermediate between fine and coarse-textured soils; includes loam, fine sandy loams, silt loam, and silt.
meltdown: Uncontrolled chain reaction in a nuclear reactor.
meristematic tissue: Plant tissue responsible for plant growth.
methane digester: Device used to produce methane gas from decomposing organic wastes.
methane gas: Flammable gas, also known as swamp gas, that is the main component of natural gas fuel; also a potent greenhouse gas. Produced by biological processes in anaerobic soils.
mineral matter: Nonliving items such as rocks.
monitoring: Watching or checking on.
monogamy: Mating system by which individuals pair up and have only one mate at a time.
mortality: The number of deaths in a given period; death, particularly from disease or on a large scale.
multiple use: A management strategy for natural resources that considers the needs of the different groups of people who use or desire to use the resources.
municipal solid waste (MSW): Waste generated by households, although a major source of municipal solid waste is from industry.
mutagenic effect: Long-term exposure to toxic substance resulting in permanent DNA changes that can be passed along to successive generations.
mutualism: Two types of wildlife living together for the mutual benefit of both.

N

natality: Number of births in comparison with the number of individuals per year.
natural attenuation landfill: Landfill designed to allow for normal percolation of water through the

buried materials and into the surrounding soil with the intention that the leachate will be diluted or otherwise neutralized with minimal impact on the environment.

natural regeneration: Reforestation of an area from seeds or vegetation without intervention.

natural wetlands: Wetlands that have not resulted from human activities.

neritic zone: Area of the ocean beyond the intertidal zone that extends to the outer edge of the continental shelf.

niche: A specific role or function within a habitat that is performed by an organism; niches enable different organisms to occupy different places in the same habitat.

nitrogen cycle: The circular flow of nitrogen from free atmospheric gas to nitrogen compounds in soil, water, and organisms and then back to atmospheric nitrogen.

nitrogen fixation: A process in which nitrogen gas from the atmosphere is converted to nitrates by soil microorganisms.

nitrogen-fixing bacteria: Bacteria that live in soil, water, or in nodules or colonies on the roots of certain plants. These bacteria are capable of changing nitrogen gas to nitrates.

nitrous oxides: Compounds that contain nitrogen and oxygen; they make up approximately 5 percent of the pollutants in automobile exhaust.

node: Joint in an underground plant root where a new plant can develop.

nonadaptive behavior: Failure of an organism to adapt to a changing environment.

nonexhaustible resource: A natural resource that, for all practical purposes, never runs out, such as sunlight.

nonhazardous waste: All waste materials that are not otherwise classified as hazardous or radioactive.

non–point source pollution: Pollution from several sources that cannot be traced to a single point of origin.

nonrenewable resource: A resource such as minerals or oil that cannot be replaced when lost because of excessive use or abuse.

no-till: Method of growing crops that involves no tillage of the soil; seeds are planted in slits in soil, and chemicals are used to control weeds.

noxious weed: Plant that is highly damaging to an environment that is controlled under the authority of state law.

nuclear energy: Energy from a source that uses the nuclear fission process.

nuclear waste: Radioactive waste materials, including spent fuel rods, that are by-products of controlled nuclear reactions.

O

oceanic zone: Area beginning at the outer edge of the continental shelf that includes the deep ocean region extending to the continental shelf of the opposite shore.

offal: Waste parts of a slaughtered animal.

O horizon: Surface soil layer composed of organic matter and a small amount of mineral matter.

oil shale: Stone (shale) that contains crude oil.

omnivore: An animal that eats both plants and other animals.

organic compound: A chemical compound that contains carbon.

organic matter: Material of plant or animal origin that decays in the soil to form humus.

organism: An individual plant, animal, or other life form with organs and parts that function together.

original tissue: As used in this text, partially decomposed organic matter that still retains recognizable characteristics of the original plant or animal.

overgrazing: A condition in which domestic or wild animals destroy the vegetative cover in an area by harvesting or trampling the plants beyond their ability to recover; occurs when the carrying capacity of the area is exceeded.

ozone: Compound that exists in limited quantities approximately 15 miles above the Earth's surface.

P

parasitism: Relationship in which one type of wildlife lives by feeding on another without killing it.

parent material: The unconsolidated mineral or organic matter from which the solum (A, E, and B horizons) has developed.

particulate matter: Tiny particles of dust and waste materials that are suspended in air or smoke.

particulates: A form of pollution consisting of small particles that are suspended in the air.

pathogen: A disease-causing agent.

percolate: To seep; water percolates through soil and leachate seeps through the soil.

perennial weed: Weed that lives for more than two years.

pesticide: A chemical that is used to kill insects, weeds, rodents, fungi, or other pests. Pesticides include herbicides, insecticides, rodenticides, and fungicides.

pesticide resistance: The ability of a pest to tolerate lethal levels of a pesticide.
pest population equilibrium: Condition that occurs when the number of pests stabilizes or remains steady.
pest resurgence: A pest's ability to repopulate after control measures have been eliminated or reduced.
petroleum: A naturally occurring oily, flammable liquid found in large underground deposits; the basic material from which a large variety of products are manufactured, including gasoline, diesel fuels, and heating oil.
pH: An expression of the acid–base relationship designated as the logarithm of the reciprocal of the hydrogen ion activity; the value of 7.0 expresses neutral solutions, values below 7.0 represent increasing acidity, and those above 7.0 represent increasingly basic solutions.
pheromone: A chemical substance that is used by animals and insects to attract mates through their sense of smell.
phloem: Cells that carry manufactured food to areas of the plant where it is stored or used.
photosynthesis: The formation of carbohydrates from carbon dioxide and water that takes place in the chlorophyll-containing tissues of plants exposed to light; oxygen is produced as a by-product.
phytoplankton: Minute plants suspended in water with little or no ability to control their position in the water mass; frequently referred to as *algae.*
pioneers: The first plants to grow naturally in an area that has been cleared or burned or has been newly formed.
plankton: Microscopic plants and animals that float on surface waters and are food for fish and other aquatic animals.
plant disease: Any abnormal plant growth.
point source pollution: Environmental pollution that can be traced to its point of origin.
pollution: Addition of any substance not normally found in or occurring in a material or ecosystem.
polygamy: Mating system in which both males and females have more than one mate.
polygyny: System in which males attract and mate with several females.
population: A group of similar organisms that is found in the same area; a coexisting and interbreeding group of individuals of the same species in a particular locality.
population ecology: Concern with the total number of individuals of a species and their relationships with the factors within their environment.
population level: The number of a given animal in a specific geographic area.
potable: Drinkable—that is, free of harmful chemicals and organisms.
prairie pothole: Wetland that relies on periodic rainfall for its water supply. Prairie potholes are usually full in the spring and early summer before water levels start to drop off and the potholes start to disappear for the rest of the year. Prairie potholes are found mainly in North Dakota, South Dakota, Minnesota, and Nebraska.
precipitation: Water falling to Earth from the atmosphere as rain, snow, mist, or hail.
predation: The type of relationship when one animal eats another animal.
predator: An animal that survives by killing and eating other animals.
preservation: An attempt to prevent the use of some natural resource or the modification of an environment simply for the sake of keeping it intact.
preservationist: A person or group who seeks to maintain some resource in its current or natural state.
primary consumer: An animal that eats plants.
primary succession: The development of an ecological community in an area where living organisms were not previously found, such as on a newly formed volcanic island.
producer: A green plant that converts solar energy and other plant nutrients to starches and sugars.
production: Number of offspring of a population.
promiscuity: Mating system in which males and females mate with numerous members of the opposite sex.
pulpwood: Wood used for fiber in manufacturing paper products.
purify: To remove all foreign material.

Q

quarantine: The isolation of pest-infested material.

R

radial growth: Growth resulting in increased diameter in a tree.
radiant energy: Energy that comes from the sun.
radiation: Emission of potentially harmful nuclear particles from atoms and molecules of certain elements such as plutonium, uranium, and radium.
radioactive material: Matter that emits radiation.
radioactive waste: Waste materials that are characterized by the active and measurable emission of subatomic particles.

radon: Colorless, radioactive gas formed by the decay of radium.
range: The geographic area in which an organism lives and moves about—for example, the range of a specific animal; a term used to describe an open, usually grassed area over which livestock move about to feed.
rangeland: Land in arid or semiarid areas with permanent plant cover used for grazing.
range of tolerance: The limiting environmental conditions within which an organism can survive and function.
reactivity: Tendency of a material to be unstable at normal temperatures and pressures or to explode or react violently when mixed with air, water, or other chemicals.
recycled animal wastes: Animal waste products such as manure and animal body parts from meat-processing plants.
recycled nutrients: Undigested materials obtained from animal wastes and used as feed to provide nutrients in an animal's ration.
recycling: Process of reusing materials that would otherwise be disposed of as waste.
renewable resource: A resource such as a forest or wild animal population that is capable of being replaced as it is used.
respiration: The process by which an animal or plant takes in oxygen from the air and gives off carbon dioxide and other products of oxidation.
rhizome: A horizontal plant stem that grows underground.
riparian zone: The land adjacent to the bank of a stream, river, or other waterway.
risk assessment: Means of evaluating the degree of danger to an environment and its organisms. When a source of contamination is identified, the risk is assessed by evaluating several characteristics of the contaminated material as well as pathways that might lead to exposure of plants, humans, and other animals.
rodenticide: A chemical poison that is used to kill rodents such as mice or rats.
room and pillar: Mining method by which coal is removed from a mine that leaves pillars of coal to help support the ceiling. Seams of coal 40 feet to 80 feet wide are cut. Once the parallels are cut, the blocks of coal are left to support the ceiling.
routes of entry: Ways in which hazardous or toxic materials are able to enter the bodies of humans and other animals.
ruminant: Animal such as deer or bison that has a series of four stomach compartments that are capable of digesting high-fiber food.

S

salinity: A measurement of the salt concentration in water; affects the survival of organisms living in aquatic environments.
salt: Compound resulting from the interaction of an acid and a base.
saltation: The process of wind-driven erosion from small grains of sand.
saltwater: Water with a salinity of 30 to 35 parts per 1,000.
sand: The largest of the soil separates—between 0.05 and 2.00 millimeters in diameter; the coarsest textural class.
sapwood: Light-colored wood through which water and dissolved plant nutrients flow through the tree.
saturated soil: Water added until all of the spaces or pores are filled.
saturation: In solutions, the maximum amount of a substance that can be dissolved in a liquid without the excess precipitating from the solution or being released into the air.
secondary consumer: A carnivorous animal that obtains its nutrition by eating primary consumers and other carnivores.
secondary succession: The gradual change in plant species that live in an area as a damaged ecosystem returns to its original stage of ecological development.
second law of energy: States that every time energy is converted from one form to another, some energy is lost in the form of heat.
seedlings: Young plant grown from seed.
seed tree method: Timber harvest method in which mature trees of the desired species are protected from cutting in locations scattered throughout the forest for the purpose of producing seeds.
seismograph: Device that uses sound waves to find oil.
semelparous species: Animals that reproduce only once in their lifetime.
semiconductor: Material that is a poor conductor until it is acted on by heat, light, or electricity.
semisubmersible rig: Partially submerged rig used in offshore oil drilling.
sex ratio: Number of males compared to number of females.

shaft mine: Type of main underground mine where the access passageways are vertical to the coal seam and are equipped with an elevator to bring machinery and personnel to and from the coal.
shelterwood method: A timber harvest method in which mature trees are left in the harvested area in sufficient numbers to shade and protect seedlings. The mature trees are harvested once the seedlings have become established.
shrink-swell potential: How much a mass of soil swells when wet and shrinks when dry; a function of the amount of swelling clays; important engineering property of soil.
shrub: A woody perennial plant that normally produces many stems or shoots from the base and does not reach more than 15 feet in height.
shrub layer: Vegetation consisting of short woody plants that occupies the stratum between the herb layer and the understory of a forest.
silt: Medium-sized soil particles carried or deposited by moving water or wind; particles between 0.05 millimeter and 0.002 millimeter in diameter.
silt load: The amount of eroded soil that is carried in the flowing waters of streams and rivers.
silvics: The study of forests and forest relationships.
silviculture: The scientific management of forests.
simple stomach: A digestive system in which digestion is accomplished in a single stomach compartment.
slope: Incline from the level or slant; change in elevation for a given horizontal distance of the Earth's surface, often expressed as a percentage.
slope mine: Coal mine with its access tunnel on a slope from the surface to the coal seam.
smog: Pollution of the atmosphere caused by ultraviolet light from the sun reacting with atmospheric pollutants; the term originally described urban mixtures of smoke and fog.
softwood: Wood from conifers.
soil: Loose mineral and organic material on the Earth's surface that serves as a medium for the growth of land plants.
soil aeration: Process by which exchanges are made between soil and atmospheric air to maintain adequate oxygen for plant roots; aeration varies according to soil condition.
soil conservation: Practice of protecting soil from erosion caused by strong winds or flowing water.
soil solution: Liquid phase of soil; consists of water and dissolved ions.
solar cell: Device that uses silicon to capture and convert light energy to electricity.
solar energy: Using the sun as a heat energy source.
solar panel: Device that traps energy from the sun in the form of heat or light.
solid waste: Waste materials that are normally disposed of in landfills, incinerators, or composting facilities.
solubility: The degree to which a substance can be dissolved in a liquid, usually expressed as milligrams per liter or percent.
spodosol: One of 12 soil orders; mainly acidic, coarse-textured forest soils of cool, humid regions; has subsoil layer with illuvial accumulation of humus, aluminum, and sesquioxides.
stand: A population of trees in a given location.
steward: An administrator or supervisor who manages land resources.
stewardship: The management or care of a resource or property that belongs to someone else.
stolon: A stem that grows above ground.
strata: Several levels or layers of vegetation or water; levels or layers of plant growth in a forest or other ecosystem.
strip mine: Process of removing coal near the Earth's surface by first removing the soil that covers it; place where strip mining occurs.
subsoil: Soil below the plow layer; generally the B horizon.
subsoil thickness: In a soil profile, the distance measured from the bottom edge of the topsoil to the top edge of the parent material.
sulfur: Pale yellow element that occurs widely in nature.
surface creep: Movement of sand particles along a soil surface by being rolled in the wind.
surface drainage: Method of artificially draining wet soils by digging a system of ditches that collect water and carry it off the field.
surface water: Water in natural or human-made bodies of water on the Earth's surface such as lakes or reservoirs.
suspension: A system of tiny particles hanging in a liquid or gas; movement of clay or silt particles in the wind by being suspended in the air.
sustainable agriculture: A philosophy and collection of practices that seek to protect resources while ensuring adequate productivity, strive to minimize off-farm inputs such as fertilizers and pesticides, and maximize on-farm resources such as livestock manure and nitrogen fixation by legumes. Soil and water management are central components.
swamp: Land area covered continuously or nearly continuously by standing water with trees or shrubs growing in the water. The water tends to be stagnant and usually dark and nontranslucent.

T

targeted pest: Identified pest that, if introduced, poses a major economic threat.
tar sand: Sand that contains a low-grade oil.
Taylor Grazing Act of 1934: The federal law that authorized the U.S. government to establish and enforce regulations regarding the grazing of public lands, eventually leading to the management of millions of acres of public lands by the Bureau of Land Management.
temperate forest biome: Group of ecosystems in which broadleaf trees are abundant and annual precipitation exceeds 30 inches.
teratogenic effect: One of three effects as a result of long-term exposure to toxic materials resulting in an increase in the risk of physical defects in a developing embryo.
terminal growth: Vertical growth in a tree.
terrestrial: Growing or living on land.
terrestrial biome: Large community of plants, animals, and other organisms living on land.
tetraethyl lead: Colorless, poisonous, and oily liquid that improved the burning qualities of gasoline and helped control engine knocking.
texture: With soil, refers to relative percentages of sand (largest), silt (midsized), and clay (smallest) particles; course-textured soil is mostly sand and is gritty to the touch; fine-textured soil is mostly clay and feels silky to the touch, becoming slick or sticky when wet.
thermal energy: Energy that is released as heat when fuels or nutrients are burned or digested.
thermal stratification: Differences in water temperatures at various depths with deep water being colder than water near the surface.
thermal stress: Stress resulting from rapid temperature change or extreme high or low temperature.
threatened species: Species that are at risk because of declining numbers in their population. They can reasonably be expected to survive if immediate steps are taken to protect the remaining populations and their habitats. It is a legal status that is established through the Endangered Species Act, which provides extra protection and management.
tidal power: Using the force of the tides to run turbines and thus produce electricity.
tidewater: The water that flows up the mouth of a river with rising or incoming ocean tides.
tillable: Soil that is workable with tools and equipment.
timberland: Forest land capable of producing more than 20 cubic feet of industrial wood per year.
topsoil: The upper layer of soil; the richest, most productive soil layer.
topsoil thickness: Depth to which topsoil extends in the soil surface.
toxic: Poisonous; a pesticide or other material that is poisonous or harmful to an organism other than the target organism.
toxic substance: Defined by government safety regulations as material that contains one or more of 39 specific contaminants such as heavy metals or organic carbon-based compounds known as *volatile organic compounds (VOCs).*
toxic waste: Waste products that are poisonous to living organisms.
transformer: Takes primary source of food, incorporates other chemicals and energy forms, and changes it into more complex organic compounds, foods, and tissue.
transpiration: The loss of water to the atmosphere from plant leaves.
trap crop: A pest-susceptible crop planted to attract a pest into a localized area.
tree: A woody plant that produces a main trunk and has a more or less distinct and elevated head (a height of 15 feet or more).
tundra biome: Group of ecosystems located in the frozen northern regions of the northern hemisphere where evaporation rates are low, precipitation is minimal, and swamplike conditions exist during summer because water cannot penetrate the frozen soil.
turbid: Muddy or cloudy water conditions, usually from having sediment stirred up.
turbine: Machine that converts the energy of falling water to rotary motion that is then used to turn electrical generators.

U

ultisol: One of 12 soil orders; the leached soils of warm climates.
undergrazed: A condition that occurs in a range area when too little grazing is done to keep the grass cover healthy.
understory: Short trees in a forest that fill an intermediate stratum of vegetation beneath the canopy created by the crowns, branches, and foliage of the tallest trees.
universal solvent: Substance that dissolves or otherwise changes most other materials—for example, water.

V

vapor drift: Movement of pesticide vapors caused by chemical volatilization of the product.
vector: A living organism that carries an infectious agent directly or indirectly from an infected to an uninfected individual.
vegetative cover: Cover made of plants or their vegetation.
vernal pool: Special type of wetland that may only last for a few months each year. Like prairie potholes, vernal pools rely on periodic rainfall to form in the spring. They disappear in early summer.
volatile organic compound (VOC): Substance that contains hydrogen and carbon and may contain oxygen, nitrogen, and other elements.

W

warm-blooded animal: An animal with the ability to regulate its body temperature.
water: Clear, colorless, tasteless, and nearly odorless liquid.
water cycle: The cycling movement of water in the form of vapor from oceans to clouds to the Earth as precipitation and back to the oceans through rivers and streams.
water hardness: Measure of the total concentration of primarily calcium and magnesium expressed in milligrams per liter (mg/l) of equivalent calcium carbonate ($CaCO_3$).
waterlogged soil: Soil with pores that are filled with water and thus are low in oxygen; caused by high water tables, poor drainage, or excess moisture from rain, irrigation, or flooding.
watershed: An area bounded by geographic features in which precipitation is absorbed in the soil to form groundwater, which eventually emerges to become surface water and ultimately drains to a particular water course or body of water.
water table: Level below which the ground is saturated with water.
weathering: The process of breaking down rocks and minerals through the actions of weather, such as by rain and by freezing and thawing.
weed: An unwanted plant.
wetland: Land areas that are flooded during all or part of the year.
wind turbine: Structure equipped with propeller blades that turn in the wind and thus drive generators that produce electricity.

X

xylem layer: Cell structures that transport water, minerals, and nutrients within plants from the roots to the leaves.

Z

zoology: The branch of biology that deals with animal life.
zooplankton: Microscopic animals that live on or near the surface of a body of water.
zygote: A fertilized egg.

INDEX

abiotic disease, 174
abiotic subsystems, 21
absorption, 370
acid, 94–95
acid precipitation, 13–14
acid rain, 13–14, 55, 63, 66
acorns, 153
adaptability, 295, 298
adaptive behavior, 224
adsorption, 328
aeration
  of soil, 122
  of water, 101
age structure, population, 292
agriculture, 251–265
  chemical use in, 255, 263, 373–374
  conflicts with wildlife, 253–254
  contour farming, 134, 136
  cover crop, 134
  crop rotation, 135
  domestic plants and animals, 252–253
  economic profitability in, 260
  environmental health and, 258–259
  erosion prevention, 134–137
  farmers and ranchers, 253–255
  fertilizer use, 258, 260, 263
  foundations for, 252–257
  improving plant/animal performance, 262–264
  intensive farming practices, 263
  land use, 255–257
  no-till, 134
  social and economic equity, 260
  stewardship in, 255, 261
  sustainable, 257–261
  tillage practices, 130, 131–132, 134, 135, 257
  wetlands and, 206–207
agriscience, 261–264
A horizon, 118
air, 61–77
  Clean Air Acts, 74–75
  composition of, 62
  definition of, 62
  greenhouse effect/global warming, 67–72
  living organisms and, 72–73
  necessity for human life, 62
  as nonexhaustible resource, 45
air pollution, 44, 55, 62–67
  natural selection as result of, 64
  overcoming effects of, 73
  pollutants, 62–67, 71, 72
  spread of pollutants, 66
air quality, 62–67, 73–75
  emission-control equipment, 63
  maintenance and improvement of, 55, 66–67, 73–74
  regulation of/Clean Air Acts, 74–75
  threats to, 62–67
air quality control, career in, 382
alcohol, as energy source, 356–357
alfisols, 177
algal blooms, 93, 100
alien species, 237, 296
alkaline, 94–95
alluvial deposits, 116
alluvial fans, 128–129
alternative fuels. *See* energy and alternative fuels
ammonia, 103–104
anaerobic decomposition, 103
anchorage, 121
animal equivalent units (AEU), 194
animals. *See also* livestock; wildlife
  basic needs of, 217–228
  behaviors and habits, 216–217
  domestic, 252–253
  growth of, 226–227
  reproduction, 227, 292–293
animal waste, recycled, 331–333
anions, 82

annual crops, 123
annual rings, 153
annual weeds, 271
anthracite coal, 339
aquatic biologist, 383
aquatic food chains/webs, 83, 85
aquatic organisms, 8
aquifers, 13, 90–91
arachnid, 270
arborist, 384
artificial regeneration, 155
asbestos, 66
asexual reproduction, 154
ash, 338
ash trees, 152
aspen trees, 149–150
atom, 354
attenuation, 328–330
Attwater's prairie chicken, 223
automobiles. *See also* energy and alternative fuels
  alternate fuels for, 345–346
  catalytic converters in, 44, 63
  exhaust gases from, 44, 55, 72
  fuel economy/efficiency of, 345

bacteria, 32
balance of nature, 46–47
bald eagle, 297–298
barbed wire, 191
bedrock, 119
beech trees, 152
bees, 274, 281
B horizon, 118
bicarbonates, 96
biennial weeds, 272
bighorn sheep, 36–37
biochemical oxygen demand (BOD), 106
biodiesel, 357
biodiversity, conserving, 164
biogas, 357
biological control of pests, 278, 280
biological succession, 33–38
  climax communities, 34
  competitive exclusion principle, 35–37
  definition of, 33–34
  pioneer plants, 34
  primary succession, 34, 35
  secondary succession, 34–35, 36
biological synthesis, 21–22
biological value, 165–166
biologist: aquatic, 382
biologist: fish and wildlife, 382
biologist: marine, 382, 383
biology, 216
biomass, 170, 362–363
biomass power, 170
biomes, 184, 300–315
  coniferous forest, 314–315
  desert, 310
  ecosystems and, 300
  freshwater, 300–305
  grassland, 312–313
  marine, 305–308
  temperate forest, 313–314
  terrestrial, 308–315
  tundra, 310–312
biosphere, 7
Biosphere II, 9
biotechnology, 53
  environmental cleanup through, 236
biotic diseases, 174
biotic potential, 238
biotic subsystems, 21
birch trees, 149
birds
  digestive system of, 220
  ornithologist, 389
  seeding by, 155
birthrate, 292, 293
bison, 188–189, 297, 313
bituminous coal, 339
black walnut, 152
black willow, 152
body heat, 24–25
bogs, 205
boron (B), 98
botanist, 383
botany, 216
bottomland hardwoods forests, 145
Bridger, Jim, 311
brown blood disease, 104–105
buffers, 96
building foundations, 125
Bureau of Land Management (BLM),184, 192

calcium, 97
California condor, 233, 299
cambium, 153, 154
canopy, 166, 313
capillary water, 92
carbonates, 96
carbon cycle, 25–27
carbon dioxide
  in carbon cycle, 25–26
  dissolved, 101–102

global warming and, 26–27, 69, 71–72
oceans and, 338
release by respiration, 22, 72
use in photosynthesis, 22, 73
carbon monoxide, 64
carcinogenic effect, 371
careers in environmental science, 379–392
air quality control, 382
biologist: aquatic, 382
biologist: fish and wildlife, 382
biologist: marine, 383
botanist, 383
dendrology, silviculture, and forestry, 383–384
ecologist, 384
educator: forestry, 384–385
entomologist, 385
environmental analyst, 385
environmental engineer, 385–386
environmental quality technician, 386
environmental scientist, 386–387
forester, 387
game bird farm manager, 387–388
herpetologist, 388
ichthyologist, 388–389
inspector: environmental safety, 398
mitigation specialist, 398
oceanologist, 398
ornithologist, 398
predatory animal control officer, 398–399
science teacher, 390
selecting a career, 380–381
soil conservationist, 390
soil specialist, 390
taxonomist, 390
technology specialist, 391
water/wastewater treatment specialist, 391
wildlife conservation officer, 391–392
wildlife technician, 392
carnivores, 31, 83, 220
carrying capacity, 47–48, 189, 296
cars. *See* automobiles
Carson, Rachel, 269
catalytic converters, 44, 63
cations, 82
causal agent, 275
cavefish, 302
cell division, 226–227
central broad-leaved forest, 145
chain reaction, 354
chemical control, 279–282
chemical energy, 24
chemical oxygen demand (COD), 106
chemical precipitation, 330
chemicals. *See also* pesticides
agricultural, 255, 263, 373–374
disposal of, 374–375
cherry trees, 152
chloride, 96, 105
chlorine, 105
chlorofluorocarbons (CFCs), 65
global warming and, 69, 72
ozone layer damage by, 65, 68, 72
chlorophyll, 22, 73, 106
C horizon, 119
clay, 118, 328
Clean Air Act(s), 74–75, 373
clean culture, 277
Clean Water Act, 53, 210–211, 373
clear-cutting, 159–160
climate, 300
climate change. *See* global warming
climax communities, 34
Clinton, Bill, 15
coal, 338–341
formation of, 338–339
mining procedures, 339–341
mining safety legislation, 341
mining waste production, 321
types of, 339
coarse-textured (sandy) soil, 119–120
commensalism, 230
communication skills, 381
community, 7
competition, 230–231
competitive advantage, 35
competitive exclusion principle, 35–37
compound, 269
computer jobs, 381, 391
condor, California, 233, 299
conductivity, 95
coniferous forest, 143
coniferous forest biome, 314–315
conifers, 314
conservation, 14–15, 50–51
of biodiversity, 164
definition of, 50
of fossil fuels, 345–346
goal, national attitude toward, 53
of matter, 14, 20
of soil, 50–51, 128–137, 178, 254
of water, 107–108
of wetlands, 200, 208–209
of wildlife and habitats, 231–233
conservationists, 15, 50
soil conservationist career, 390

consumers, 85
containment landfills, 330–331
continental shelf, 308
contour farming, 134, 136
coppice method, 161
Corps of Engineers. *See* U.S. Army Corps of Engineers
corrosivity, 325, 369
cotton boll weevil, 270
cover crop, 134
crop (bird digestive organ), 220
crop ecosystem, 284
cropland, 123
crop rotation, 135
crops. *See* agriculture
crude oil, 12
cultivars, 278
cultural control, 276–278
cut, 340
cycles. *See* natural cycles

dams, 361
DDT, 269, 297–298
deciduous forest, 313
deciduous trees, 142, 313
decomposers, 21, 127
decomposition, 22, 103
decreasers, 192
defoliation, 270
dendrology, career in, 383–384
denitrification, 28
density, population, 292
derrick, 342
desert biome, 310
developed vs. undeveloped nations, population characteristics of, 294–295
dichlorodiphenyltrichloroethane. *See* DDT
dilution, 329–330
direct exposure pathway, 371
direct seeding, 155–156
disease
  definition of, 174
  in forests, 174
  insect/arachnid vectored, 270
  plant diseases, 275–276
  triangle, 276
dissolved oxygen (DO), 99–101
distribution, population, 292
domestic plants and animals, 252–253
drift mine, 341
drift, pesticide, 276, 277
drillship, 342
Ducks Unlimited, 209, 305

ecological resources, 235
ecological wetlands, 198
ecologist, 6
  career as, 384
  vs. environmentalist, 6
ecology
  definition of, 6
  principles of, 7–8
  ultimate concept in, 23
economic threshold level, 285
economic value, 234–235
economy of scale, 258
ecosphere, 7
ecosystem management, 41–57
  balance of nature, 46–47
  carrying capacity, 47–48
  conservation, 50–51
  human population and, 48–49
  multiple use, 52
  national attitude, 53
  nature of resources, 44–48
  preservation, 51–52
  reclaiming damaged/polluted resources, 53–54
  resource management, 42–44
ecosystems, 21–23, 284. *See also* biomes; ecosystem management
  balanced vs. unbalanced, 8
  biomes and, 300
  biotic and abiotic subsystems, 21
  biotic processes in, 21–22
  components of, 21
  damaged, resource management in, 44
  definition of, 7
  energy flow in, 8, 23
  manipulation of, 284–285
education, careers in
  forestry education, 384–385
  science teacher, 390
electrical energy, 24
electrostatic precipitators, 66
elemental cycles, 25
elements, 25, 269
emigration, 295–296
emissions. *See* air quality; automobiles; global warming
encroachment, 234
endangered species, 232–233, 239
  causes of endangerment, 296–299
  Endangered Species Act, 232–233, 299, 315
  recovery plans, 299–300
energy and alternative fuels, 23–25, 349–365
  alcohol, 356–357
  biodiesel, 357

first law of, 24
flow in ecosystems, 8, 23
geothermal energy, 355–356
hydropower, 360–361
methane, 358–360
nuclear power, 353–355
renewable vs. nonrenewable sources, 350
second law of, 24–25
solar energy, 350–353
tidal power, 361–362
types of, 24
wind, 362
wood (biomass), 362–363
entomologist, career as, 385
entomophagous, 274
environment, 42, 55
environmental analyst, 385
environmental engineer, 385–386
environmental health, 258–259
environmental impact statements, 163, 178
environmentalist, 6
Environmental Protection Agency (EPA), 74–75, 164, 209
environmental quality technician, 386
environmental safety inspector, 389
environmental science
definition of, 6
major principles in, 19–39
environmental scientist, 386–387
environmental stewardship, 236, 261
EPA. *See* Environmental Protection Agency
eradication, 282
erodibility, 130
erosion, 10, 50–51, 120, 128–137
agriculture and, 254
preventing and controlling, 143–147
process of, 129–133
saltation, 132
statistics on, 128
steps of, 129
water, 129–131
wind, 128, 131–133
estuary, 308
ethanol. *See* alcohol
ethylenediura (EDU), 73
evaporation, 86
Everglades, 198, 203
evergreen trees, 142, 143, 146–149, 314–315
evolution, 298
exhaustible resources, 45–46
exoskeleton, 273–274
extinction, 232, 236–238, 296–300
causes of, 296–300
evolution and, 298
Exxon Valdez oil spill, 13

farming. *See* agriculture
farm wildlife, 239–240, 256–257
fecundity, 238, 293
fencing, 191
fertility, 90, 293
fertilization, 227
fertilizers, 258, 260, 263
filtration, 330
fine-textured (clay) soil, 120
fish
dams and, 361
freshwater vs. marine, 105
habitat for, 88
ichthyologist, 388–389
metal poisoning, 98–99
oxygen stress, 99–101, 303
pH and, 94–95
as renewable resource, 45
silt and, 129
water pollution and, 37
water quality needs, 99–105, 303–304
fishing, 244
fission, 354
fixed platform, 342
flood hazard, 120
floodplains, 203–206
fluoride, 96
fly ash, 338
food, 217–220
food chains, 8, 31–32, 85
aquatic, 83
carrying capacity and, 47–48
definition of, 83
marine, 302
food pyramid, 32–33
food webs, 32, 83, 85
forage, 218
forbs, 184, 192
forester, 158
career as, 387
forest fires, 36, 175–176
forestland, 142
forest management, 154–163
clear-cutting, 159–160
coppice method, 161
environmental impact statements, 163
planning harvests, 158–159
reproduction and regeneration, 154–157

forest management *(Cont'd.)*
- salvage harvesting, 162
- seeding, 155–157, 158
- seed tree method, 160
- selection cutting, 161–162
- shelterwood method, 161
- sustained yield, 163
- timber management, 158–162
- wetlands and, 207
- woodlot management, 157–158

forestry, 142
- careers in, 383–385, 387
- educator in, 384–385

forests, 141–181. *See also* forest fires; forest management; trees
- acid precipitation and, 13–14, 55
- biodiversity and, 164
- biological value of, 165–166
- biomass, 170
- biomes, 313–315
- canopy, 166, 313
- climate moderation value of, 169
- deciduous, 313
- definition of, 142
- diseases, 174
- economic value of, 165
- energy value, 169–170
- floor, 167, 314
- habitat value of, 166–168
- herb layer, 167, 314
- insects and, 174–175
- livestock value of, 170–172
- pests, 173–174
- protecting, 134, 173–176
- recreation value of, 172
- regions of North America, 143–146
- reproduction and regeneration of, 154–157
- as resource, 45, 165–173, 315
- shrub layer, 167, 314
- soil management in, 126
- soils in, 176–178
- strata, 166, 313
- tree growth and physiology, 152–154
- trees, important types and species in, 146–152
- understory, 167, 313
- uses of, 50, 315
- watersheds in, 86–88
- watershed value of, 172–173
- wildlife management in, 240–241

fossil fuels, 23, 24, 26, 337–347
- carbon dioxide from, 338
- coal, 338–341
- conservation, 345–346
- formation of, 338–339, 341–342
- natural gas, 343–344
- oil (petroleum), 341–343
- oil shale and tar sands, 344–345

foundations for buildings, 125
free water, 92
freshwater, 82
freshwater biomes, 300–305
fuels. *See* energy and alternative fuels; fossil fuels
fungicides, 280

game bird farm manager, 387–388
gametes, 227
gas bubble disease, 105
gasohol, 357
genetic engineering, 53, 260, 278–279
geothermal energy, 355–356
germination, 156
gestation length, 293–294
gizzard, 220
glacial deposits, 115–116
global warming, 26–27, 68–72
- precautions against, 71–72

governmental agencies, careers with, 381–382. *See also specific agencies*

grasslands, 183–196, 211
- agencies managing, 184, 192
- biome, 312–313
- carrying capacity of, 189
- definition of, 184
- fencing and, 191
- fire and, 188–189, 312–313
- grazing capacity, 194
- grazing management, 189–192, 194–195, 255–257
- history of, 187–192
- management techniques, 193–196
- names for, 186–187
- overgrazing of, 190–191
- restoration of, 195–196
- settlement and, 190–191
- Taylor Grazing Act of 1934, 192
- types of, 185–187
- undergrazed, 193
- vegetation of, 185–187, 192–193

gravimeter, 342
gravitational water, 92
grazing capacity, 194
grazing land, 123–124, 193–196. *See also* grasslands
- management of, 189–192, 194–195, 255–257
- overgrazing, 190–191
- restoration of, 195–196

Taylor Grazing Act of 1934, 192
undergrazing, 193
Great Basin, 311
Great Dismal Swamp, 198–199
greenhouse effect, 67–68
gregarious animals, 225
groundwater, 13, 90–91, 92
contamination, 258, 276
safe-guarding, 91
growth impact, 175

habitat. *See also* biomes
definition of, 62, 232
destruction of, 223, 237, 296
fragmentation of, 228
in freshwater biomes, 304
lentic, 304
loss, 224–226
lotic, 304
preservation/restoration of, 231–233
value of forests, 166–168
wildlife and, 223–224, 228
hardwoods, 142, 144–145, 149–152
hardwoods forests, 145
Hawaiian forest, 146
hazardous materials, 10, 12, 92–93. *See also* toxic and hazardous substances
harm from, 370–371
hazardous waste, 126, 324–326
HAZMAT, 368–369
heartwood, 152, 154
heat, 24–25
heat capacity, 95
herbicides, 11, 107, 280
herbivores, 31, 83, 220
herb layer, 167, 314
herpetologist, 388
hickory trees, 152
honeybees, 274, 281
humans
food pyramid dominance, 32–33
impacts on wildlife, 233–236
interaction with environment, 42, 55
population characteristics in developed vs. undeveloped nations, 294–295
resource use and, 48–49, 52
world population of, 48, 258
humus, 117
hydric soils, 197, 201, 202
hydrocarbons, 63
hydrogen sulfide, 103, 358
hydrologic cycle. *See* water cycle
hydrology, 196
hydrophytes, 196
hydropower, 360–361
hygroscopic water, 92

ice ages, 115
ichthyologist, 388–389
igneous rocks, 115
ignitability, 325, 368
immediate effect, 371
immigration, 296
imprinting, 233
increasers, 192–193
indicator species, 93
indirect exposure pathway, 371
industrial solid waste, 321
industrial waste, 10
industry, 206
waste from, 10, 321–324
ingestion, 370–371
inhalation, 370
injection, 370
inner bark, 153
inorganic compounds, 269
insecticides, 11, 93–94, 106, 280
insects. *See also* integrated pest management (IPM); pesticides; pests
anatomy of, 273–274
beneficial, 273, 274
control using sterile males, 279
damage from, 175, 283–284
definition of, 270
disease and, 270
in forests, 174–175
as pests, 273–274, 283–284
pheromones, 279
inspector: environmental safety, 389
integrated pest management (IPM), 174, 267–289
alternatives for pest-control, 276–282
biological control, 278, 280
chemical control, 279–282
components of, 286–287
crop and biology ecosystem, 284
cultural control, 276–278
description of, 268–269
ecosystem manipulation, 284–285
environmental concerns in, 276
eradication of pests, 282
genetic control, 278–279
history of, 269
identification of pests, 287
implementation of, 287
insect pheromones, 279
key pests, 282–284

integrated pest management (IPM) *(Cont'd.)*
  monitoring in, 286, 287
  need for, 270
  pesticide resistance, 279
  pesticides, 280–281
  pest population equilibrium, 284–285
  plant diseases, 275–276
  principles and concepts, 282–286
  quarantines, 281–282
  regulatory control, 281–282
  resurgence of pests, 279
  threshold levels in, 285, 287
  types of pests, 270–275
intertidal zone, 307–308
invaders, 193, 196
ion exchange, 329
ions, 82, 96–98
iron, algae growth and, 27
iterparous species, 293

jack-up rig, 342
jobs. *See* careers in environmental science
jurisdictional wetlands, 197

key pests, 282–284
kinetic energy, 24

lakes wildlife, 244–245
land, 113–139. *See also* soil
  definition of, 114
  relationships with water, 86–89
  as reservoir for water, 90–91
  uses, uses in agriculture, 123–127, 255–257
landfills, 10–11, 126, 327–331
  containment, 330–331
  definition/description, 327–328
  design, 328
  natural attenuation, 328–330
  sanitary, 326
lava flow, biological succession over, 34
law of conservation of matter, 14, 20
laws of energy, 23–25
leachate, 327
lead, 63–64
lemmings, 312
lentic habitat, 304
lignite, 339
lime requirement test, 98
limnologist, 382
liner, 330
liquified natural gas (LNG), 344
livestock, 190–196. *See also* grazing capacity; grazing land
livestock value (of forests), 170–172
load-bearing capacity, 125
loess deposits, 116
logging, 158–159, 163. *See also* forest management
  waste, 321–322
long-term effect, 371
longwall system, 341
lotic habitat, 304
Love Canal, 369

magnesium, 97
magnetometer, 342
management of ecosystems. *See* ecosystem management
manure, 332
maple trees, 149
marine biologist, 382, 383
marine biome, 305–308
marine deposits, 116
marshes, 88–90, 202–203
mass transit systems, 67
mast, 218
mating systems, 292
matter, conservation of, 14, 20
McDonald's restaurants, waste generation from, 322–323
medium-textured (loamy) soil, 120
meltdown, 355
meristematic tissue, 272
metal poisoning, 98–99
metamorphic rocks, 115
methane, 72, 358–360
  digester, 358–360
  gas (energy), 358–360
  global warming and, 69, 72
  in water, 105
methemoglobinemia, 105
mice, 274–275
mineral matter, in soil, 115, 118
mining
  coal mining, 339–341
  industrial solid waste from, 321
  Utah Crandall Canyon mine collapse, 340
  wetlands and, 207
Mississippi River floodplain, 203
mitigation specialist, 389
mitosis (cell division), 226–227
mobility, 295–296
monitoring, 286, 287

monogamy, 292
mortality, 223, 293
moths, evolution of, 64
mottling, 202
mulch, 130, 321–322
multiple use of resources, 52
municipal solid waste (MSW), 321–324
musk ox, 312
mutagenic effect, 371
mutualism, 229–230

natality, 292
National Parks (U.S.), establishment of, 51
National Resource Conservation Service (NRCS), 201
National Wetlands Inventory (NWI), 200–201
Native Americans, 188, 189
natural attenuation landfills, 328–330
natural cycles, 25–30
  carbon cycle, 25–27
  elemental cycle, 25
  nitrogen cycle, 27–28
  water cycle, 28–31
natural gas, 343–344
  vehicles powered by, 345
natural regeneration, 155
natural wetlands, 90, 208
neritic zone, 308
New Orleans, wetlands and, 198
niche, 35
nitrate, 96
nitrite, 104–105
nitrogen, 44, 105
nitrogen cycle, 27–28
nitrogen fixation, 27–28
nitrogen-fixing bacteria, 27
nitrous oxide (N_20), 63–64, 72
  global warming and, 69, 72
node, 272
nonadaptive behavior, 238
nonexhaustible resources, 44–45
nonhazardous waste, 326
non-native species. *See* alien species
non-point source pollution, 373
nonrenewable resources, 45–46
northern coniferous forest, 143
northern hardwoods forest, 144
no-till, 134
noxious weeds, 272
nuclear energy, 353
nuclear power, 64, 353–355
nuclear waste, 355
nutrients
  for animals, 217–220
  for plant growth, 121–122
  recycled, 332

oak trees, 149, 313
ocean environments, 305–308
oceanic zone, 308
oceanologist, 389
offal, 331–333
off-site identification, 200–201
O horizon, 118
oil (petroleum), 12–13, 106, 341–343. *See also* fossil fuels
  conservation, 345–346
  estimated oil supply, 343
  as exhaustible resource, 46
  exploration and drilling, 342
  oil spills, 12–13
  OPEC, 345
  underground fuel tank leakage, 12, 13
oil shale/tar sands, 343, 344–345
omnivores, 220
on-site identification, 201–202
OPEC (Organization of Petroleum Exporting Countries), 345
organic compounds, 280
organic matter
  in soil, 117–118, 176–177
  soil formation and, 116
  in water, 82, 105–107
organism, definition of, 7
original tissue, 117
ornithologist, 389
overgrazing, 190
oxygen, 121–122
  biochemical oxygen demand (BOD), 106
  chemical oxygen demand (COD), 106
  consumption during respiration, 72
  dissolved, 99–101, 303–304
  necessity for life, 121
  release by photosynthesis, 72–73, 121
  saturation, 101
  solubility, temperature and, 95
  stress, 99–101
  use during respiration, 121
ozone, 65
ozone layer, hole in, 65, 67–68

Pacific Coast forest, 146
parasites, use in pest control, 278
parasitism, 229

parent material, 114–115
particulate matter, from wood burning, 169
particulate organic carbon, 106
particulates, 65–66
passenger pigeon, 238, 297, 314
pathogens, 271
  use in pest control, 278
peat moss mining, 207
peat soils, 116
penicillin, 299
percolate, 327
peregrine falcon, 297–298
perennial forages, 123
perennial weeds, 272
permafrost, 311
pesticides, 11–12, 66, 106–107, 280–281
  air quality and, 66
  drift/vapor drift, 276, 277
  environmental concerns, 276
  improper use of, 253–254
  pest resurgence and, 279
  resistance, 279
  safe-guarding groundwater from, 91
  species endangerment and, 297
  types of, 11–12, 106–107
pests. *See also* integrated pest management (IPM)
  damage from, 175, 283–284
  definition of, 66
  in forests, 173–174
  insects, 273–274, 283–284
  integrated pest management (IPM), 174, 267–289
  key, 282–284
  plant diseases, 275–276
  population equilibrium, 284–285
  rodents, 274–275
  sensible pest control, 93–94
  targeted, 282
  threshold levels, 285, 287
  types of, 270–275
  weeds, 271–272
petroleum. *See* oil
pH, 94–95, 97
  buffers, 96, 97
pheromones, 279
phloem, 153
phosphate, 96
photosynthesis, 22, 26, 72–73
  research to improve, 74
phytoplankton, 301–302
pine trees, 146–148
pioneer (plant), 34
plankton, 301–302
plant diseases, 275–276
plant growth, 121–122
point source pollution, 373
polar bear, 9
pollination, 274, 281
pollution, 10–14. *See also* air pollution; water pollution
  definition of, 10
  fish survival and, 37
  non-point source, 373
  point source, 373
  reclaiming polluted resources, 53–54
  wetlands and, 206
polygamy, 292
polygyny, 292
ponds, 204, 244–245
poplar trees, 149
population, 7
  characteristics, 292–295
  definition of, 292
  in developed vs. undeveloped nations, 294–295
  level, 47
  world population (human), 48, 258
population ecology, 291–317
  biomes, 300–315
  definition of, 292
  extinction and its causes, 296–300
  population characteristics, 292–293
  population growth factors, 293–296
pores
  in leaves, 30
  in soil, 92
potable water, 80
potassium, 97
potato blight disease, 270
prairie chicken, 223
prairie potholes, 205–206, 235
prairies, 184, 185–186, 312. *See also* grasslands
precipitation, 86
predation, 230
predators, 31–32, 220
  pest control using, 278
predatory animal control officer, 398–399
preservation, 51–52
preservationists, 14–15, 52
primary consumers, 31
primary succession, 34, 35
producers, 21, 83, 85
production, 293
promiscuity, 292
pronghorn antelope, 189, 313
pulpwood, 143
purify (water), 92

quarantines, 281–282

radial growth, 175
radiant energy, 24
radiation, 354, 355
radioactive, definition, 354
radioactive dust and materials, 64
radioactive waste, 325–326
radon, 64, 73
ranchers, 253–255. *See also* agriculture; grazing land; livestock
range, 184
rangelands, 124, 184, 255–257. *See also* grasslands
  management techniques, 193–196
  objectives of management, 193–194
  restoration of, 195–196
range of tolerance, 37–38
rats, 274–275
reactivity, 325, 369
recreation
  uses of soil, 126–127
  value of forests, 172
recycled animal wastes, 331–333
recycled nutrients, 332
recycling, 45–46, 324, 333
  of petroleum products, 93
renewable resources, 45
reproduction
  animal, 227, 292–293
  forest, 154–157
  sexual and asexual, 154
  vegetative, 154
reproductive growth, 272
resource management, 42–44. *See also* ecosystem management
resources
  conservation of, 50–51
  ecological, 235
  exhaustible (nonrenewable), 45–46
  forests as, 165–173, 315
  human population, use by, 48–49
  multiple use of, 52–53
  nature of, 44–48
  nonexhaustible, 44–45
  preservation of, 51–52
  reclaiming damaged/polluted, 53–54
  renewable, 45
respiration, 22, 25, 72, 122
rhizome, 272
riparian zone, 171, 243
risk assessment, 372
rivers, 204. *See also* streams
rocks, 115
Rocky Mountain forest, 146
rodenticide, 11–12
rodents, 274–275, 312
room and pillar system, 341
roots, 121–122
routes of entry, 370–371
ruminant, 219–220

safety training and precautions, 381
salinity, 95–96, 306
saltation, 132
salts/salt content, 82
saltwater, 82
sand, 118
sapwood, 153
saturated soil, 92
saturation of oxygen, 101
science teacher, 390
scientist, environmental, 386–387
scrubbers/scrubbing, 66, 358
secondary consumers, 31
secondary succession, 34–35, 36
second law of thermodynamics, 352
sedimentary rocks, 115
seeding (of forests), 155–157, 158
seedlings, 158
seed tree method, 160
seismograph, 342
semelparous species, 292–293
semiconductor, 351
semisubmersible rig, 342
septic systems, 94, 125–126
sewer gas, 358
sex ratio, 293
sexual reproduction, 154
shaft mines, 341
shelterwood method, 161
shrink-swell potential, 125
shrub layer, 167, 314
shrubs, 142, 192
Silent Spring (Carson), 269
silica, 96
silt, 10, 13, 51
  definition of, 51, 118
  load, 171
  water quality and, 114
  wind-blown (loess), 116
silvics, 154
silviculture, 154
  career in, 383–384
simple stomach, 219
slaughterhouses, waste from, 332–333

slope, 119
  water erosion and, 130
slope mine, 341
smog, 55
snowmobiles in Yellowstone Park, 15
sodium absorption ratio, 97
sodium percentage, 97
softwoods, 142, 146–149
soil, 113–139. *See also* soil erosion
  aeration, 122
  alfisols, 177
  alluvial and marine deposits, 116
  careers in, 390
  characterization of, 118–121
  composition of, 114
  conservation of, 50–51, 128–137, 178, 254
  definition of, 62, 114, 176
  drainage, 120
  as exhaustible resource, 46
  fertility, 90
  flood hazard likelihood, 120
  forest soils, 176–178
  formation of, 114–116
  glacial deposits, 115–116
  hydric, 197, 201, 202
  as living environment, 127–128
  loess deposits, 116
  as medium for plant growth, 121–122
  minerals and rocks, 115, 118
  orders, 177
  organic deposits, 116
  organic matter, 117–118, 176–177
  parent materials, 114–115
  physical properties of, 119–121
  profile, 118–119
  saturated, 92
  slope, 119
  solution of nutrients, 122
  spodosols, 177
  subsoil, 118
  texture, 119–120, 129, 130, 176
  thickness, 121
  tillable, 118
  tilling practices, 130, 131–132, 134, 135
  topsoil, 118
  uses, 123–127
  utisols, 177
  waterlogged, 122
  weathering, 116–117
soil conservationist, 390
soil erosion, 10, 50–51, 120, 128–137
  of forest soils, 177–178
  preventing and controlling, 143–147
soil specialist, 390
solar energy, 350–353
  active vs. passive, 352
  solar cell, 350–351
  solar panel, 351–352
solid wastes, 10–11, 320–327
  definitions/description, 320–324
  types of, 324–326
solubility, 95
southern cavefish (*Trphlichthys subterraneus*), 302
southern forest, 145
specific conductivity, 95
Spencer, Roy, 70
spodosols, 177
stack scrubbers, 63
stand, 154
stewards, 255
stewardship, 236, 261
stolon, 272
strata, 166, 313
streams, 204, 242–244
strip mines, 339–340
subsoil, 118, 121
succession, biological, 33–38
sulfate, 96
sulfur, 63
Superfund, 372–373
surface creep, 133
surface water, 10
  as nonexhaustible resource, 44–45
  pollution, 20
suspension, 133
sustainable agriculture, 257–261
Swampbuster provision, 200, 201, 207
swamp gas. *See* methane
swamps, 198–199, 204–205
sweetgum trees, 149
sycamore trees, 152

targeted pests, 282
tar sands, 343, 344–345
Taxol, 299
taxonomic groups, 7
taxonomist, 390
Taylor Grazing Act of 1934, 192
technology specialist, 391
temperate forest biome, 313–314
temperature. *See also* global warming
  body heat, 24–25
  Earth's surface temperature, 68–70
  of water, 95
teratogenic effect, 371
terminal growth, 175

terrestrial biomes, 308–315
terrestrial organisms, 8
tetraethyl lead, 63–64
texture of soil, 119–120, 129, 130, 176
thermal energy, 24
thermal stratification, 304
thermal stress, 95
thermodynamics, second law of, 352
threatened species, 232–233, 239. *See also* endangered species
tidal power, 361–362
tidewater, 82
tillable soil, 118
tilling practices, 130, 131–132, 134, 135, 257
timberland, 142
timber management, 158–162
topsoil, 118
  erosion, 10, 120, 128–137
  thickness, 121
total dissolved solids (TDS), 96
total organic carbon test, 106
toxic and hazardous substances, 367–377, 369
  harm from, 370–371
  HAZMAT, 368–369
  legislation, 372–375
  Love Canal, 369
  risk assessment, 372
  routes of entry, 370–371
  Superfund, 372–373
  toxic substance, 368
toxic materials, 10–11, 66, 98–99
toxic waste, 255, 325
trace elements, 98–99
transformers, 21
transpiration, 30, 89
transportation. *See* automobiles; fossil fuels; oil (petroleum)
trap crop, 277
trees, 146–152. *See also* forests; *specific trees*
  annual rings, 153
  cambium, 153, 154
  commercial products from, 143
  definition of, 142
  evergreens, 142, 143, 146–149, 314–315
  growth and physiology of, 152–154
  growth impact, 175
  hardwoods, 142, 144–145, 149–152
  phloem (inner bark), 153
  radial growth, 175
  seeding of, 155–157
  seedlings, 158
  softwoods, 142, 146–149
  stand of, 154
  terminal growth, 175
  types of, 142, 146–152
  xylem layer (sapwood), 153
tropical forest, 146
tundra biome, 310–312
turbid water, 98, 303
turbine, 360, 362

Udall, Stewart, 53
undergrazed areas, 193
understory, 167, 313
United Nations Framework Convention on Climate Change, 71
universal solvent, 83–85
upwelling currents, 307
uranium, 354–355
urban forester, 384
urbanization, 124–125, 206
U.S. Army Corps of Engineers, 197, 198, 201, 209–210
U.S. Endangered Species Act, 232–233, 299, 315
U.S. Fish and Wildlife Service (USFWS),197–198, 209
  endangered species and, 232, 245
  jobs with, 381–382
U.S. Forest Service, 184
U.S. National Park Service, 184
Utah Crandall Canyon mine collapse, 340
utisols, 177

vapor drift, 276, 277
vector, 270
vegetative cover, 223
vegetative growth, 272
vegetative reproduction, 154
ventilation systems, 73–74
volatile organic compound (VOC), 368

warm-blooded animal, 229
waste/waste management, 10–11, 20–21, 318–335. *See also* landfills; toxic and hazardous substances
  conservation of matter and, 20
  disposal of waste, 125–126, 326–327
  hazardous waste, 126, 324–325
  industrial solid waste, 321
  industrial waste, 10
  landfills, 326, 327–331
  municipal solid waste, 321–324
  nonhazardous waste, 326
  nuclear waste, 355
  percolate/leachate, 327
  radioactive waste, 325–326
  recycled animal waste, 331–333
  recycling, 45–46, 324, 333

waste/waste management *(Cont'd.)*
  solid wastes, 10–11, 320–327
  total waste generation, 323
  toxic waste, 255, 325
wastewater
  disposal, 125–126
  treatment, 53–54
wastewater treatment specialist, 391
water, 62, 79–111. *See also* surface water; wastewater; water pollution; water purification; watersheds; wetlands
  alkalinity, total, 97–98
  capillary, 92
  Clean Water Act, 53
  composition/statistics, 62, 82, 94
  conservation, 107–108
  cycle, 28–30, 45, 85–86
  dissolved gases, 99–105
  dissolved ions, 82, 96–98
  dissolved minerals, 303
  dissolved oxygen, 99–101, 303–304
  dissolved solids, total, 96
  environment, limiting factors in, 302–303
  food chains and webs, 83–84
  free, 92
  fresh vs. salt water, 82
  freshwater biomes, 300–305
  functions of, 81
  gravitational, 92
  groundwater, 13, 90–91, 92
  hardness, 97–98
  hygroscopic, 92
  importance in living organisms, 28, 80, 121
  importance in the environment, 28–29, 80–81
  infiltration of, 129–130
  land, relationships with, 86–89
  land as reservoir for, 90–91
  lime requirement test, 98
  as living environment, 94–107
  marine biome, 305–308
  nature of, 81–86, 94
  organic material in, 82, 105–107
  pH, 94–95, 96, 97
  potable, 80
  precipitation and evaporation, 86
  quality, improving, 91–94
  salinity, 95–96
  in soil, 90–92, 114, 121
  table, 90
  temperature, 95, 100
  thermal stratification, 304
  trace elements in, 98–99
  transpiration, 30
  turbid, 98, 303
  as universal solvent, 83–85
  water-saving tips, 107–108
  wildlife and, 221–223
water cycle, 28–30, 45, 85–86
water erosion, 129–131
waterlogged soil, 122
water pollution, 53–54, 84–85
  indicator species, 93
  reducing, 91–94
  safe-guarding groundwater, 90–91
  wastewater treatment, 53–54
water purification, 92
  by marshes and wetlands, 54, 88–90, 208–209
Watershed Protection and Flood Prevention Act, 305
watersheds, 29, 86–89, 172–173
  definition of, 86, 172
  value of forests in, 172–173
water treatment specialist, 391
weathering, 116–117
weeds, 271–272
wetlands, 196–211, 234–236, 304–305
  agencies managing, 197–198, 200, 201, 209
  causes of loss, 206–207, 210
  characteristics of, 196–197
  constructed, 90
  definitions of, 196, 197–198
  ecological, 198
  as ecological resources, 235
  economic value of, 234–235
  floodplains, 203–206
  governmental programs and laws, 209–211, 305
  history in United States, 198–200
  identification of, 200–202
  jurisdictional, 197
  management of, 209–211, 305
  marshes, 88–90, 202–203
  National Wetlands Inventory (NWI), 200–201
  natural, 90, 208
  preservation of, 200, 208–209
  restoration of, 209, 210, 305
  status in United States, 206–207
  Swampbuster provision, 200, 201, 207
  types of, 202, 203–206
  water purification by, 54, 88–90, 208–209
  wildlife management in, 241–242
whooping crane, 297
wildlife, 215–247
  animal behaviors and habits, 216–217
  animal growth, 226–227
  animal reproduction, 227
  basic needs of, 217–228

endangered/threatened species, 232–233, 239
extinction and, 232, 236–238
food needs of, 217–220
habitat arrangement for, 228
habitat needs of, 223–224
habitat preservation/restoration, 231–233
human impacts on, 233–236
management of. *See* wildlife management
relationships, 229–231
as resource, 45, 50
safety training/precautions for working with, 381
shelter needs of, 223
space needs of, 224–226
stewardship for, 236
water needs of, 221–223
wildlife conservation officer, 391–392
wildlife management, 239–245
endangered/threatened species, 239
farm wildlife, 239–240, 256–257
forest wildlife, 240–241
lakes and ponds wildlife, 244–245
stream wildlife, 242–244
wetlands wildlife, 241–242
wildlife technician, 392
windbreaks, 134, 136
wind energy, 362
wind erosion, 128, 131–133
wind turbine, 362
wood, as energy source, 362–363. *See also* forests; trees
woodlot management, 157–158

xylem layer, 153

Yellowstone National Park, 313
bison in, 46
snowmobiles in, 15
succession after forest fires in, 36
Young, Brigham, 311

zoology, 216
zooplankton, 301–302
zygote, 227